T0201388

HYBRID CONTROL AND MOTION PLANNING OF DYNAMICAL LEGGED LOCOMOTION

HYBRID CONTROL AND MOTION PLANNING OF DYNAMICAL LEGGED LOCOMOTION

Nasser Sadati
Guy A. Dumont
Kaveh Akbari Hamed
William A. Gruver

IEEE PRESS

A JOHN WILEY & SONS, INC., PUBLICATION

Library of Congress Cataloging-in-Publication Data:

Hybrid control and motion planning of dynamical legged locomotion / Nasser Sadati. . . [et al.].
    p. cm.
  ISBN 978-1-118-31707-5 (hardback)
 1. Mobile robots. 2. Robots–Motion. 3. Walking.   I. Sadati, Nasser.
  TJ211.415.H93   2012
  629.8′932–dc23

                                                                    2012002035

ISBN: 978-1-118-31707-5

CONTENTS

During the last three decades, enormous advances have occurred in robot control of dynamical legged locomotion. The desire to study legged locomotion has been motivated by the desire to assist people with disabilities to walk and replace humans in hazardous environments. The control of dynamical locomotion is complicated by (i) limb coordination, (ii) hybrid nature of walking and running due to presence of impact and takeoff, (iii) underactuation, (iv) overactuation, (v) inability to apply the Zero Moment Point criterion during dynamic walking and running, (vi) lack of algorithms to achieve feasible period-one orbits, and (vii) conservation of angular momentum about the robot's center of mass during flight phases of running. New applications of complex legged robots also require the use of system engineering approaches to resolve these issues that are beyond any single traditional engineering discipline. As new problems in legged locomotion require multidisciplinary methodologies, there is a critical need for a comprehensive book covering motion planning algorithms and hybrid control. This book fills that gap for researchers, professionals, and students who are versed in robotics and control theory. This book serves as a reference and essential guide for researchers and engineers to perform future research and development in order to advance various topics of hybrid control of legged locomotion. This volume also provides a comprehensive overview of hybrid models describing the evolution of planar and 3D legged robots during dynamical legged locomotion, and hybrid control schemes to asymptotically stabilize periodic orbits for the resulting closed-loop systems. The major topics of this book include hybrid systems, systems with impulse effects, offline and online motion planning algorithms to generate periodic walking and running motions and two-level control schemes including within-stride feedback laws to reduce the dimension of hybrid systems, continuous-time update laws for online minimization of a general cost function, and event-based update laws to asymptotically stabilize the generated desired orbits. This volume can be viewed as a handbook in this important field, as well as a reference book for researchers and practicing engineers.

Chapter 2 introduces basic ideas, definitions, and results from the literature of hybrid systems. Chapter 3 shows how to design a continuous-time-invariant feedback law that asymptotically stabilizes a feasible periodic trajectory using an extension of hybrid zero dynamics for a hybrid model of walking. The main objective is to develop a continuous-time-invariant control law for walking of a planar biped robot during the double support phase.

A number of control problems for reconfiguration of a planar multilink robot during flight phases have been considered in the literature. However, these methods cannot be employed online to solve the reconfiguration problem for monopedal running. For this reason, Chapters 4 and 5 present online reconfiguration algorithms that provide a solution to this latter problem for given flight times and angular momenta. The algorithms proposed in this book are expressed using the methodology of reachability and optimal control for time-varying linear systems with input and state constraints. In addition, a two-level control scheme based on the online reconfiguration algorithms and hybrid zero dynamics is proposed in Chapter 4 to asymptotically stabilize a desired period-one orbit for a hybrid model describing running by planar monopedal robots. Chapter 6 presents a time-invariant control scheme to asymptotically stabilize a desired feasible periodic orbit for running by a 3D legged robot along a straight line. A systematic algorithm to generate desired feasible periodic orbits for 3D running is also presented. Chapter 6 extends the results of Chapters 4 and 5 to 3D running robots.

In order to reduce the number of actuated joints for walking on a flat surface and restore walking motion for persons with disabilities, a motion planning algorithm is developed in Chapter 7 for walking with passive knees. In addition, a time-invariant two-level control scheme is presented to stabilize the desired motions that are generated. In Chapter 8, an analytical approach for designing a class of continuous-time update laws to update the parameters of stabilizing controllers during continuous phases is proposed such that (i) a general cost function, such as the energy of the control input over single support, can be minimized online, and (ii) the exponential stability behavior of the limit cycle for the closed-loop system is not affected.

Book Webpage: Supplemental materials are available at the following URL:

$$\texttt{http://booksupport.wiley.com.}$$

This webpage includes MATLAB codes for motion planning algorithms and hybrid control schemes of several legged robots studied in this book, an erratum, and a link to submit errors found in this book.

NASSER SADATI
GUY A. DUMONT
KAVEH AKBARI HAMED
WILLIAM A. GRUVER

November 14, 2011

CHAPTER 1

Introduction

1.1 OBJECTIVES OF LEGGED LOCOMOTION AND CHALLENGES IN CONTROLLING DYNAMIC WALKING AND RUNNING

The most effective type of locomotion in rough terrains is legged locomotion. During the past three decades enormous advances have occurred in robot control and motion planning of dynamic walking and running locomotion. In particular, hundreds of walking mechanisms have been built in research laboratories and companies throughout the world. The desire to study legged locomotion has been motivated by the need to assist people with disabilities to walk and replace humans in hazardous environments. Underactuation, impulsive nature of the impact between the lower limbs and the environment, the existence of foot structure and the large number of degrees of freedom (DOF) are basic problems in controlling legged robots. Underactuation is naturally associated with dexterity. For example, headstands are considered dexterous [1]. In this case, the contact point between the body and the ground is acting as a pivot without actuation. The nature of the impact between the lower limbs of legged robots and the environment causes the dynamics of the system to be hybrid and impulsive. The impact between the foot and the ground is one of the main difficulties in designing control laws for walking and running robots. Unlike robotic manipulators, legged robots are always free to detach from the walking surface, thereby leading to various types of motions. Finally, the existence of many degrees of freedom in the mechanism of legged robots causes the coordination of the links to be difficult. As a result of these latter issues, the design of practical controllers for legged robots remains to be a challenging problem. Also, these features complicate the application of traditional stability margins. Consequently, the major issues in the control of dynamic walking and running are as follows:

1. *Limb coordination.* Legged robots are high degree of freedom mechanisms, and consequently, coordination of their links to achieve dynamic walking and running locomotion is complex.

Hybrid Control and Motion Planning of Dynamical Legged Locomotion, Nasser Sadati, Guy A. Dumont, Kaveh Akbari Hamed, and William A. Gruver.

2. *Hybrid nature of locomotion due to presence of impact and liftoff.* The presence of impact, foot touchdown, and liftoff leads to *models with impulse effects* and *hybrid systems* consisting of multiple *continuous* and *discrete phases*. In particular, mathematical models describing the evolution of legged robots during walking and running include both discrete and continuous phenomena. Instantaneous discrete phases arise when feet impact the ground or feet liftoff the ground, whereas ordinary differential equations based on classical Lagrangian mechanics describe the evolution of legged robots during continuous phases of locomotion.

3. *Underactuation.* During certain phases of walking and running such as *single support* (one leg on the ground) in walking and *flight* (no leg on the ground) in running, legged robots have fewer actuators than degrees of freedom.

4. *Overactuation.* During *double support phase* of bipedal walking (both legs on the ground), biped robots have fewer degrees of freedom than actuators. Due to overactuation, the control input corresponding to a specific trajectory in the state space is not unique.

5. *Inability to apply the Zero Moment Point criterion.* Most past work in the literature of legged robots emphasizes the *quasi-static stability criteria* and *flat-footed walking* based on the *Zero Moment Point (ZMP)* [2–16] and the *Foot Rotation Indicator (FRI) point* [17]. The ZMP is defined as the point on the ground where the net moment generated from ground reaction forces has zero moment about two axes that lie in the plane of ground [3]. The ZMP is contained in the robot's *support polygon*, where the support polygon is defined as the convex hull formed by all contact points with the ground. The ZMP criterion states that when the ZMP is contained within the *interior* of the support polygon, the robot is stable so that it will not topple. Thus, in this kind of stability, as long as the ZMP lies strictly inside the support polygon of the foot the trajectories are feasible. If the ZMP lies on the edge of the support polygon, then the trajectories may not be feasible. The center of pressure (COP) is a standard notion in mechanics which was renamed as the ZMP by Vukobratovic [3]. The FRI point is a concept defined when the foot is in rotation with respect to the ground [17]. The FRI is the point on the ground where the net ground reaction force would have to act to keep the foot stationary. Thus, if FRI is within the convex hull of the stance foot, the robot can walk and it does not roll over its extremities, such as the heel or the toe. This type of walking is called as *fully actuated walking*. If the FRI is not in the projection of the foot on the ground, the stance foot rotates about the extremities. Such an event is also known as *underactuated walking*. As long as the foot does not rotate about its extremities, the ZMP, COP, and FRI points are equivalent [15] (see Fig. 1.1). In the literature of legged locomotion, a *statically stable gait* is a periodic locomotion in which the robot's center of mass (COM) does not leave the support polygon. A *quasi-statically stable gait* is a periodic locomotion in which the COP of the robot is within the *interior* of the support polygon. Moreover, a *dynamically stable gait* is a periodic locomotion where the robot's COP is on the boundary of

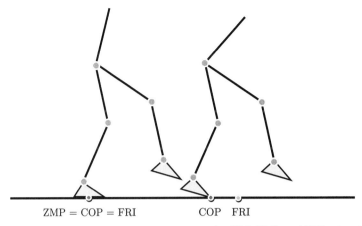

ZMP = COP = FRI   COP   FRI

Figure 1.1   Two planar bipedal models to compare the COP, ZMP, and FRI points. The FRI point is a point on the ground contact surface, within or outside the convex hull of the foot support area, at which the resultant moment of the force/torque impressed on the foot is normal to the surface [17]. In the left figure, the foot does not rotate about its extremities, thus, the ZMP, COP, and FRI points are equivalent. At the right figure, the foot is starting to rotate since the FRI point is outside the convex hull of the stance foot. We note that the COP is at the tip of the stance foot about which the foot rotates.

the support polygon for at least part of the walking cycle [18]. Thus, during dynamic walking and running cycles, the location of the robot's COP is on the boundary of the support polygon and, as a result, this will prohibit the use of the ZMP criterion. To make this notion more precise, the ZMP criterion is a sufficient and necessary condition for the stance foot not to rotate. However, this does not imply that the walking motion is asymptotically stable in the sense of Lyapunov [18, Chapter 11].

6. *Lack of algorithms to achieve feasible period-one orbits and limit cycles.* The main problem in control of legged locomotion is how to design a feedback law that guarantees the existence of a stable *limit cycle* for the closed-loop system. Underactuation and *unilateral constraints* must be included in order to design a *feasible* periodic orbit for legged locomotion. Unilateral constraints are constraints on the state and control inputs of the mechanical system that represent feasible contact conditions between the leg ends and the ground. In particular, leg ends, whether they are terminated with feet or points, are not attached to the ground. Hence, the ground reaction forces must lie in the *friction cone* to prevent slippage and foot liftoff. Thus, normal forces at the leg ends can only act in one direction, and are unilateral. In addition, if the foot is to remain flat on the ground and not rotate about its extremities, then the FRI must be between the heel and toe, a condition that can be expressed as a pair of unilateral constraints. These facts combined with underactuation during the single support and flight phases complicate the design of motion planning algorithms to generate feasible periodic locomotion.

7. *Conservation of angular momentum about the robot's COM during flight phases.* During flight phases of running, conservation of angular momentum about the robot's COM is a *nonintegrable Pfaffian constraint* which complicates the path planning and control of the robot's configuration during landing (flight to stance phase).

1.2 LITERATURE OVERVIEW

1.2.1 Tracking of Time Trajectories

Most existing control algorithms in the literature of legged robots are *time-dependent* approaches based on tracking of *predetermined* time trajectories generated by the ZMP criterion, *inverted pendulum model* and *nonlinear oscillators* as central pattern generators of the spinal cord. The ZMP criterion [2–16] has been used for trajectory tracking in ASIMO [2] and WABIAN [5, 6]. The *Linear Inverted Pendulum Model (LIPM)* [19, 20] and ZMP criterion-based approaches for stable walking reference generation have been reported in the literature. In these techniques, generally, the ZMP reference during a stepping motion is kept fixed in the middle of the supporting foot. Erbatur and Kurt [9] proposed a reference generation algorithm based on the LIPM and moving support foot ZMP. In addition, they made use of a simple inverse kinematics-based joint space controller to test the reference trajectory for simulation of a 3D, 12-DOF biped robot model. By allowing a variation of ZMP over the convex hull of foot polygon, Bum-Joo et al. [10] proposed an algorithm to modify the walking period and the step length in both the sagittal and lateral planes of the humanoid robot HanSaRam-VII. Motoi et al. [11] presented a real-time gait planning algorithm based on ZMP for pushing motion of humanoid robots to deal with an object with unknown mass. Kajita et al. [12] presented a ZMP-based running pattern generation algorithm for running of the humanoid robot HRP-2LR. ZMP-based online jumping pattern generation for running of a monopedal robot has also been reported in Ref. [13]. Sato et al. [14] proposed walking trajectory planning on stairs for biped robots using the method of virtual slope and the ZMP criterion. Sardain and Bessonnet [15] proved the coincidence of COP and ZMP and they examined related control aspects. In this latter reference, the virtual COP–ZMP was also defined to extend the concept for walking on uneven terrain. Ref. [21] approximated the biped model as an *inverted pendulum* and made use of trajectory tracking to control dynamic walking locomotion in Biper-3. Katoh and Mori used PID controllers to track reference trajectories generated by a van der Pol oscillator in the control of BIPMAN [22]. To control walking in Kenkyaku, Furusho and Masubuchi applied PID controllers for tracking joint reference trajectories [23]. Furusho and Sano also applied a decoupled control approach for control of motions in the frontal and sagittal planes during walking of the three-dimensional bipedal prototype BLR-G2 [24, 25]. PID controllers were employed to track the time trajectories generated by a length-varying inverted pendulum during walking of Meltran II by Kajita et al. [26, 27]. In these latter references, to maintain the biped's COM at a constant height, the pendulum's length is assumed to vary in a proper manner. In Ref. [28], PID control was also used to track predetermined trajectories to improve the stability behavior. A computed torque method was used to control a planar, 5-DOF bipedal robot in Ref. [29]. The performance of three control techniques including PID, computed torque, and sliding mode control in the tracking of joint trajectories during walking by a planar, 5-DOF biped was compared by Raibert et al. [30]. Tracking of time trajectories, based on computed torque with gravity compensation and a length-varying inverted pendulum model, has also been applied during walking by a three-dimensional bipedal robot in Refs. [8, 31, 32]. Trajectories generated by an inverted pendulum was also used by Kajita et al., to control walking of HRP-2 [33, 34].

The ZMP criterion has become a very powerful tool for trajectory generation in walking of legged robots. However, it needs a stiff joint control of the prerecorded trajectories that leads to poor robustness in unknown rough terrains whereas humans and animals show exceptional robustness when walking on irregular terrains. It is well known in biology that there are *Central Pattern Generators (CPG)* in spinal cord coupling with the musculoskeletal system [35–37]. The CPG and feedback networks can coordinate the body links of the vertebrates during locomotion. There are several mathematical models that have been proposed for a CPG. Among them, Matsuoka [38–41] has studied an approach in which a CPG is modeled by a *Neural Oscillator* consisting of two mutually inhibiting neurons. Each neuron is represented by a nonlinear differential equation. Matsuoka's approach has been used by Taga [36, 37] and Miyakoshi et al. [42] for biped robots. Kimura has also used this approach at the hip joints of quadruped robots [43, 44]. Ref. [45] presented a hybrid CPG–ZMP controller for the real-time balance of a simulated flexible spine humanoid robot. The CPG component of the controller allows the mechanical spine and feet to exhibit rhythmic motions using two control parameters. By monitoring the measured ZMP location, the control scheme modulates the neural activity of the CPG to allow the robot to maintain balance on the sagittal and frontal planes in real time.

1.2.2   Poincaré Return Map and Hybrid Zero Dynamics

As stated in Section 1.1, the main problem in controlling legged locomotion is how to design a controller that guarantees the existence of an *asymptotically stable limit cycle* for the closed-loop mechanical system. A classical technique for analyzing stability of periodic orbits for *time-invariant* dynamical systems described by ordinary differential equations is the *method of Poincaré sections*. This method establishes an equivalence between the stability analysis of the *periodic orbit* for an nth-order *continuous-time system* and that of the corresponding *equilibrium point* for an $(n - 1)$th-order *discrete-time system*. Grizzle et al. [46] extended the method of Poincaré sections to *systems with impulse effects*. A system with impulse effects consists of a continuous phase described by an ordinary differential equation and a discrete phase described by an instantaneous reinitialization rule for the differential equation. To simplify the application of the Poincaré sections method in the design of *time-invariant controllers* for walking by an underactuated three-link biped robot and *instantaneous* double support phase, Grizzle et al. [46] created *zero dynamics manifolds* that are *forward invariant* under the flow of the continuous phase of walking (i.e., single support phase). However, since the zero dynamics manifolds are not invariant under the flow of the discrete phase (i.e., impact model), the approach of Ref. [46] resulted in a *restricted Poincaré return map* (i.e., the Poincaré return map restricted to the zero dynamics manifolds) that cannot be expressed in a closed-form. The outputs in Ref. [46], corresponding to the zero dynamics manifolds, were selected as *holonomic* functions referred to as *virtual constraints*. Virtual constraints are a set of holonomic output functions defined on the configuration space of the mechanical system. They are forced to be zero by within-stride feedback laws to reduce the dimension of the Poincaré return map and to coordinate the links of biped robots during walking [47]. The method of virtual constraints for designing time-invariant controllers in walking of planar biped robots with one degree of underactuation, point feet and instantaneous double support has been studied in Refs. [18, Chapter 6, 46–52]. Westervelt et al. [52] created virtual constraints to ensure that the corresponding zero dynamics manifold is *hybrid invariant* under the closed-loop hybrid model of walking and introduced the notion of *hybrid zero dynamics* (HZD). The zero dynamics manifolds are said to be hybrid invariant if they are both forward invariant (i.e., invariant under the flow of the continuous phase) and *impact invariant* (i.e., invariant under the flow of the discrete phase). During walking of a planar biped robot with an underactuated *cyclic* variable [1, 53], HZD results in a two-dimensional zero dynamics manifold, and consequently, a one-dimensional restricted Poincaré return map that can be expressed in closed form. This approach was also extended by Choi and Grizzle [54] for creating a two-dimensional zero dynamics manifold during walking of a planar fully actuated biped robot in fully actuated and underactuated phases. To reduce the dimension of the full-order hybrid model of running, which in turn simplifies the stabilization problem of the desired orbit, Ref. [55] proposed that the configuration of the mechanical system should be transferred from a specified initial pose (immediately after the takeoff) to a specified final pose (immediately before the landing) during flight phases. This problem is referred to as *landing in a fixed configuration* or *configuration determinism at landing* [18, p. 252]. By using the virtual constraints approach and the configuration determinism at landing, Ref. [55] obtained a closed-form expression for the one-dimensional restricted Poincaré return map of running by the five-link, four-actuator planar bipedal robot, RABBIT [47]. Moreover, to ensure that the stance phase zero dynamics manifold is hybrid invariant under the closed-loop hybrid model of running, an additional constraint was imposed on the vector of generalized velocities at the end of flight phases. To satisfy the configuration determinism at landing and hybrid invariance, Ref. [55] utilized the approach of *parameterized* HZD. In particular, using the Implicit Function Theorem and a *numerical* nonlinear optimization problem with an equality constraint, the parameters of the virtual constraints of the *flight phase* were updated in a *step-by-step fashion* during the discrete transition from stance to flight (i.e., takeoff). However, the stance phase controller was assumed to be fixed. For running of RABBIT, an alternative parameterized control law was proposed by Morris et al. [56]. However, their approach did not create HZD. The use of *event-based* control laws to update the parameters of time-invariant controllers for stabilization of periodic orbits in systems with impulse effects was presented in Refs. [57–59]. When the amount of underactuation during locomotion of biped robots is increased, it becomes difficult to create hybrid invariant manifolds. Morris and Grizzle [60] proposed a method to generate an open-loop *augmented* system with impulse effects, a new holonomic output function for the resultant system and an event-based update law for the parameters of the output such that the zero dynamics manifold associated with this output is hybrid invariant under the closed-loop augmented system. This latter approach has been used in design of time-invariant controllers for walking of a 3D biped robot in Refs. [61, 62] and also for walking and running of planar bipedal robots with springs, MABEL [63–65] and ATRIAS [66]. Hürmüzlü also applied the method of Poincaré sections to a planar, five-link bipedal robot and imposed a mix of holonomic and nonholonomic constraints on the mechanical system to obtain a closed-form expression for the robot's trajectory [67].

1.3 THE OBJECTIVE OF THE BOOK

In this book we provide a comprehensive overview of hybrid models describing the evolution of planar and 3D legged robots during dynamical legged locomotion and also propose hybrid control schemes to asymptotically stabilize desired periodic orbits for the closed-loop systems. The topics include (i) hybrid systems, (ii) systems with impulse effects, (iii) offline and online motion planning algorithms to generate desired feasible periodic walking and running motions, (iv) two-level control schemes, including within-stride feedback laws to reduce the dimension of the hybrid systems, (v) continuous-time update laws to minimize a general cost function online, and (vi) event-based update laws to asymptotically stabilize the desired periodic orbits. This book also provides a comprehensive presentation of issues and challenges faced by researchers and practicing engineers in motion planning and hybrid control of dynamical legged locomotion. Furthermore, we describe the current state of the art and future directions across all domains of dynamical legged locomotion so that readers can extend the proposed motion planning algorithms and hybrid control methodologies to other planar and 3D legged robots. The main objectives of this book are as follows.

1.3.1 Hybrid Zero Dynamics in Walking with Double Support Phase

There has been little attention given to control of biped robots during the double support phase with unilateral constraints. Such constraints present challenges for the design of controllers. The objective of Chapter 3 is to develop an analytical approach for designing a continuous feedback law that realizes a desired period-one trajectory as an asymptotically stable orbit for a planar biped robot. The robot is assumed to be a five-link, four-actuator planar mechanism in the sagittal plane with point feet. The fundamental assumption is that the double support phase is *not* instantaneous. Hence, bipedal walking can be represented by a hybrid model with two continuous phases, including a single support phase and a double support phase, and discrete transitions between the continuous phases. In the single support phase, the mechanical system has one degree of underactuation, whereas it is overactuated in the double support phase. Chapter 3 shows how to design a continuous time-invariant feedback law that asymptotically stabilizes a feasible periodic trajectory using an extension of HZD for a hybrid model of walking [68, 69]. The main contribution is to develop a continuous time-invariant control law for walking of a planar biped robot during the double support phase. Since the mechanical system in the double support phase has three degrees of freedom and four actuators, a *constrained dynamics approach* [70, p. 157] is used to describe the reduced-order dynamics of the system. Then, two virtual constraints are proposed as holonomic outputs for the constrained system and an output zeroing problem with two control inputs is solved. This results in a *nontrivial* two-dimensional zero dynamics manifold corresponding to the virtual constraints in the state manifold of the constrained system. Moreover, the corresponding zero dynamics has two control inputs that are not employed for output zeroing. Instead, they are used to satisfy the unilateral constraints. Furthermore, these inputs are obtained such that the control has *minimum norm* on the desired periodic trajectory. It can be shown that the constrained dynamics of the double support phase is completely feedback linearizable on an open subset of the state manifold. However, since our objective is to design a continuous time-invariant controller based on nontrivial HZD, in contrast to Ref. [71] we do not use input-state linearization nor a discontinuous time optimal control for tracking trajectories. An analogous approach is used in Refs. [54, 72] for creating a two-dimensional zero dynamics manifold in the state space of a fully actuated phase of walking where the fully actuated dynamics is completely feedback linearizable. The control strategy is presented at the following two levels. At the first level, we employ *within-stride controllers* including single and double support phase controllers. These are continuous time-invariant and parameterized feedback laws that create a *family of two-dimensional finite-time attractive and invariant submanifolds* on which the dynamics of the mechanical system is restricted. At the second level, the parameters of the within-stride controllers are updated at the end of the single support phase (in a stride-to-stride manner) by an event-based update law to achieve *hybrid invariance* and *stabilization*. As a consequence, the stability properties of the desired periodic orbit can be analyzed using a one-dimensional restricted Poincaré return map.

1.3.2 Hybrid Zero Dynamics in Running with an Online Motion Planning Algorithm

Chapter 4 presents an analytical approach for designing a *two-level control law* to asymptotically stabilize a desired period-one orbit during running by a planar monopedal robot. The monopedal robot is a three-link, two-actuator planar mechanism in the sagittal plane with point foot. The desired periodic orbit is generated by the method developed in Ref. [73]. It is assumed that the model of monopedal running can be expressed by a hybrid system with two continuous phases, including stance phase and flight phase, and discrete transitions between the continuous phases, including takeoff and landing (impact). The configuration of the mechanical system is specified by the *absolute orientation* with respect to an inertial world frame and by the joint angles determining the shape of the robot. During the flight phase, the angular momentum of the mechanical system about its COM is conserved. To reduce the dimension of the full-order hybrid model of running, which in turn simplifies the stabilization problem of the desired orbit, as proposed in Ref. [55], the configuration determinism at landing should be solved. However, the flight time and angular momentum about the COM may differ during consecutive steps. Consequently, the *reconfiguration problem* must be solved online. A number of control problems for reconfiguration of a planar multilink robot with zero angular momentum have been considered in the literature, for example, Refs. [74–78]. For the case that the angular momentum is not necessarily zero, a method based on the Averaging Theorem [79, Theorem 2.1] was presented in Ref. [80] such that for any value of the angular momentum, joint motions can reorient the multilink arbitrarily over an arbitrary time interval. However, when the angular momentum is not zero, this method cannot be employed online for solving the reconfiguration problem for monopedal running. For this reason, Chapter 4 presents an *online reconfiguration algorithm* that solves this problem for given flight times and angular momenta [81, 82]. The algorithm proposed in Chapter 4 is expressed using the methodology of *reachability* and *optimal control* for time-varying linear systems with input and state constraints. The main contribution of this chapter is to present an *analytical* approach for online generation of twice continuously differentiable (C^2) *modified reference trajectories* during flight phases of running to satisfy the configuration determinism at landing [81]. Moreover, by relaxing the constraint of Ref. [55] on the vector of generalized velocities at the end of the flight phases, Chapter 4 presents a two-level control scheme based on the reconfiguration algorithm to asymptotically stabilize a desired periodic orbit. In this scheme, within-stride controllers, including stance and flight phase controllers, are employed at the first level. The stance phase controller is chosen as a time-invariant and *parameterized* feedback law to generate a *family* of finite-time attractive zero dynamics manifolds. An alternative approach based on continuous feedback law is employed here to track the modified reference trajectories generated by the reconfiguration algorithm during the flight phases. To generate a family of hybrid invariant manifolds, an event-based controller updates the parameters of the *stance phase controller* during the transition from flight to stance (i.e., impact) [81]. Consequently, the stability properties of the desired periodic orbit can be analyzed and modified by a one-dimensional discrete-time system defined on the basis of a restricted Poincaré return map.

1.3.3 Online Motion Planning Algorithms for Flight Phases of Running

Following the results of Chapter 4, to asymptotically stabilize the desired periodic orbit for the hybrid model of running using a one-dimensional restricted Poincaré return map and HZD approach, the configuration of the mechanical system should be transferred from a predetermined initial pose (immediately after takeoff) to a predetermined final pose (immediately before landing) during the flight phases of running. The objective of Chapter 5 is to present *modified* online motion planning algorithms for generation of continuous (C^0) and continuously differentiable (C^1) open-loop trajectories in the body configuration space of the mechanical system such that the reconfiguration problem is solved [82, 83]. The algorithms presented in Chapter 5 are extensions of that presented in Chapter 4. In particular, the generated trajectories in Chapter 4 were C^2 while the reachable sets associated with the algorithms of Chapter 5 are larger than that of Chapter 4. We address the motion planning problem for general planar open kinematic chains composed of $N \geq 3$ rigid links interconnected with frictionless and rotational joints. The main contribution of Chapter 5 is to present online motion planning algorithms based on *virtual time* for generation of joint motions to satisfy configuration determinism at transitions. In particular, it is assumed that the time trajectory of a desired joint motion, precomputed offline, solves the reconfiguration problem. By replacing the time argument of the desired motion by a strictly increasing function of time called the *virtual time*, Chapter 5 shows how to determine continuous and continuously differentiable joint motions in an online manner during consecutive steps of running so that they solve the reconfiguration problem.

1.3.4 Hybrid Zero Dynamics in 3D Running

Chapter 6 presents a motion planning algorithm to generate periodic time trajectories for running by a 3D monopedal robot. In order to obtain a *symmetric* gait along a straight line, the overall open-loop model of running can be expressed as a hybrid system with four continuous phases consisting of two stance phases and two flight phases and discrete transitions among them (takeoff and impact). The robot is assumed to be a 3D, three-link, three-actuator, monopedal mechanism with a point foot. During the stance phases, the robot has three degrees of underactuation, whereas it has six degrees of underactuation in the flight phases. The motion planning algorithm is developed on the basis of a *finite-dimensional nonlinear optimization problem* with equality and inequality constraints and extends the results of Refs. [73, 84] for planar bipedal robots. The main objective of Chapter 6 is to develop time-invariant feedback scheme to exponentially stabilize a desired periodic orbit generated by the motion planning algorithm for the hybrid model of running.

Chapter 6 shows how to create hybrid invariant manifolds during 3D running [85]. By assuming that the control inputs of the mechanical system have discontinuities during discrete transitions between continuous phases, the takeoff switching hypersurface can be expressed as a zero level set of a scalar holonomic function. In other words, takeoff occurs when a scalar quantity, a strictly increasing function of time on the desired gait, passes through a threshold value. The virtual constraints during stance phases are defined as the summation of two terms including a *nominal* holonomic output function vanishing on the periodic orbit and an *additive* parameterized *Bézier polynomial*, both in terms of the latter strictly increasing scalar. By properties of Bézier polynomials, an update law for the parameters of the stance phase virtual constraints is developed, which in turn results in a *common* intersection of the parameterized stance phase zero dynamics manifolds and the takeoff switching hypersurface. By this approach, creation of hybrid invariance can be easily achieved by updating the other parameters of the Bézier polynomial. Consequently, a parameterized restricted

Poincaré return map can be defined on the common intersection for studying the stabilization problem. Thus, the overall feedback scheme can be considered at two levels. At the first level, within-stride controllers including stance and flight phase controllers, which are continuous time-invariant and parameterized feedback laws, are employed to create a family of attractive zero dynamics manifolds in each of the continuous phases. At the second level, the parameters of the within-stride controllers are updated by event-based update laws during discrete transitions between continuous phases to achieve hybrid invariance and stabilization. By this means, the stability analysis of the periodic orbit for the full-order hybrid system can be treated in terms of a reduced-order hybrid system with a five-dimensional Poincaré return map.

1.3.5 Hybrid Zero Dynamics in Walking with Passive Knees

In Chapter 7, a motion planning algorithm to generate time trajectories of a periodic walking motion by a five-link, two-actuator planar bipedal robot is presented. In order to reduce the number of actuated joints for walking on a flat ground and restore the walking motion in the disabled, it is assumed that the robot has passive point feet and unactuated knee joints. In other words, only the hip joints of the robot are assumed to be actuated. The motion planning algorithm is developed on the basis of a finite-dimensional nonlinear optimization problem with equality and inequality constraints. The equality constraints are necessary and sufficient conditions by which the impulsive model of walking has a period-one orbit. Whereas the inequality constraints are introduced to guarantee (i) the feasibility of the periodic motion and (ii) capability of applying the proposed two-level control scheme for stabilization of the orbit. The main objective of Chapter 7 is to present a time-invariant two-level feedback law based on the notion of virtual constraints and HZD to exponentially stabilize a desired periodic motion generated by the motion planning algorithm [86]. The studied mechanical system has three degrees of underactuation during single support. Chapter 7 presents a control methodology for creation of hybrid invariant manifolds and stabilization of a desired periodic orbit for the impulsive model of walking. In particular, for a given integer number $M \geq 2$, we introduce $M - 1$ *within-stride switching hypersurfaces* and thereby split the single support phase into M *within-stride phases*. The within-stride switching hypersurfaces are defined as level sets of a scaler holonomic quantity that is strictly increasing function of time on the desired walking motion. To stabilize the desired orbit, the overall controller is chosen as a two-level feedback law. At the first level, during a within-stride phase, a parameterized holonomic output function is defined for the dynamical system and imposed to be zero by using a continuous-time feedback law. The output function is expressed as the difference between the actual values of the angle of hip joints and their desired evolutions, in terms of the latter increasing holonomic quantity. At the second level, the parameters of continuous-time feedback laws are updated during within-stride transitions by event-based update laws. The purpose of updating the parameters is (i) achieving hybrid invariance, (ii) continuity of continuous-time feedback laws during within-stride transitions, and (iii) stabilization of the desired orbit. From the construction procedure of the parameterized output functions and event-based update laws, it is shown that the intersections of the corresponding zero dynamics manifolds and within-stride switching hyper-surfaces are independent of the parameters. Consequently, by choosing one of these common intersections as the Poincaré section, stabilization can be addressed on the basis of a five-dimensional restricted Poincaré return map.

1.3.6 Hybrid Zero Dynamics with Continuous-Time Update Laws

To improve the convergence rate, the idea of updating the parameters of time-invariant stabilizing controllers by event-based update laws has been described in Refs. [57–59]. The contribution of Chapter 8 is to develop a method for designing a class of continuous-time update laws to update the parameters of stabilizing controllers *during continuous phases* of locomotion such that (i) a general cost function (such as the energy of the control input over single support) can be minimized in an online manner, and (ii) the exponential stability behavior of the limit cycle for the closed-loop system is not affected [87]. In addition, Chapter 8 introduces a class of continuous-time update laws with *radial basis step length* to minimize a desired cost function in terms of the controller parameters and initial states.

CHAPTER 2

Preliminaries in Hybrid Systems

In this chapter, we will introduce some of the basic ideas, definitions, and results from the literature of hybrid systems. We will concentrate on the ideas used most often for generalization of the Poincaré return map. Our treatment is primarily intended as a review for the reader's convenience, with some additional focus on the geometric aspects of the subject. This chapter may be skipped in the first reading, or by the reader familiar with the results. We will refer to texts such as Ref. [18] for more complete developments and details.

2.1 BASIC DEFINITIONS

To define a hybrid model with two continuous phases, consider the differential equations $\dot{x}_1 = f_1(x_1)$ and $\dot{x}_2 = f_2(x_2)$ which are defined on the state spaces $\mathcal{X}_1$ and $\mathcal{X}_2$, respectively. It is assumed that $\mathcal{X}_1$ and $\mathcal{X}_2$ are embedded submanifolds of $\mathbb{R}^{n_1}$ and $\mathbb{R}^{n_2}$ for some $n_1, n_2 \in \mathbb{N}$. Let $\mathcal{S}_1^2$ and $\mathcal{S}_2^1$ be switching hypersurfaces in the state spaces $\mathcal{X}_1$ and $\mathcal{X}_2$ on which the transition from $\mathcal{X}_1$ to $\mathcal{X}_2$ and the transition from $\mathcal{X}_2$ to $\mathcal{X}_1$ occur, according to the continuously differentiable transition maps $\Delta_1^2 : \mathcal{S}_1^2 \rightarrow \mathcal{X}_2$ and $\Delta_2^1 : \mathcal{S}_2^1 \rightarrow \mathcal{X}_1$, respectively. It is assumed that $\mathcal{S}_1^2$ and $\mathcal{S}_2^1$ can be expressed as

$$\mathcal{S}_1^2 = \{x_1 \in \mathcal{X}_1 | H_1^2(x_1) = 0\}$$

$$\mathcal{S}_2^1 = \{x_2 \in \mathcal{X}_2 | H_2^1(x_2) = 0\},$$

where $H_1^2 : \mathcal{X}_1 \rightarrow \mathbb{R}$ and $H_2^1 : \mathcal{X}_2 \rightarrow \mathbb{R}$ are C^1 such that $\forall x_1 \in \mathcal{S}_1^2$, $\frac{\partial H_1^2}{\partial x_1}(x_1) \neq 0$ and $\forall x_2 \in \mathcal{S}_2^1$, $\frac{\partial H_2^1}{\partial x_2}(x_2) \neq 0$. By these assumptions, $\mathcal{S}_1^2$ and $\mathcal{S}_2^1$ are embedded submanifolds of $\mathcal{X}_1$ and $\mathcal{X}_2$. Moreover, suppose that $\mathcal{S}_1^2 \cap \Delta_2^1(\mathcal{S}_2^1) = \phi$ and $\mathcal{S}_2^1 \cap \Delta_1^2(\mathcal{S}_1^2) = \phi$. The autonomous hybrid model with two continuous phases denoted by the 8-tuple

Hybrid Control and Motion Planning of Dynamical Legged Locomotion, Nasser Sadati, Guy A. Dumont, Kaveh Akbari Hamed, and William A. Gruver.

13

$\Sigma(\mathcal{X}_1, \mathcal{X}_2, \mathcal{S}_1^2, \mathcal{S}_2^1, \Delta_1^2, \Delta_2^1, f_1, f_2)$ is defined as follows

$$
\Sigma_1 : \begin{cases}
\mathcal{X}_1 \subset \mathbb{R}^{n_1} \\
\mathcal{F}_1 : \quad \dot{x}_1 = f_1(x_1) \\
\mathcal{S}_1^2 = \{x_1 \in \mathcal{X}_1 \mid H_1^2(x_1) = 0\} \\
\mathcal{T}_1^2 : \quad x_2^+ = \Delta_1^2(x_1^-)
\end{cases}
$$

$$
\Sigma_2 : \begin{cases}
\mathcal{X}_2 \subset \mathbb{R}^{n_2} \\
\mathcal{F}_2 : \quad \dot{x}_2 = f_2(x_2) \\
\mathcal{S}_2^1 = \{x_2 \in \mathcal{X}_2 \mid H_2^1(x_2) = 0\} \\
\mathcal{T}_2^1 : \quad x_1^+ = \Delta_2^1(x_2^-),
\end{cases}
$$

$$(2.1)$$

where for every $i \in \{1, 2\}$, $\mathcal{F}_i$ represents the flow of the autonomous differential equation $\dot{x}_i = f_i(x_i)$. Moreover, $x_i^-(t) := \lim_{\tau \nearrow t} x_i(\tau)$ and $x_i^+(t) := \lim_{\tau \searrow t} x_i(\tau)$ are the left and right limits of the trajectory $x_i(t) \in \mathcal{X}_i$, respectively.

Definition 2.1 (Continuously Differentiable Hybrid Model) *The autonomous hybrid model* $\Sigma(\mathcal{X}_1, \mathcal{X}_2, \mathcal{S}_1^2, \mathcal{S}_2^1, \Delta_1^2, \Delta_2^1, f_1, f_2)$ *is said to be continuously differentiable or* C^1 *if for every* $i \in \{1, 2\}$, $f_i : \mathcal{X}_i \to T\mathcal{X}_i$ *is* C^1.

As in Ref. [18, p. 92], a solution of the hybrid model (2.1) is constructed by piecing together the trajectories of the flows $\mathcal{F}_i$, $i = 1, 2$ such that the transitions take place when these flows intersect the switching hypersurfaces $\mathcal{S}_i^j$, $i, j = 1, 2, i \neq j$. The new initial conditions for the differential equations $\dot{x}_j = f_j(x_j)$, $j = 1, 2$ are also determined by the transition maps Δ_i^j, $i, j = 1, 2, i \neq j$ (i.e., $x_j^+ = \Delta_i^j(x_i^-)$). To make this notion precise, we need to define exactly what we mean by a "solution" of the hybrid model.

Definition 2.2 (Solutions of the Hybrid Model) *Let* $\mathcal{X} := \mathcal{X}_1 \cup \mathcal{X}_2$ *be the union of the state spaces. Assume that* $(t_0, x_0) \in \mathbb{R} \times \mathcal{X}$ *and* $t_f \in \mathbb{R} \cup \{\infty\}$ *are a given initial pair and final time, respectively. Suppose that there exists a closed discrete subset* $\mathcal{T} := \{t_0 < t_1 < \cdots < t_j < \cdots\} \subset [t_0, t_f)$ *representing the switching times, and a function* $i : \mathcal{T} \to \{1, 2\}$ *determining the continuous phases of the hybrid model such that* $i(j) \neq i(j+1)$ *for all* $j \geq 0$, *where* $i(j) := i(t_j)$. *A function* $\varphi : [t_0, t_f) \times \{x_0\} \to \mathcal{X}$ *is said to be a solution of the hybrid model (2.1) if*

1. *for all* $j \geq 0$,
   (a) $\varphi(t, x_0)$ *restricted to the interval* $[t_j, t_{j+1})$ *is right continuous, and* $\varphi(t, x_0) \in \mathcal{X}_{i(j)}$ *for every* $t \in [t_j, t_{j+1})$;
   (b) *for every point* $t \in [t_j, t_{j+1})$, *the left limit* $\varphi^-(t, x_0) := \lim_{\tau \nearrow t} \varphi(\tau, x_0)$ *exists and is finite. Furthermore,* $\varphi^-(t, x_0) \notin \mathcal{S}_{i(j)}^{i(j+1)}$;

(c) *for every point* $t \in (t_j, t_{j+1})$, $\frac{\partial}{\partial t}\varphi(t, x_0) = f_{i(j)}(\varphi(t, x_0))$; *and*

2. *for all* $j \geq 1$ *and* $t_j < \infty$, $\varphi^+(t_j, x_0) = \Delta_{i(j)}^{i(j+1)}(\varphi^-(t_j, x_0))$.

Next let the autonomous hybrid model $\Sigma(\mathcal{X}_1, \mathcal{X}_2, \mathcal{S}_1^2, \mathcal{S}_2^1, \Delta_1^2, \Delta_2^1, f_1, f_2)$ satisfy the following hypothesis:

(H1) The vector fields $f_i : \mathcal{X}_i \to T\mathcal{X}_i$, $i = 1, 2$ are continuous and the solutions of the differential equations $\dot{x}_i = f_i(x_i)$, $i = 1, 2$ for a given initial condition in the state spaces $\mathcal{X}_i$, $i = 1, 2$ are unique and depend continuously on the initial conditions.

Then, the solutions of Σ are unique. By definition, the solution is also right continuous, and there does not exist a value of t such that $\varphi(t, x_0) \in \mathcal{S}_i^j$ for $i, j = 1, 2, i \neq j$. Consequently, if $x_0 \in \mathcal{S}_i^j$ for some $i, j = 1, 2, i \neq j$, then, $\varphi(t, x_0)$ is defined as $\varphi(t, \Delta_i^j(x_0))$ [18, p. 86]. A solution $\varphi(t, x_0)$ of the hybrid model (2.1) defined on the interval $[t_0, \infty)$ is said to be *periodic* if there exists a finite $T > 0$ such that $\varphi(t + T, x_0) = \varphi(t, x_0)$ for all $t \geq t_0$.

Now we will direct our attention to the definitions of *orbital stability* from the literature of hybrid systems. For this purpose, let $\mathcal{O}$ be a periodic orbit of the hybrid model corresponding to the periodic solution $\varphi(t, x_0)$, that is, $\mathcal{O} = \{\varphi(t, x_0) | t \geq t_0\}$. The following definitions adopted from Ref. [18, p. 86] will be used for orbital stability throughout this book.

Definition 2.3 (Stability) *The periodic orbit $\mathcal{O}$ of the hybrid model (2.1) is said to be stable if for every $\varepsilon > 0$, there is an open neighborhood $\mathcal{N}$ of $\mathcal{O}$ such that for every $x \in \mathcal{N}(\mathcal{O})$ there exists a solution $\varphi(t, x)$ of the hybrid model defined on the interval $[0, \infty)$ such that*

$$dist(\varphi(t, x), \mathcal{O}) < \varepsilon$$

for all $t \geq 0$.

Definition 2.4 (Asymptotic Stability) *The periodic orbit $\mathcal{O}$ of the hybrid model (2.1) is said to be asymptotically stable if it is stable, and there is an open neighborhood $\mathcal{N}(\mathcal{O})$ such that for every $x \in \mathcal{N}(\mathcal{O})$ there exists a solution $\varphi(t, x)$ of the hybrid model defined on the interval $[0, \infty)$ where*

$$\lim_{t \to \infty} dist(\varphi(t, x), \mathcal{O}) = 0.$$

Definition 2.5 (Exponential Stability) *The periodic orbit $\mathcal{O}$ of the hybrid model (2.1) is said to be exponentially stable if there exist an open neighborhood $\mathcal{N}(\mathcal{O})$ and positive scalar numbers N and γ such that for every $x \in \mathcal{N}(\mathcal{O})$ there exists a solution*

$\varphi(t, x)$ *of the hybrid model defined on the interval* $[0, \infty)$ *where*

$$dist(\varphi(t, x), \mathcal{O}) \leq N \, dist(x, \mathcal{O}) \, exp(-\gamma t)$$

for all $t \geq 0$.

2.2 POINCARÉ RETURN MAP FOR HYBRID SYSTEMS

This section reviews some of the key concepts that will be useful in the existence and stability analysis of periodic solutions of closed-loop hybrid systems. A classical approach for analyzing dynamical systems is due to Poincaré. In the method of Poincaré sections, the flow of an nth-order autonomous continuous-time system is replaced with an $(n - 1)$th-order autonomous discrete-time system. Thus, the Poincaré return map reduces the order of the dynamical system. Moreover, it establishes an equivalence between the existence and stability analysis of limit cycles of continuous-time systems and those of corresponding equilibrium points of discrete-time systems. In this section, the method of the Poincaré sections for autonomous hybrid systems with two continuous phases is introduced. To do this, assume that hypothesis H1 is satisfied. Let $\varphi_i(t, x_0)$ for $i \in \{1, 2\}$ represent the unique integral curve of the differential equation $\dot{x}_i = f_i(x_i)$ with the initial condition x_0, that is, there exists $t_{f,i}(x_0) > 0$ such that $\frac{\partial}{\partial t}\varphi_i(t, x_0) = f_i(\varphi_i(t, x_0))$ for all $t \in [0, t_{f,i}(x_0))$ and furthermore, $\varphi_i(0, x_0) = x_0$. Define the function $T_2 : \mathcal{X}_2 \to \mathbb{R} \cup \{\infty\}$ as the first time at which the flow $\varphi_2(t, x_0)$ intersects the switching manifold $\mathcal{S}_2^1$. This is made precise in the following definition:

$$T_2(x_0) := \begin{cases} \inf\{t \geq 0 | \varphi_2(t, x_0) \in \mathcal{S}_2^1\} & \text{if } \exists t \text{ such that } \varphi_2(t, x_0) \in \mathcal{S}_2^1 \\ \infty & \text{otherwise.} \end{cases} \quad (2.2)$$

Next, introduce $\tilde{\mathcal{S}}_1^2$ as the set of all $x_1 \in \mathcal{S}_1^2$ for which the flow $\varphi_2(t, \Delta_1^2(x_1))$ intersects the manifold $\mathcal{S}_2^1$ transversally in a positive finite time. In a similar manner, the function $T_1 : \mathcal{X}_1 \to \mathbb{R} \cup \{\infty\}$ is defined as follows:

$$T_1(x_0) := \begin{cases} \inf\{t \geq 0 | \varphi_1(t, x_0) \in \tilde{\mathcal{S}}_1^2\} & \text{if } \exists t \text{ such that } \varphi_1(t, x_0) \in \tilde{\mathcal{S}}_1^2 \\ \infty & \text{otherwise.} \end{cases} \quad (2.3)$$

Moreover, define $\tilde{\mathcal{S}}_2^1$ as the set of all $x_2 \in \mathcal{S}_2^1$ for which the flow $\varphi_1(t, \Delta_2^1(x_2))$ intersects the manifold $\tilde{\mathcal{S}}_1^2$ transversally in a positive finite time. We refer the reader to Ref. [18, p. 94] for more details. A formal definition of the Poincaré return map can be expressed as follows. Let the maps $P_2 : \tilde{\mathcal{S}}_1^2 \to \mathcal{S}_2^1$ and $P_1 : \tilde{\mathcal{S}}_2^1 \to \tilde{\mathcal{S}}_1^2$ can be introduced in the following forms:

$$P_2(x_1) := \varphi_2(T_2 \circ \Delta_1^2(x_1), \Delta_1^2(x_1))$$
$$P_1(x_2) := \varphi_1(T_1 \circ \Delta_2^1(x_2), \Delta_2^1(x_2)). \tag{2.4}$$

Then, the *Poincaré return map* $P : \tilde{S}_2^1 \to S_2^1$ is defined as

$$P(x_2) := P_2 \circ P_1(x_2). \tag{2.5}$$

The following theorem is an important result which allows us to consider P as the Poincaré return map for a system with impulse effects. Consequently, the results developed for the existence and stability analysis of systems with impulse effects can be applied to hybrid systems with two continuous phases.

Theorem 2.1 (P as the Poincaré Map for an Impulsive System) *[18, p. 95]*
Assume that hypothesis H1 is satisfied and P denotes the Poincaré return map for the autonomous hybrid model Σ. Then, P is well-defined and continuous. If Σ is also C^1, P is continuously differentiable. Moreover, P is the Poincaré return map for the autonomous system with impulse effects $\Sigma_{ie}(\mathcal{X}_2, \mathcal{S}, \Delta, f_2)$, where $\mathcal{S} := \tilde{S}_2^1$ and $\Delta(x_2) := \Delta_1^2 \circ P_1(x_2)$.

It is remarkable that the system with impulse effects $\Sigma_{ie}(\mathcal{X}_2, \mathcal{S}, \Delta, f_2)$ can be expressed as

$$\Sigma_{ie} : \begin{cases} \dot{x}_2 = f_2(x_2) & x_2^- \notin \mathcal{S} \\ x_2^+ = \Delta(x_2^-) & x_2^- \in \mathcal{S}. \end{cases} \tag{2.6}$$

The definition of solutions for the system Σ_{ie} is similar to that presented for the hybrid system Σ. Following the results of Ref. [18, p. 88], it can be shown that under hypothesis H1 the switching map Δ is continuous. Moreover, if the hybrid system Σ is C^1, Δ is also C^1. The geometric description of Theorem 2.1 is illustrated in Fig. 2.1. Let $\mathcal{O}$ be a periodic orbit of the hybrid system Σ transversal to $\tilde{S}_1^2$ and $\tilde{S}_2^1$. To define what we mean by the notion of "transversality," let $\mathcal{O}_1 := \mathcal{O} \cap \mathcal{X}_1$ and $\mathcal{O}_2 := \mathcal{O} \cap \mathcal{X}_2$. In addition, for the later purposes, define $\{x_1^*\} := \overline{\mathcal{O}_1} \cap \tilde{S}_1^2$, $\{x_2^*\} := \overline{\mathcal{O}_2} \cap \tilde{S}_2^1$, $T_1^* := T_1 \circ \Delta_2^1(x_2^*)$, and $T_2^* := T_2 \circ \Delta_1^2(x_1^*)$, where $\overline{\mathcal{O}_1}$ and $\overline{\mathcal{O}_2}$ represent the closure sets of $\mathcal{O}_1$ and $\mathcal{O}_2$, respectively. The periodic orbit $\mathcal{O} = \mathcal{O}_1 \cup \mathcal{O}_2$ is said to be *transversal* to $\mathcal{S}_1^2$ and $\mathcal{S}_2^1$ if (i) $\{x_1^*\}$ and $\{x_2^*\}$ are singletons and (ii) $L_{f_1} H_1^2(x_1^*), L_{f_2} H_2^1(x_2^*) \neq 0$, where $L_{f_i} H_i^j(x_i^*) := \frac{\partial H_i^j}{\partial x_i}(x_i^*) f_i(x_i^*)$ for $i, j \in \{1, 2\}$ and $i \neq j$.

By the construction of the system Σ_{ie}, $\mathcal{O}_2$ is a periodic orbit of the system Σ_{ie}. Conversely, if $\mathcal{O}_2$ is the periodic orbit of the system Σ_{ie}, hypothesis H1 implies that there exists a unique solution of the differential equation $\dot{x}_1 = f_1(x_1)$ with the initial condition $\Delta_2^1(x_2^*)$ and, as a consequence, $\mathcal{O}$ is the only periodic orbit of the system Σ such that $\mathcal{O} \cap \mathcal{X}_2 = \mathcal{O}_2$.

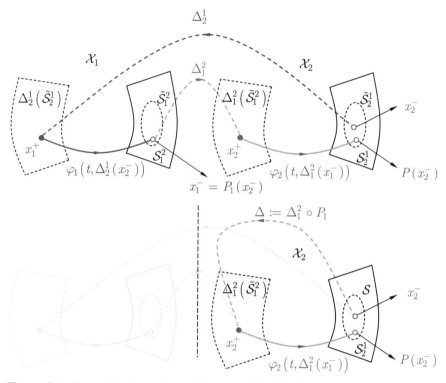

Figure 2.1 Geometric description of Theorem 2.1. The Poincaré return map of the autonomous hybrid system $\Sigma(\mathcal{X}_1, \mathcal{X}_2, \mathcal{S}_1^2, \mathcal{S}_2^1, \Delta_1^2, \Delta_2^1, f_1, f_2)$, $P : \tilde{\mathcal{S}}_2^1 \rightarrow \mathcal{S}_2^1$, is also the Poincaré return map for the autonomous system with impulse effects $\Sigma_{ie}(\mathcal{X}_2, \mathcal{S}, \Delta, f_2)$, where $\mathcal{S} := \tilde{\mathcal{S}}_2^1$ and $\Delta(x_2) := \Delta_1^2 \circ P_1(x_2)$. (See the color version of this figure in color plates section.)

Now we are in a position to introduce one of the fundamental theorems of this chapter. This theorem is basically a generalization of the method of Poincaré sections for systems with impulse effects. It also establishes an equivalence between the stability analysis of periodic orbits of Σ_{ie} and corresponding equilibrium points of the discrete-time system $x_2[k + 1] = P(x_2[k])$ in the state space $\mathcal{S}$.

Theorem 2.2 (Method of Poincaré Sections) *[18, p. 89]*[1] *Assume that hypothesis H1 is satisfied. Then, the following statements are true:*

1. *x_2^* is a stable (asymptotically stable) equilibrium point of $x_2[k + 1] = P(x_2[k])$ if and only if the orbit $\mathcal{O}_2$ is stable (asymptotically stable) for the impulsive system Σ_{ie}.*

---

[1] This theorem is a restatement of Theorem 4.1 of Ref. [18, p. 89].

2. *If the hybrid system Σ is also continuously differentiable, then, x_2^* is an exponentially equilibrium point of $x_2[k+1] = P(x_2[k])$ if and only if the orbit $\mathcal{O}_2$ is exponentially stable for the impulsive system Σ_{ie}.*

Using Theorem 2.2, the stability analysis of the transversal periodic orbit $\mathcal{O}_2$ for the system with impulse effects Σ_{ie} can be translated into the stability analysis of the equilibrium point x_2^* for the discrete-time system $x_2[k+1] = P(x_2[k])$ defined on the state space $\mathcal{S}$. Next, suppose that the periodic orbit $\mathcal{O}_2$ is stable for the system with impulse effects Σ_{ie}. What can we say about the stability behavior of the periodic orbit $\mathcal{O}$ for the hybrid system Σ? To investigate the stability properties of the periodic orbit $\mathcal{O}_2$ on the basis of stability of the periodic orbit $\mathcal{O}$, we present the following theorem.

Theorem 2.3 (Equivalence of the Stability Behavior) *Assume that hypothesis H1 is satisfied. Then, the following statements are true:*

1. *$\mathcal{O}$ is stable (asymptotically stable) for the hybrid system Σ if and only if $\mathcal{O}_2$ is stable (asymptotically stable) for the system with impulse effects Σ_{ie}.*
2. *If the hybrid system Σ is also continuously differentiable, then $\mathcal{O}$ is exponentially stable for the hybrid system Σ if and only if $\mathcal{O}_2$ is exponentially stable for the system with impulse effects Σ_{ie}.*

Proof. The first statement is immediate, so only the second statement is proved. Moreover, since the necessity of the second statement is straightforward, the sufficiency is proved.

From Definition 2.5, if the periodic orbit $\mathcal{O}$ is exponentially stable,

$$\text{dist}(\varphi(t, x), \mathcal{O}) \leq N \, \text{dist}(x, \mathcal{O}) \exp(-\gamma t) \tag{2.7}$$

for every $t \geq 0$ and $x \in \mathcal{N}(\mathcal{O})$. Also from the definition of the distance function on $\mathcal{X}$ [18, p. 93], inequality (2.7) can be expressed as

$$\begin{cases} \text{dist}(\varphi(t, x), \mathcal{O}_2) \leq Nd \, \exp(-\gamma t), & \varphi(t, x) \in \mathcal{X}_2 \\ \text{dist}(\varphi(t, x), \mathcal{O}_1) \leq Nd \, \exp(-\gamma t), & \varphi(t, x) \in \mathcal{X}_1 \end{cases} \tag{2.8}$$

for all $t \geq 0$, where $d := \text{dist}(x, \mathcal{O})$. Without loss of generality, assume that $x \in \tilde{\mathcal{X}}_2 \subset \mathcal{X}_2$, where $\tilde{\mathcal{X}}_2$ is the set of all points $x_2 \in \mathcal{X}_2$ for which there exists a closed discrete subset $\mathcal{T}(x_2) := \{0 = t_0 < t_1 < ... < t_j < ...\} \subset [0, \infty)$ determining the switching times corresponding to the initial condition x_2, and a solution $\varphi(., x_2) : [0, \infty) \to \mathcal{X}$ such that $\varphi(t, x_2) \in \mathcal{X}_2$ for all $t \in [t_{2k}, t_{2k+1}), k = 0, 1, 2, \cdots$ and $\varphi(t, x_2) \in \mathcal{X}_1$ for all $t \in [t_{2k+1}, t_{2k+2}), k = 0, 1, 2, \cdots$. Hence, inequality (2.8) can be rewritten as follows

$$\begin{cases} \text{dist}(\varphi(t, x), \mathcal{O}_2) \leq Nd \, \exp(-\gamma t), & t \in [t_{2k}, t_{2k+1}) \\ \text{dist}(\varphi(t, x), \mathcal{O}_1) \leq Nd \, \exp(-\gamma t), & t \in [t_{2k+1}, t_{2k+2}). \end{cases} \tag{2.9}$$

Since in the construction of the system with impulse effects Σ_{ie}, the time duration of phase 1 is omitted, the set of switching times for the corresponding solution of the system Σ_{ie} can be expressed as $\mathcal{T}_{ie}(x) := \{\bar{t}_k\}_{k=0}^{\infty}$, where

$$\bar{t}_0 := t_0 = 0$$
$$\bar{t}_1 := t_1$$
$$\bar{t}_2 := \bar{t}_1 + t_3 - t_2 = t_1 + t_3 - t_2$$
$$\bar{t}_3 := \bar{t}_2 + t_5 - t_4 = t_1 + t_3 + t_5 - t_2 - t_4$$
$$\vdots$$
$$\bar{t}_k := \bar{t}_{k-1} + t_{2k-1} - t_{2k-2} = t_1 + \sum_{i=1}^{k-1}(t_{2i+1} - t_{2i}).$$

In addition, let us define

$$\psi(\bar{t}, x) := \varphi(\bar{t} - \bar{t}_k + t_{2k}, x), \qquad \bar{t}_k \leq \bar{t} < \bar{t}_{k+1} \tag{2.10}$$

for $k = 0, 1, 2, \cdots$. By construction, if $\varphi(t, x), t \geq 0$ is the solution of the hybrid system Σ, then $\psi(\bar{t}, x), \bar{t} \geq 0$ is the corresponding solution of the impulsive system Σ_{ie}. Due to the fact that $\mathcal{O}_2$ is an exponentially stable periodic orbit of Σ_{ie}, there exist positive scalers N_2 and γ_2 such that

$$\text{dist}(\psi(\bar{t}, x), \mathcal{O}_2) \leq N_2 d \, \exp(-\gamma_2 \bar{t}) \tag{2.11}$$

for every $\bar{t} \geq 0$. Inequality (2.11) in combination with equation (2.10) results in

$$\text{dist}(\varphi(t, x), \mathcal{O}_2) \leq N_2 d \, \exp(-\gamma_2(t + \bar{t}_k - t_{2k}))$$
$$= N_2 d \, \exp(-\gamma_2 t) \, \exp\left(\gamma_2 \sum_{i=1}^{k}(t_{2i} - t_{2i-1})\right) \tag{2.12}$$
$$\leq N_2 d \, \exp(-\gamma_2 t) \, \exp(k\gamma_2 T_{1,\max})$$

for any $t \in [t_{2k}, t_{2k+1}), k = 0, 1, 2, \cdots$, where

$$T_{1,\max}(x) := \sup_{i \geq 1}(t_{2i} - t_{2i-1})$$

is the supremum of all time durations of phase 1 corresponding to the trajectory $\varphi(t, x)$. Furthermore, $t_{2k} \geq k T_{\min}$ for $k = 0, 1, 2, \cdots$, where

$$T_{\min}(x) := \inf_{i \geq 1}(t_{2i} - t_{2i-2})$$

is the infimum of all time durations corresponding to the steps of $\varphi(t, x)$. Thus, for every $t \in [t_{2k}, t_{2k+1})$, $k \leq \frac{t}{T_{\min}}$ and consequently, $t - kT_{1,\max} \geq (1 - \frac{T_{1,\max}}{T_{\min}})t$. From Lemma C.1 of Ref. [18, p. 439], hypothesis H1 implies that the functions T_1 and T_2 are continuous on the sets $\Delta_2^1(\tilde{S}_2^1)$ and $\Delta_1^2(\tilde{S}_1^2)$, respectively. This fact in combination with the following inequality which holds on the periodic orbit $\mathcal{O}$,

$$T_{1,\max} = T_1^* < T^* = T_{\min}$$

implies that there exists $\bar{\varepsilon} > 0$ such that for any $0 < \varepsilon < \bar{\varepsilon}$ and $x \in \mathcal{N}_\varepsilon(\mathcal{O}) \cap \tilde{X}_2$, $T_{1,\max}(x) < T_{\min}(x)$, where $\mathcal{N}_\varepsilon(\mathcal{O})$ is an ε-neighborhood of $\mathcal{O}$. Thus, for every $0 < \varepsilon < \bar{\varepsilon}$ and $x \in \mathcal{N}_\varepsilon(\mathcal{O}) \cap \tilde{X}_2$, (2.12) can be expressed as

$$\begin{aligned} \mathrm{dist}(\varphi(t, x), \mathcal{O}_2) &\leq N_2 d \, \exp(-\gamma_2(t - kT_{1,\max})) \\ &\leq \bar{N}_2 d \, \exp(-\bar{\gamma}_2 t) \end{aligned} \tag{2.13}$$

for all $t \in [t_{2k}, t_{2k+1})$, $k = 0, 1, 2, \cdots$, where $\bar{N}_2 := N_2$ and $\bar{\gamma}_2 := \gamma_2(1 - \frac{T_{1,\max}}{T_{\min}})$.
Next, let us define $\{x_2[k]\}_{k=1}^\infty$ and $\{x_1[k]\}_{k=1}^\infty$ by

$$\begin{aligned} x_2[k] &:= \overline{\varphi(t_{2k-1}, x)} \cap \mathcal{S}_2^1 \\ x_1[k] &:= \overline{\varphi(t_{2k}, x)} \cap \mathcal{S}_1^2, \end{aligned} \tag{2.14}$$

where $\overline{\varphi}$ is the set closure of φ. Since the hybrid system Σ is C^1, f_1 and Δ_2^1 are Lipschitz continuous with Lipschitz constants L_1 and L_2^1 on some *convex* subsets $\mathbf{X}_1 \subset \mathcal{X}_1$ and $\mathbf{S}_2^1 \subset \mathcal{S}_2^1$, respectively such that $\mathcal{O}_1 \subset \mathbf{X}_1$ and $x_2^* \in \mathbf{S}_2^1$ [88, Lemma 3.1, p. 89]. Thus, using the standard results for continuous dependence on initial states of the solutions of $\dot{x}_1 = f_1(x_1)$ [88, Theorem 3.4, p. 96], for every $t \in [t_{2k+1}, t_{2k+2})$, $k = 0, 1, 2, \cdots$,

$$\begin{aligned} \mathrm{dist}(\varphi(t, x), \mathcal{O}_1) &\leq \left\| \varphi_1\left(t, \Delta_2^1(x_2[k+1])\right) - \varphi_1\left(t, \Delta_2^1(x_2^*)\right) \right\| \\ &\leq \left\| \Delta_2^1(x_2[k+1]) - \Delta_2^1(x_2^*) \right\| \exp(L_1(t - t_{2k+1})) \\ &\leq L_2^1 \left\| x_2[k+1] - x_2^* \right\| \exp(L_1(t_{2k+2} - t_{2k+1})) \\ &\leq L_2^1 \left\| x_2[k+1] - x_2^* \right\| \exp(L_1 T_{1,\max}). \end{aligned} \tag{2.15}$$

Furthermore, since Σ is C^1 and $\mathcal{O}_2$ is an exponentially stable periodic orbit of Σ_{ie} transversal to $\mathcal{S}$, Theorem 2.2 implies that there exist scalers $\tilde{N}_2 > 0, 0 < \tilde{\gamma}_2 < 1$ and $r > 0$ such that for any $x_2[1] \in B_r(x_2^*) \cap \mathcal{S}$,

$$\| x_2[k+1] - x_2^* \| \leq \tilde{N}_2 \| x_2[1] - x_2^* \| (\tilde{\gamma}_2)^k, \qquad k = 1, 2, \cdots, \tag{2.16}$$

where $B_r(x_2^*) := \{x_2 \in \mathbb{R}^{n_2} | \|x_2 - x_2^*\| \leq r\}$. Next, let $T_{\max} := T_{2,\max} + T_{1,\max}$, where

$$T_{2,\max} := \sup_{i \geq 1}(t_{2i-1} - t_{2i-2})$$

is the supremum of all time durations of phase 2 corresponding to $\varphi(t, x)$. Then, $t_{2k+1} \leq t < t_{2k+2} \leq (k + 1)T_{\max}$ and consequently from inequality (2.16),

$$\|x_2[k + 1] - x_2^*\| \leq \tilde{N}_2 \|x_2[1] - x_2^*\| \exp\left(\ln \tilde{\gamma}_2\left(\frac{t}{T_{\max}} - 1\right)\right). \tag{2.17}$$

This fact in combination with inequality (2.15) also results in

$$\mathrm{dist}(\varphi(t, x), \mathcal{O}_1) \leq \bar{N}_1 d \exp(-\bar{\gamma}_1 t)$$

for all $t \in [t_{2k+1}, t_{2k+2})$, $k = 0, 1, 2, \cdots$, where $\bar{\gamma}_1 := -\frac{\ln \tilde{\gamma}_2}{T_{\max}}$ and $\bar{N}_1$ is such that

$$\bar{N}_1 \geq \eta \, \frac{\tilde{N}_2 \, L_2^1 \, \exp(L_1 T_{1,\max})}{\tilde{\gamma}_2},$$

where $\eta > 0$ is an arbitrary scalar. Choosing $N = \max_{x \in \mathcal{N}(\mathcal{O}_2)}(\bar{N}_1, \bar{N}_2)$, $\gamma = \min_{x \in \mathcal{N}(\mathcal{O}_2)}(\bar{\gamma}_1, \bar{\gamma}_2)$ and $\mathcal{N}(\mathcal{O}_2)$ as the following compact set

$$\begin{aligned}\mathcal{N}(\mathcal{O}_2) &= \{x \in \mathcal{X}_2 | \|x_2[1] - x_2^*\| \leq \eta \, \mathrm{dist}(x, \mathcal{O}_2)\} \\ &= \{x \in \mathcal{X}_2 | \|\varphi_2(T_2(x), x) - x_2^*\| \leq \eta \, \mathrm{dist}(x, \mathcal{O}_2)\}\end{aligned}$$

completes the proof. ∎

Theorems 2.2 and 2.3 establish an analytical approach to investigate the stability behavior of the transversal periodic orbit $\mathcal{O}$. However, this approach is particularly useful when we have a closed-form expression for the Poincaré return map P. For example, it is often convenient to check the exponential stability in terms of eigenvalues of the Jacobian matrix $DP(x_2^*)$. Under hypothesis H1, if the hybrid system Σ is C^1, the Poincaré return map is continuously differentiable. To obtain a closed-form expression for the Jacobian matrix $DP(x_2^*)$, define

$$\begin{aligned}\Phi_1(t, x) &:= D_x\varphi_1(t, x) \\ \Phi_2(t, x) &:= D_x\varphi_2(t, x)\end{aligned}$$

as the *trajectory sensitivity matrices*. From Ref. [89, p. 316], since the periodic orbit $\mathcal{O}$ is transversal to $\mathcal{S}_1^2$ and $\mathcal{S}_2^1$, the functions T_1 and T_2 are differentiable at the points

$\Delta_2^1(x_2^*)$ and $\Delta_1^2(x_1^*)$, respectively, and

$$DT_1\left(\Delta_2^1(x_2^*)\right) = \frac{-1}{L_{f_1} H_1^2(x_1^*)} \frac{\partial H_1^2}{\partial x_1}(x_1^*) \, \Phi_1\left(T_1^*, \Delta_2^1(x_2^*)\right)$$

$$DT_2\left(\Delta_1^2(x_1^*)\right) = \frac{-1}{L_{f_2} H_2^1(x_2^*)} \frac{\partial H_2^1}{\partial x_2}(x_2^*) \, \Phi_2\left(T_2^*, \Delta_1^2(x_1^*)\right),$$

(2.18)

which in turn result in

$$DP(x_2^*) = \left[I_{n_2 \times n_2} - \frac{f_2(x_2^*)\frac{\partial H_2^1}{\partial x_2}(x_2^*)}{L_{f_2} H_2^1(x_2^*)} \right] \Phi_2\left(T_2^*, \Delta_1^2(x_1^*)\right) \frac{\partial \Delta}{\partial x_2}(x_2^*), \qquad (2.19)$$

where

$$\frac{\partial \Delta}{\partial x_2}(x_2^*) = \frac{\partial \Delta_1^2}{\partial x_1}(x_1^*) \left[I_{n_1 \times n_1} - \frac{f_1(x_1^*)\frac{\partial H_1^2}{\partial x_1}(x_1^*)}{L_{f_1} H_1^2(x_1^*)} \right] \Phi_1\left(T_1^*, \Delta_2^1(x_2^*)\right) \frac{\partial \Delta_2^1}{\partial x_2}(x_2^*).$$

We observe that the domain of the Jacobian matrix is the $(n_2 - 1)$-dimensional tangent space $T_{x_2^*}S$. Therefore, the proper notation for the Jacobian matrix is $DP|_{T_{x_2^*}S}(x_2^*)$. However, for clarity, we will not use it. From equation (2.19), we require the trajectory sensitivity matrices $\Phi_1(T_1^*, \Delta_2^1(x_2^*))$ and $\Phi_2(T_2^*, \Delta_1^2(x_1^*))$ to compute the Jacobian $DP(x_2^*)$. To obtain the trajectory sensitivity matrix $\Phi_i(T_i^*, \Delta_j^i(x_j^*))$, $i, j = 1, 2, j \neq i$, the well-known *variational equation* [89, p. 305] is appended to the original differential equation during the continuous phase, that is,

$$\begin{bmatrix} \dot{x}_i \\ \dot{\Phi}_i \end{bmatrix} = \begin{bmatrix} f_i(x_i) \\ D_{x_i} f_i(x_i)\Phi_i \end{bmatrix}$$

which is integrated over the time interval $[0, T_i^*]$ with the initial condition

$$\begin{bmatrix} \Delta_j^i(x_j^*) \\ I_{n_i \times n_i} \end{bmatrix}.$$

Thus, the computations required to check the stability are complex. This situation motivates us to look for ways to simplify the stability analysis.

2.3 LOW-DIMENSIONAL STABILITY ANALYSIS

Theorems 2.2 and 2.3 present a natural and analytical approach to investigate the stability behavior of the transversal periodic orbit $\mathcal{O}$ for the hybrid system Σ. As mentioned in the previous section, when the Poincaré return map can be written down in a closed-form expression, this approach is useful. However, determining the Poincaré return map requires the solutions of the differential equations $\dot{x}_i = f_i(x_i)$, $i = 1, 2$ which in general can only be computed by applying numerical integration algorithms. Therefore, in the general case, the Poincaré return map cannot be expressed in a closed-form expression. In order to simplify the stability analysis, this section presents special circumstances where the stability behavior of the periodic orbit $\mathcal{O}$ can be tested by low-dimensional tools such as the restricted Poincaré return maps. The main ideas and results in developing the notion of restricted Poincaré return maps, which are employed for inducing asymptotically stable periodic solutions in walking and running of the biped robots, are due to Grizzle et al. [18, 46, 52]. We first present the following definitions.

Definition 2.6 (Hybrid Invariance for Impulsive Systems) *The set $Z_2 \subset \mathcal{X}_2$ is said to be hybrid invariant for the system with impulse effects $\Sigma_{ie}(\mathcal{X}_2, \mathcal{S}, \Delta, f_2)$ if*

1. *Z_2 is forward invariant under the flow of the differential equation $\dot{x}_2 = f_2(x_2)$, that is, for every $x_2 \in Z_2$, there exists $t_f(x_2) > 0$ such that $\varphi_2(t, x_2) \in Z_2$ for $t \in [0, t_f)$; and*
2. *Z_2 is impact invariant, that is, $\mathcal{S} \cap Z_2 \neq \phi$ and $\Delta(\mathcal{S} \cap Z_2) \subset Z_2$.*

Definition 2.7 (Hybrid Invariance for Hybrid Systems) *The sets $Z_1 \subset \mathcal{X}_1$ and $Z_2 \subset \mathcal{X}_2$ are said to be hybrid invariant for the hybrid system*

$$\Sigma\left(\mathcal{X}_1, \mathcal{X}_2, \mathcal{S}_1^2, \mathcal{S}_2^1, \Delta_1^2, \Delta_2^1, f_1, f_2\right)$$

if

1. *for every $i \in \{1, 2\}$, Z_i is forward invariant under the flow of the differential equation $\dot{x}_i = f_i(x_i)$ and*
2. *for every $i, j \in \{1, 2\}$ and $i \neq j$, $\mathcal{S}_i^j \cap Z_i \neq \phi$ and $\Delta_i^j(\mathcal{S}_i^j \cap Z_i) \subset Z_j$.*

Definition 2.8 (Finite-Time Attractiveness) *For $i \in \{1, 2\}$, the settling time to the set Z_i, $T_{Z_i} : \mathcal{X}_i \to \mathbb{R} \cup \{\infty\}$, is defined to be the infimum of all times for which the trajectory $\varphi_i(t, x_i)$ arrives at the set Z_i and remain there until the maximal time of existence. Note that if there is not such a time, $T_{Z_i}(x_i)$ is defined to be ∞. Moreover, the set Z_i is said to be locally continuously finite-time attractive if (i) Z_i is forward invariant under the flow of the differential equation $\dot{x}_i = f_i(x_i)$ and (ii) there exists an open set $\mathcal{N}_i$ containing Z_i such that for every $x_i \in \mathcal{N}_i$, the function T_{Z_i} at x_i is well-defined (i.e., finite) and continuous [18, p. 96].*

Now assume that the sets $Z_i \in \mathcal{X}_i$, $i = 1, 2$ are embedded submanifolds of the state spaces $\mathcal{X}_i$, $i = 1, 2$ satisfying the following hypotheses:

(H2) For every $i, j \in \{1, 2\}$ and $i \neq j$, $S_i^j \cap Z_i$ is an embedded submanifold of $\mathcal{X}_i$ such that $\dim(S_i^j \cap Z_i) = \dim(Z_i) - 1$;

(H3) For every $i \in \{1, 2\}$, Z_i is locally continuously finite-time attractive;

(H4) Z_1 and Z_2 are hybrid invariant for the autonomous hybrid system Σ; and

(H5) For every $i \in \{1, 2\}$, $\mathcal{O}_i \subset Z_i$, i.e., $\mathcal{O}_i$ is an integral curve of the restriction dynamics $\dot{x}_i = f_i|_{Z_i}(x_i)$.

Hypothesis H3 implies finite-time convergence to the manifolds Z_1 and Z_2. Due to the fact that finite-time convergence implies nonuniqueness of solutions in reverse time, if the set $Z_i, i = 1, 2$ is finite-time attractive, it is not possible that the vector field $f_i : \mathcal{X}_i \to T\mathcal{X}_i, i = 1, 2$ is Lipschitz continuous. By hypothesis H4, solutions of the hybrid system Σ initialized in $Z := Z_1 \cup Z_2$ remain in Z until the maximal time of existence. Therefore, we can construct the following reduced-order hybrid model to study the stability behavior of the transversal periodic orbit $\mathcal{O}$

$$\Sigma_1|_{Z_1} : \begin{cases} & Z_1 \subset \mathcal{X}_1 \\ \mathcal{F}_1|_{Z_1} : & \dot{z}_1 = f_1|_{Z_1}(z_1) \\ & S_1^2 \cap Z_1 = \{z_1 \in Z_1 \mid H_1^2|_{Z_1}(z_1) = 0\} \\ \mathcal{T}_1^2|_{Z_1} : & z_2^+ = \delta_1^2(z_1^-) \end{cases}$$

(2.20)

$$\Sigma_2|_{Z_2} : \begin{cases} & Z_2 \subset \mathcal{X}_2 \\ \mathcal{F}_2|_{Z_2} : & \dot{z}_2 = f_2|_{Z_2}(z_2) \\ & S_2^1 \cap Z_2 = \{z_2 \in Z_2 \mid H_2^1|_{Z_2}(z_2) = 0\} \\ \mathcal{T}_2^1|_{Z_2} : & z_1^+ = \delta_2^1(z_2^-), \end{cases}$$

where for $i, j \in \{1, 2\}$ and $i \neq j$, $f_i|_{Z_i}$ and $H_i^j|_{Z_i}$ denote the restrictions of f_i and H_i^j to the manifold Z_i, respectively. Also, $\mathcal{F}_i|_{Z_i}$ represents the flow of the restriction dynamics $\dot{z}_i = f_i|_{Z_i}(z_i)$. Moreover, $\mathcal{T}_i^j|_{Z_i}$ denotes the restricted switching maps $\delta_i^j : S_i^j \cap Z_i \to Z_j$ defined by the continuously differentiable laws $\delta_i^j := \Delta_i^j|_{S_i^j \cap Z_i}$. For simplicity, the reduced-order hybrid model (2.20), which is called the *hybrid restriction dynamics*, is denoted by the 8-tuple

$$\Sigma|_Z(Z_1, Z_2, S_1^2 \cap Z_1, S_2^1 \cap Z_2, \delta_1^2, \delta_2^1, f_1|_{Z_1}, f_2|_{Z_2}).$$

If the hybrid system Σ satisfies hypothesis H1, so does the hybrid restriction dynamics $\Sigma|_Z$, that is, the vector fields $f_i|_{Z_i} : Z_i \to TZ_i, i = 1, 2$ are continuous. Also, the solutions of the restriction dynamics $\dot{z}_i = f_i|_{Z_i}(z_i), i = 1, 2$ for every initial condition in the state space $Z_i, i = 1, 2$ exist and are unique. Moreover, these solutions depend continuously on the initial conditions. Thus, for every initial condition in the state space Z, there exists a unique solution of the hybrid restriction dynamics $\Sigma|_Z$. From hypothesis

H2, for every i, $j \in \{1, 2\}$ and $i \neq j$, $\mathcal{S}_i^j \cap Z_i$ is an embedded submanifold of Z_i with dimension one less than the dimension of Z_i. In addition, $\mathcal{S}_j^i \cap \delta_i^j(\mathcal{S}_i^j \cap Z_i) = \phi$, which in turn implies that a switching does not occur immediately after another switching. Figure 2.2 illustrates the geometry of the hybrid restriction dynamics $\Sigma|_Z$.

The solution of $\Sigma|_Z$ initialized from $z \in Z$ can be expressed as $\varphi|_Z(t, z)$, where $\varphi(t, z)$ represents the corresponding solution of Σ. Thus, the *restricted Poincaré return map* for $\Sigma|_Z$ can be defined as $\rho : \tilde{\mathcal{S}}_2^1 \cap Z_2 \to \mathcal{S}_2^1 \cap Z_2$ by

$$\rho(z_2) := \rho_2 \circ \rho_1(z_2), \qquad (2.21)$$

where $\rho_2 : \tilde{\mathcal{S}}_1^2 \cap Z_1 \to \mathcal{S}_2^1 \cap Z_2$ and $\rho_1 : \tilde{\mathcal{S}}_2^1 \cap Z_2 \to \tilde{\mathcal{S}}_1^2 \cap Z_1$ are given by

$$\begin{aligned}
\rho_2(z_1) &:= \varphi_2|_{Z_2}\left(T_2|_{Z_2} \circ \delta_1^2(z_1), \delta_1^2(z_1)\right) \\
\rho_1(z_2) &:= \varphi_1|_{Z_1}\left(T_1|_{Z_1} \circ \delta_2^1(z_2), \delta_2^1(z_2)\right).
\end{aligned} \qquad (2.22)$$

By the construction of $\Sigma|_Z$, $\rho(z_2) = P|_Z(z_2)$, where $P|_Z$ is the restriction of the Poincaré return map of the full-dimensional hybrid system Σ to Z. Applying Theorem 2.1 implies that ρ is also the Poincaré return map for the reduced-order system with impulse effects $\Sigma_{ie}|_{Z_2}(Z_2, \mathcal{S} \cap Z_2, \delta, f_2|_{Z_2})$, where $\delta(z_2) := \delta_1^2 \circ \rho_1(z_2)$. By hypothesis H5 and the construction procedure, the transversal periodic orbit $\mathcal{O}$ of the hybrid system Σ is the periodic orbit of the hybrid restriction dynamics $\Sigma|_Z$ which is also transversal to $\mathcal{S}_1^2 \cap Z_1$ and $\mathcal{S}_2^1 \cap Z_2$ (see Fig. 2.2). Now we are in a position to present the fundamental theorem of this section. This theorem establishes an equivalence between the stability analysis of the transversal periodic orbit

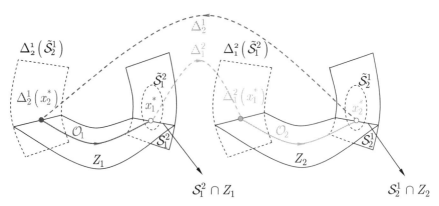

Figure 2.2   Geometric description of the hybrid restriction dynamics. By hypotheses H2–H5 and the construction of the hybrid restriction dynamics, the transversal periodic orbit $\mathcal{O}$ of the hybrid system Σ is also the periodic orbit of the hybrid restriction dynamics $\Sigma|_Z$ which is transversal to $\mathcal{S}_1^2 \cap Z_1$ and $\mathcal{S}_2^1 \cap Z_2$. (See the color version of this figure in color plates section.)

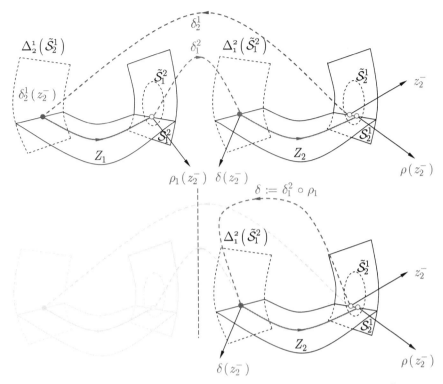

Figure 2.3 Geometric description of the restricted Poincaré return map $\rho : \tilde{\mathcal{S}}_2^1 \cap Z_2 \to \mathcal{S}_2^1 \cap Z_2$. By hypotheses H2–H5 and the construction of $\Sigma|_Z$, $\rho(z_2) = P|_Z(z_2)$, where $P|_Z$ is the restriction of the Poincaré return map of the full-dimensional hybrid system Σ to Z. By applying Theorem 2.1, it follows that ρ is also the Poincaré return map for the reduced-order system with impulse effects $\Sigma_{ie}|_{Z_2}(Z_2, \mathcal{S} \cap Z_2, \delta, f_2|_{Z_2})$, where $\delta(z_2) := \delta_1^2 \circ \rho_1(z_2)$. (See the color version of this figure in color plates section.)

$\mathcal{O}$ of the full-dimensional hybrid system Σ satisfying hypotheses H1–H5 and the stability analysis of the equilibrium point x_2^* of the reduced-order discrete-time system $z_2[k + 1] = \rho(z_2[k])$ with the state space $\mathcal{S} \cap Z_2$ (see Fig. 2.3). Consequently, the stability behavior of the periodic orbit of the full-order hybrid model can be determined by low-dimensional tools which require less computation.

Theorem 2.4 (Low-Dimensional Stability Analysis) *[18, p. 99][2] Assume that the full-dimensional autonomous hybrid system $\Sigma(\mathcal{X}_1, \mathcal{X}_2, \mathcal{S}_1^2, \mathcal{S}_2^1, \Delta_1^2, \Delta_2^1, f_1 \, f_2)$ satisfies hypothesis H1. Furthermore, suppose that there exist embedded submanifolds $Z_i, i = 1, 2$ of the state spaces $\mathcal{X}_i, i = 1, 2$ satisfying hypotheses H2–H5. Then, the following statements are true:*

---

[2] This theorem is a restatement of Theorem 4.5 of Ref. [18, p. 99].

1. x_2^* is a stable (asymptotically stable) equilibrium point of $z_2[k+1] = \rho(z_2[k])$ if and only if the orbit $\mathcal{O}_2$ is stable (asymptotically stable) for the corresponding full-dimensional impulsive system $\Sigma_{ie}(\mathcal{X}_2, \mathcal{S}, \Delta, f_2)$.

2. If both $f_i|_{Z_i}, i = 1, 2$ are also continuously differentiable, then x_2^* is an exponentially stable equilibrium point of $z_2[k+1] = \rho(z_2[k])$ if and only if the orbit $\mathcal{O}_2$ is exponentially stable for the corresponding full-dimensional impulsive system $\Sigma_{ie}(\mathcal{X}_2, \mathcal{S}, \Delta, f_2)$.

Theorem 2.4 in combination with Theorem 2.3 immediately implies the following result.

Corollary 2.1 *Assume that the hybrid system $\Sigma(\mathcal{X}_1, \mathcal{X}_2, \mathcal{S}_1^2, \mathcal{S}_2^1, \Delta_1^2, \Delta_2^1, f_1, f_2)$ satisfies hypothesis H1. Furthermore, suppose that there exist embedded submanifolds $Z_i, i = 1, 2$ of the state spaces $\mathcal{X}_i, i = 1, 2$ satisfying hypotheses H2–H5. Then, the following statements are true:*

1. *x_2^* is a stable (asymptotically stable) equilibrium point of $z_2[k+1] = \rho(z_2[k])$ if and only if the orbit $\mathcal{O}$ is stable (asymptotically stable) for the full-dimensional hybrid system Σ.*

2. *If both $f_i|_{Z_i}, i = 1, 2$ are also continuously differentiable, then x_2^* is an exponentially stable equilibrium point of $z_2[k+1] = \rho(z_2[k])$ if and only if the orbit $\mathcal{O}$ is exponentially stable for the full-dimensional hybrid system Σ.*

2.4 STABILIZATION PROBLEM

Consider the open-loop hybrid system Σ_{ol} taking the following form

$$
\Sigma_{1,ol} : \begin{cases} \mathcal{X}_1 \subset \mathbb{R}^{n_1} \\ \mathcal{F}_1 : \quad \dot{x}_1 = f_1(x_1) + g_1(x_1)u \\ \quad\; \mathcal{S}_1^2 = \{x_1 \in \mathcal{X}_1 \mid H_1^2(x_1) = 0\} \\ \mathcal{T}_1^2 : \quad x_2^+ = \Delta_1^2(x_1^-) \end{cases}
$$

$$(2.23)$$

$$
\Sigma_{2,ol} : \begin{cases} \mathcal{X}_2 \subset \mathbb{R}^{n_2} \\ \mathcal{F}_2 : \quad \dot{x}_2 = f_2(x_2) + g_2(x_2)u \\ \quad\; \mathcal{S}_2^1 = \{x_2 \in \mathcal{X}_2 \mid H_2^1(x_2) = 0\} \\ \mathcal{T}_2^1 : \quad x_1^+ = \Delta_2^1(x_2^-), \end{cases}
$$

where $u \in \mathcal{U}$ is the control input vector. Moreover, $\mathcal{U} \subset \mathbb{R}^m$ called the *admissible control input region* is defined to be the set of all piecewise continuous functions

$t \mapsto u(t)$ with the property $\|u\|_{\mathcal{L}_\infty} := \sup_{t \geq 0} \|u(t)\| < u_{\max}$, where $u_{\max}$ is a positive scalar. Suppose that $\mathcal{O} = \mathcal{O}_1 \cup \mathcal{O}_2$ is an orbit corresponding to a period-one solution of the open-loop hybrid system Σ_{ol}. This section addresses the problem of asymptotic stabilization of $\mathcal{O}$ for the system Σ_{ol}. The main idea of this section has been taken from Ref. [57–59].

To asymptotically stabilize the orbit $\mathcal{O}$ for Σ_{ol}, a two-level control scheme is presented. At the first level of the control scheme, parameterized and time-invariant continuous feedback laws are employed during continuous phases $i = 1, 2$ to create a family of parameterized, finite-time attractive and forward invariant manifolds on which the differential equation $\dot{x}_i = f_i(x_i) + g_i(x_i)u$ is restricted. As mentioned in Section 2.3, this will reduce the complexity of the calculations required for obtaining the Poincaré return map. Let $Z_{1,\alpha}$ and $Z_{2,\beta}$ represent the parameterized manifolds created in the phases 1 and 2, respectively. Also, α and β denote the parameters of the controllers during phases 1 and 2 which takes values in the open sets $\mathcal{A}$ and $\mathcal{B}$. To show explicitly the dependence on the parameters α and β, the time-invariant feedback laws in phases 1 and 2 are denoted by $u_1(x_1; \alpha)$ and $u_2(x_2; \beta)$, respectively. With these control laws, the closed-loop dynamics of phases 1 and 2 can be given by $\dot{x}_1 = f_{1,\text{cl}}(x_1; \alpha)$ and $\dot{x}_2 = f_{2,\text{cl}}(x_2; \beta)$, where

$$f_{1,\text{cl}}(x_1; \alpha) := f_1(x_1) + g_1(x_1)u_1(x_1; \alpha)$$
$$f_{2,\text{cl}}(x_2; \beta) := f_2(x_2) + g_2(x_2)u_2(x_2; \beta).$$

Now assume that the following hypotheses are satisfied:

(H6) For every $\alpha \in \mathcal{A}$ and $\beta \in \mathcal{B}$,
  (a) the sets $\mathcal{S}_1^2 \cap Z_{1,\alpha}$ and $\mathcal{S}_2^1 \cap Z_{2,\beta}$ are independent of α and β, respectively. The common intersections are also denoted by $\mathcal{S}_1^2 \cap Z_1$ and $\mathcal{S}_2^1 \cap Z_2$. Furthermore, $\mathcal{S}_1^2 \cap Z_1$ and $\mathcal{S}_2^1 \cap Z_2$ are embedded submanifolds of $\mathcal{X}_1$ and $\mathcal{X}_2$ with the properties $\dim(\mathcal{S}_1^2 \cap Z_1) = \dim(Z_{1,\alpha}) - 1$ and $\dim(\mathcal{S}_2^1 \cap Z_2) = \dim(Z_{2,\beta}) - 1$;
  (b) $\Delta_1^2(\mathcal{S}_1^2 \cap Z_1) \subset Z_{2,\beta}$ and $\Delta_2^1(\mathcal{S}_2^1 \cap Z_2) \subset Z_{1,\alpha}$;
  (c) $Z_{1,\alpha}$ and $Z_{2,\beta}$ are locally continuously finite-time attractive for the closed-loop dynamics $\dot{x}_1 = f_{1,\text{cl}}(x_1; \alpha)$ and $\dot{x}_2 = f_{2,\text{cl}}(x_2; \beta)$, respectively; and

(H7) There exist $\alpha^* \in \mathcal{A}$ and $\beta^* \in \mathcal{B}$ such that $\mathcal{O}_1 \subset Z_{1,\alpha^*}$ and $\mathcal{O}_2 \subset Z_{2,\beta^*}$, that is, $\mathcal{O}_1$ and $\mathcal{O}_2$ are integral curves of the differential equations $\dot{x}_1 = f_{1,\text{cl}}(x_1; \alpha^*)$ and $\dot{x}_2 = f_{2,\text{cl}}(x_2; \beta^*)$, respectively.

Hypothesis H6 motivates us to define the parameterized restricted Poincaré return map for the closed-loop hybrid system as $\rho_{\alpha,\beta} : \mathcal{S}_2^1 \cap Z_2 \to \mathcal{S}_2^1 \cap Z_2$ by $\rho_{\alpha,\beta}(z_2) := \rho_{2,\beta} \circ \rho_{1,\alpha}(z_2)$, where $\rho_{1,\alpha} : \mathcal{S}_2^1 \cap Z_2 \to \mathcal{S}_1^2 \cap Z_1$ and $\rho_{2,\beta} : \mathcal{S}_1^2 \cap Z_1 \to \mathcal{S}_2^1 \cap Z_2$ are the parameterized versions of the maps defined in Section 2.3. Thus, to study the stabilization problem, we can define the following discrete-time system

$$z_2[k+1] = \rho(z_2[k]; \alpha[k], \beta[k]), \tag{2.24}$$

with the state space $S_2^1 \cap Z_2$ and the control inputs $\alpha[k]$ and $\beta[k]$, where $\rho(z_2; \alpha, \beta) := \rho_{\alpha,\beta}(z_2)$. Let us continue the problem of stabilizing the periodic orbit $\mathcal{O}$ for the open-loop hybrid system (2.23). To do this, assume that there exist continuous functions $\alpha_{cl} : S_2^1 \cap Z_2 \to \mathcal{A}$ and $\beta_{cl} : S_2^1 \cap Z_2 \to \mathcal{B}$ such that $\alpha_{cl}(x_2^*) = \alpha^*$ and $\beta_{cl}(x_2^*) = \beta^*$. Moreover, suppose that the equilibrium point x_2^* is asymptotically stable for the closed-loop discrete-time system

$$z_2[k+1] = \rho_{cl}(z_2[k]), \tag{2.25}$$

where $\rho_{cl}(z_2) := \rho(z_2; \alpha_{cl}(z_2), \beta_{cl}(z_2))$. Then, at the second level of the control scheme, the parameters of the feedback laws of phases 1 and 2 can be updated at the end of phase 2 by an *event-based update law*[3] in a step-to-step fashion. To make this notion more precise, the parameters α and β for the next step are updated by the following static laws

$$\alpha[k+1] = \alpha_{cl}(x_2[k]), \quad k = 1, 2, \cdots$$
$$\beta[k+1] = \beta_{cl}(x_2[k]), \quad k = 1, 2, \cdots,$$

where k denotes the step number and $x_2[k]$ was defined in equation (2.14). We observe that the parameters α and β are held constant during continuous phases, and consequently, the two-level control strategy will result in the following closed-loop hybrid system (see Fig. 2.4)

$$\Sigma_{1,cl}^{\alpha} : \begin{cases} & \mathcal{X}_1 \subset \mathbb{R}^{n_1} \\ \mathcal{F}_1 : & \dot{x}_1 = f_{1,cl}(x_1; \alpha) \\ & S_1^2 = \{x_1 \in \mathcal{X}_1 \mid H_1^2(x_1) = 0\} \\ \mathcal{T}_1^2 : & x_2^+ = \Delta_1^2(x_1^-) \end{cases}$$

$$\Sigma_{2,cl}^{\beta} : \begin{cases} & \mathcal{X}_2 \subset \mathbb{R}^{n_2} \\ \mathcal{F}_2 : & \dot{x}_2 = f_{2,cl}(x_2; \beta) \\ & S_2^1 = \{x_2 \in \mathcal{X}_2 \mid H_2^1(x_2) = 0\} \\ \mathcal{T}_2^1 : & \begin{bmatrix} x_1^+ \\ \alpha^+ \\ \beta^+ \end{bmatrix} = \begin{bmatrix} \Delta_2^1(x_2^-) \\ \alpha_{cl}(x_2^-) \\ \beta_{cl}(x_2^-) \end{bmatrix} . \end{cases} \tag{2.26}$$

---

[3] The terminology of an event-based update law is taken from Ref. [18, p. 199].

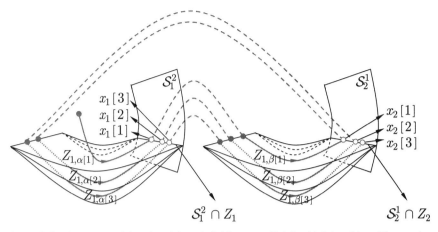

Figure 2.4 Geometry of the closed-loop hybrid system (2.26) which is achieved by employing the two-level control strategy. At the first level, the parameterized, time-invariant, and continuous feedback laws $u_1(x_1; \alpha)$ and $u_2(x_2; \beta)$ are employed to create the parameterized, finite-time attractive, and invariant manifolds $Z_{1,\alpha}$ and $Z_{2,\beta}$ during phases 1 and 2, respectively. This will result in the restriction dynamics of phases 1 and 2 as $\dot{x}_1 = f_{1,cl}|_{Z_{1,\alpha}}(x_1; \alpha)$ and $\dot{x}_2 = f_{2,cl}|_{Z_{2,\beta}}(x_2; \beta)$. At the second level of the control strategy, the parameters of the feedback laws in phases 1 and 2 are updated by an event-based update law at the end of phase 2 (i.e., in a step-to-step fashion). In particular, $\alpha[k+1] = \alpha_{cl}(x_2[k])$ and $\beta[k+1] = \beta_{cl}(x_2[k])$, where k denotes the step number. Consequently, the restricted Poincaré return map replaces the flow of the closed-loop hybrid system with the closed-loop discrete-time system $x_2[k+1] = \rho_{cl}(x_2[k])$. The graphs depict the three steps of the closed-loop hybrid system. It is assumed that the state of the system is initiated at a point off the manifold $Z_{1,\alpha[1]}$. Due to the fact that $Z_{1,\alpha}$ is finite-time attractive, the state of the system enters onto $Z_{1,\alpha}$ in a finite-time and remains in it throughout phase 1. When the state enters S_1^2 (at $x_1[1]$), a discrete event occurs, according to the transition map Δ_1^2. Hypothesis H6 implies that $\Delta_1^2(S_1^2 \cap Z_{1,\alpha[k]}) \subset Z_{2,\beta[k]}$ and consequently, during phase 2, the state of the system evolves in $Z_{2,\beta[1]}$ until it enters S_2^1 (at $x_2[1]$). At this point, a discrete event occurs, according to the transition map Δ_2^1. Moreover, event-based update law updates the parameters α and β for the next step as $\alpha[2] = \alpha_{cl}(x_2[1])$ and $\beta[2] = \beta_{cl}(x_2[1])$. These parameters are held constant during the second step and the process repeats. We observe that by hypothesis H6, $\Delta_2^1(S_2^1 \cap Z_{2,\beta[k]}) \subset Z_{1,\alpha[k+1]}$.

The following theorem is a precise statement concerning the equivalence between the stability behavior of the periodic orbit $\mathcal{O}$ for the closed-loop hybrid system (2.26) and the stability behavior of the equilibrium point x_2^* for the closed-loop discrete-time system (2.25).

Theorem 2.5 (Stabilization Policy) *Assume that the time-invariant continuous feedback laws $u_1 : \mathcal{X}_1 \times \mathcal{A} \to \mathcal{U}$ and $u_2 : \mathcal{X}_2 \times \mathcal{B} \to \mathcal{U}$ are chosen such that the families of the manifolds $\{Z_{1,\alpha}| \alpha \in \mathcal{A}\}$ and $\{Z_{2,\beta}| \beta \in \mathcal{B}\}$ satisfy hypotheses H6–H7. Furthermore, suppose that the static update laws $\alpha_{cl} : S_2^1 \cap Z_2 \to \mathcal{A}$ and*

$\beta_{cl} : \mathcal{S}_2^1 \cap Z_2 \to \mathcal{B}$ are continuous maps and the following additional hypothesis is satisfied:

(H8) For every $\alpha \in \mathcal{A}$ and $\beta \in \mathcal{B}$, the vector fields $f_{1,cl} : \mathcal{X}_1 \to T\mathcal{X}_1$ and $f_{2,cl} : \mathcal{X}_2 \to T\mathcal{X}_2$ are continuous. In addition, the solutions of the augmented differential equations

$$
\begin{bmatrix} \dot{x}_1 \\ \dot{\alpha} \end{bmatrix} = \begin{bmatrix} f_{1,cl}(x_1, \alpha) \\ 0 \end{bmatrix}
$$

$$
\begin{bmatrix} \dot{x}_2 \\ \dot{\beta} \end{bmatrix} = \begin{bmatrix} f_{2,cl}(x_2, \beta) \\ 0 \end{bmatrix}
$$

for every initial conditions in $\mathcal{X}_1 \times \mathcal{A}$ and $\mathcal{X}_2 \times \mathcal{B}$ are unique and depend continuously on the initial conditions.

Then, the following statements are true:

1. x_2^* is a stable (asymptotically stable) equilibrium point of $z_2[k+1] = \rho_{cl}(z_2[k])$ if and only if the orbit $\mathcal{O}$ is stable (asymptotically stable) for the closed-loop hybrid system (2.26).
2. If the static update laws $\alpha_{cl} : \mathcal{S}_2^1 \cap Z_2 \to \mathcal{A}$ and $\beta_{cl} : \mathcal{S}_2^1 \cap Z_2 \to \mathcal{B}$ and the restricted vector fields $f_{1,cl}|_{Z_{1,\alpha}} : Z_{1,\alpha} \to TZ_{1,\alpha}$ and $f_{2,cl}|_{Z_{2,\beta}} : Z_{2,\beta} \to TZ_{2,\beta}$ are continuously differentiable for every $\alpha \in \mathcal{A}$ and $\beta \in \mathcal{B}$, then x_2^* is an exponentially stable equilibrium point of $z_2[k+1] = \rho_{cl}(z_2[k])$ if and only if the orbit $\mathcal{O}$ is exponentially stable for the closed-loop hybrid system (2.26).

Proof. Define the augmented state spaces and switching manifolds as $\mathcal{X}_{1,a} := \mathcal{X}_1 \times \mathcal{A} \times \mathcal{B}$, $\mathcal{X}_{2,a} := \mathcal{X}_2 \times \mathcal{A} \times \mathcal{B}$, $\mathcal{S}_{1,a}^2 := \mathcal{S}_1^2 \times \mathcal{A} \times \mathcal{B}$, and $\mathcal{S}_{2,a}^1 := \mathcal{S}_2^1 \times \mathcal{A} \times \mathcal{B}$. The vector fields and switching maps corresponding to the augmented state spaces and switching manifolds can also be defined in the following forms

$$
f_{1,a}(x_1, \alpha, \beta) := \begin{bmatrix} f_{1,cl}(x_1; \alpha) \\ 0 \\ 0 \end{bmatrix} \qquad f_{2,a}(x_2, \alpha, \beta) := \begin{bmatrix} f_{2,cl}(x_2; \beta) \\ 0 \\ 0 \end{bmatrix}
$$

$$
\Delta_{1,a}^2(x_1, \alpha, \beta) := \begin{bmatrix} \Delta_1^2(x_1) \\ \alpha \\ \beta \end{bmatrix} \qquad \Delta_{2,a}^1(x_2, \alpha, \beta) := \begin{bmatrix} \Delta_2^1(x_2) \\ \alpha_{cl}(x_2) \\ \beta_{cl}(x_2) \end{bmatrix}.
$$

Next consider the following augmented hybrid system

$$
\Sigma_{1,a} : \begin{cases}
\mathcal{X}_{1,a} \subset \mathbb{R}^{n_1} \times \mathcal{A} \times \mathcal{B} \\[4pt]
\mathcal{F}_{1,a} : \begin{bmatrix} \dot{x}_1 \\ \dot{\alpha} \\ \dot{\beta} \end{bmatrix} = f_{1,a}(x_1, \alpha, \beta) \\[10pt]
\mathcal{S}_{1,a}^2 = \{(x_1, \alpha, \beta) \in \mathcal{X}_{1,a} \mid H_1^2(x_1) = 0\} \\[6pt]
\mathcal{T}_{1,a}^2 : \begin{bmatrix} x_2^+ \\ \alpha^+ \\ \beta^+ \end{bmatrix} = \Delta_{1,a}^2(x_1^-, \alpha^-, \beta^-)
\end{cases}
$$

$$\text{(2.27)}$$

$$
\Sigma_{2,a} : \begin{cases}
\mathcal{X}_{2,a} \subset \mathbb{R}^{n_2} \times \mathcal{A} \times \mathcal{B} \\[4pt]
\mathcal{F}_{2,a} : \begin{bmatrix} \dot{x}_2 \\ \dot{\alpha} \\ \dot{\beta} \end{bmatrix} = f_{2,a}(x_2, \alpha, \beta) \\[10pt]
\mathcal{S}_{2,a}^1 = \{(x_2, \alpha, \beta) \in \mathcal{X}_{2,a} \mid H_2^1(x_2) = 0\} \\[6pt]
\mathcal{T}_{2,a}^1 : \begin{bmatrix} x_1^+ \\ \alpha^+ \\ \beta^+ \end{bmatrix} = \Delta_{2,a}^1(x_2^-, \alpha^-, \beta^-).
\end{cases}
$$

From the hypotheses of Theorem 2.5 and by introducing the augmented manifolds

$$
Z_{1,a} := \{(x_1, \alpha, \beta) \in \mathcal{X}_{1,a} \mid x_1 \in Z_{1,\alpha}\}
$$
$$
Z_{2,a} := \{(x_2, \alpha, \beta) \in \mathcal{X}_{2,a} \mid x_2 \in Z_{2,\beta}\},
$$

all of the hypotheses of Corollary 2.1 are satisfied, and consequently the stability behavior of the augmented periodic orbit $\mathcal{O}_a := \mathcal{O} \times \{\alpha^*\} \times \{\beta^*\}$ for the augmented hybrid model (2.27) can be studied based on the stability analysis of the equilibrium point $(x_2^*, \alpha^*, \beta^*)$ for the following discrete-time system

$$
z_2[k+1] = \rho_{\mathrm{cl}}(z_2[k])
$$
$$
\alpha[k+1] = \alpha_{\mathrm{cl}}(x_2[k]) \qquad\qquad \text{(2.28)}
$$
$$
\beta[k+1] = \beta_{\mathrm{cl}}(x_2[k]).
$$

In addition, the stability properties of the equilibrium point $(x_2^*, \alpha^*, \beta^*)$ for the system (2.28) are equivalent to those of the equilibrium point x_2^* for the discrete-time system $z_2[k+1] = \rho_{\mathrm{cl}}(z_2[k])$. This fact in combination with the equivalence between the stability properties of $\mathcal{O}_a$ for the augmented hybrid system (2.27) and $\mathcal{O}$ for the closed-loop system (2.26) completes the proof. ∎

Asymptotic Stabilization of Periodic Orbits for Walking with Double Support Phase

3.1 INTRODUCTION

The objective of this chapter is to develop an analytical approach for designing a continuous feedback law that realizes a desired period-one trajectory as an asymptotically stable orbit for a planar biped robot. The robot is assumed to be a five-link, four-actuator planar mechanism in the sagittal plane with point feet. The fundamental assumption is that the double support phase is not instantaneous. Hence, bipedal walking can be represented by a hybrid model with two continuous phases, including a single support phase (one leg on the ground) and a double support phase (two legs on the ground), and discrete transitions between the continuous phases. In the single support phase, the mechanical system has one degree of underactuation, whereas it is overactuated in the double support phase.

Recently, the method of virtual constraints has been used to design time-invariant feedback laws for bipedal locomotion with one degree of underactuation [46–52]. Virtual constraints are a set of holonomic outputs in the configuration space of the mechanical system that coordinate the links of biped robots during walking [47]. In the case that the zero dynamics manifold corresponding to the virtual constraints is invariant under the impact map of walking, the notion of hybrid zero dynamics (HZD) was introduced in Ref. [52]. Moreover, a constructive method, based on parameterization of the virtual constraints and updating their parameters in a stride-to-stride basis, was presented in Ref. [60] for creating an augmented HZD during bipedal walking with more than one degree of underactuation. This method was used in Refs. [61, 62] to induce asymptotically stable walking by an underactuated spatial biped robot. Also,

Hybrid Control and Motion Planning of Dynamical Legged Locomotion, Nasser Sadati, Guy A. Dumont, Kaveh Akbari Hamed, and William A. Gruver.

in [55], parameterized virtual constraints have been utilized to create an HZD during running of a planar biped robot.

There has been little attention given to control of biped robots during the double support phase with unilateral constraints, which are constraints on the state and control inputs of the mechanical system that represent feasible contact conditions between the leg ends and the ground. Such constraints present challenges for the design of controllers. Moreover, due to overactuation, the control input corresponding to a specific trajectory in the state space is not unique.

In this chapter, we show how to design a continuous time-invariant feedback law that asymptotically stabilizes a feasible periodic trajectory using an extension of HZD for a hybrid model of walking [68, 69]. The main contribution is to develop a continuous time-invariant control law for walking of a planar biped robot during the double support phase. Since the mechanical system in the double support phase has three DOF and four actuators, a constrained dynamics approach [70, p. 157] is used to describe the reduced-order dynamics of the system. Then, we propose two virtual constraints as holonomic outputs for the constrained system and solve an output zeroing problem with two control inputs. This results in a nontrivial two-dimensional zero dynamics manifold corresponding to the virtual constraints in the state manifold of the constrained system. Moreover, the corresponding zero dynamics has two control inputs that are not employed for output zeroing. Instead, they are used to satisfy the unilateral constraints. Furthermore, these inputs are obtained such that the control has minimum norm on the desired periodic trajectory. It can be shown that the constrained dynamics of the double support phase is completely feedback linearizable on an open subset of the state manifold. However, since our objective is to design a continuous time-invariant controller based on HZD, in contrast to Ref. [71], we do not use input-state linearization nor a discontinuous time optimal control for tracking trajectories. An analogous approach is used in Refs. [54, 72] for creating a two-dimensional zero dynamics manifold in the state space of a fully actuated phase of walking where the fully actuated dynamics is completely feedback linearizable.

We present the control strategy at the following two levels. At the first level, we employ within-stride controllers including single and double support phase controllers. These are continuous time-invariant and parameterized feedback laws that create a family of two-dimensional finite-time attractive and invariant submanifolds on which the dynamics of the mechanical system is restricted. At the second level, the parameters of the within-stride controllers are updated at the end of the single support phase (in a stride-to-stride manner) by an event-based update law to achieve hybrid invariance and stabilization. Hybrid invariance yields a reduced-order hybrid model for walking that is referred to as HZD. As a consequence, the stability properties of the desired periodic orbit can be analyzed using low-dimensional tools developed for systems with impulse effects, such as restricted Poincaré return maps (see Section 2.3). The idea of updating parameters of the within-stride controller in an event-based manner to achieve stabilization has been described in Section 2.4.

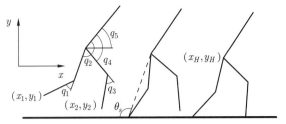

Figure 3.1 Three different phases of the biped walker motion: flight phase (left), single support phase (middle), and double support phase (right). In the single support phase, the virtual leg is depicted by the dashed line. All of the angles increase in the clockwise direction, whereas the absolute angle decreases.

3.2 MECHANICAL MODEL OF A BIPED WALKER

3.2.1 The Biped Robot

Throughout this chapter, we consider a planar biped robot, composed of five rigid links with distributed masses, walking on a flat surface. The links are connected by four revolute body joints: two knee joints and two hip joints. This robot has also point feet (see Fig. 3.1). There is an actuator located at each knee and hip joint. It is assumed that the robot cannot apply torques at the end of its legs. We also assume that a coordinate frame is attached to the flat ground called the *world frame*.

3.2.2 Dynamics of the Flight Phase

For representing the configuration of the walking robot, a convenient choice of the configuration variables consists of the body angles, the absolute orientation, and the absolute position of the robot with respect to the world frame. Unless otherwise stated, we follow the notation of Ref. [55]. The body angles are the relative angles $q_b := (q_1, q_2, q_3, q_4)'$ describing the shape of the biped robot. The absolute orientation of the biped robot is represented by q_5, whereas the absolute position is represented by the Cartesian coordinates of its COM, $p_{cm} := (x_{cm}, y_{cm})'$. Hence, the generalized coordinates for the flight phase (no leg is in contact with the ground) can be defined as $q_f := (q', p'_{cm})'$, where $q := (q'_b, q_5)'$ and prime represents matrix transpose. We remark that q_5 and p_{cm} are unactuated variables during the flight phase.

Following the notation of Ref. [55], the dynamical model during the flight phase can be expressed as

$$D_f(q_b)\ddot{q}_f + C_f(q_b, \dot{q}_f)\dot{q}_f + G_f(q_f) = B_f u, \tag{3.1}$$

where $D_f(q_b)$ is a (7×7) positive-definite mass-inertia matrix with the following form:

$$D_f(q_b) := \begin{bmatrix} A(q_b) & 0_{5 \times 2} \\ 0_{2 \times 5} & m_{\text{tot}} I_{2 \times 2} \end{bmatrix} \tag{3.2}$$

and m_{tot} represents the total mass of the biped walker. C_f is a (7×7) matrix containing the Coriolis and centrifugal terms, G_f is a (7×1) gravity vector, $u := (u_1, u_2, u_3, u_4)'$ is a vector of actuator torques, and $B_f := [I_{4 \times 4} \; 0_{4 \times 3}]'$. Using the block diagonal form of the mass-inertia matrix in equation (3.2) and equations (7.60) and (7.62) of Ref. [90, p. 256], C_f can be expressed as[1]

$$C_f(q_b, \dot{q}_f) = \begin{bmatrix} \bar{C}(q_b, \dot{q}) & 0_{5 \times 2} \\ 0_{2 \times 5} & 0_{2 \times 2} \end{bmatrix}. \tag{3.3}$$

Moreover, G_f can be given by

$$G_f(q) := m_{\text{tot}} g_0 \begin{bmatrix} 0 & 0 & 0 & 0 & 0 & 0 & 1 \end{bmatrix}', \tag{3.4}$$

in which g_0 represents the gravitational constant. Consequently, the dynamical equation of the flight phase (3.1) can be decomposed as follows:

$$A(q_b)\ddot{q} + \bar{C}(q_b, \dot{q})\dot{q} = B u \tag{3.5}$$

$$m_{\text{tot}} \ddot{p}_{\text{cm}} + m_{\text{tot}} \begin{bmatrix} 0 \\ g_0 \end{bmatrix} = 0_{2 \times 1}, \tag{3.6}$$

where $B := [I_{4 \times 4} \; 0_{4 \times 1}]'$.

Remark 3.1 (Cyclic Variables of the Flight Phase) *During the flight phase, q_5 and p_{cm} are the cyclic variables [1, 53] in the sense that $\frac{\partial \mathcal{K}_f}{\partial q_5} = \frac{\partial \mathcal{K}_f}{\partial x_{\text{cm}}} = \frac{\partial \mathcal{K}_f}{\partial y_{\text{cm}}} = 0$, where $\mathcal{K}_f(q_f, \dot{q}_f) := \frac{1}{2}\dot{q}'_f D_f(q_b)\dot{q}_f$ denotes the total kinetic energy of the mechanical system. Thus, the matrices D_f and C_f in equations (3.2) and (3.3) are independent of q_5 and p_{cm}.*

Remark 3.2 (Conservation of Angular Momentum in the Flight Phase) *From Proposition B.9 of Ref. [18, p. 427], the last row of equation (3.5) can be expressed as $\dot{\sigma}_{\text{cm}} = 0$, in which $\sigma_{\text{cm}} := A_5(q_b)\dot{q}$ denotes the angular momentum of the mechanical system about the COM and A_5 is the last row of the matrix A.*

---

[1] From equation (7.64) of Ref. [90, p. 257], the (k, j)th element of the matrix $C_f(q_b, \dot{q}_f)$ can be expressed as

$$C_{f,kj} = \sum_{i=1}^{7} \frac{1}{2} \left(\frac{\partial D_{f,kj}}{\partial q_{f,i}} + \frac{\partial D_{f,ki}}{\partial q_{f,j}} - \frac{\partial D_{f,ij}}{\partial q_{f,k}} \right) \dot{q}_{f,i}.$$

3.2.3 Dynamics of the Single Support Phase

During the single support phase, the contacting leg is called the *stance leg* and the other is called the *swing leg*. By definition, the *virtual leg* is defined by the line connecting the end of the stance leg and the hip joints [61] (see Fig. 3.1). Let $(x_1, y_1)' = p_1(q_f) := p_{cm} - f_1(q) \in \mathbb{R}^2$ and $(x_2, y_2)' = p_2(q_f) := p_{cm} - f_2(q) \in \mathbb{R}^2$ represent the Cartesian coordinates of the end of leg-1 and leg-2, respectively. Also, f_1 and f_2 are two smooth functions of the configuration variables. Since both legs of the robot are identical, without loss of generality, let leg-1 be the stance leg. In addition, assume that the end of leg-1 is on the origin of the world frame, that is, $p_{cm} = f_1(q)$. Then, using Remark B.14 of Ref. [18, p. 433], a reduced-order model for describing the evolution of the mechanical system during the single support phase can be expressed as[2]

$$D(q_b)\ddot{q} + C(q_b, \dot{q})\dot{q} + G(q) = Bu, \qquad (3.7)$$

where

$$D(q_b) := A + m_{tot}\frac{\partial f_1}{\partial q}'\frac{\partial f_1}{\partial q}$$

$$C(q_b, \dot{q}) := \bar{C} + m_{tot}\frac{\partial f_1}{\partial q}'\frac{\partial}{\partial q}\left(\frac{\partial f_1}{\partial q}\dot{q}\right)$$

$$G(q) := m_{tot}\frac{\partial f_1}{\partial q}'\begin{bmatrix} 0 \\ g_0 \end{bmatrix}.$$

By defining $x_s := (q_s', \dot{q}_s')'$ as the state vector during the single support phase, where $q_s := q$ and $\dot{q}_s := \dot{q}$, equation (3.7) can be expressed in the state space form $\dot{x}_s = f_s(x_s) + g_s(x_s)u$. Moreover, during single support, the state manifold is assumed to be[3]

$$\mathcal{X}_s := T\mathcal{Q}_s := \{x_s := (q_s', \dot{q}_s')' | q_s \in \mathcal{Q}_s, \dot{q}_s \in \mathbb{R}^5\},$$

where $\mathcal{Q}_s$ denotes the configuration space of the single support phase and it is chosen as a simply connected and open subset of $[(0, \frac{\pi}{2}) \times (0, \frac{3\pi}{2})]^2 \times (-\frac{\pi}{2}, \frac{\pi}{2})$.

---

[2] Equation (3.7) can also be obtained by applying the principle of virtual work in equation (3.5), similar to that presented in proof of Theorem 3.1 during the double support phase.

[3] In our notation, the tangent bundle of the manifold $\mathcal{M}$ is denoted by $T\mathcal{M} := \cup_{x \in \mathcal{M}} T_x\mathcal{M}$, in which $T_x\mathcal{M}$ is the tangent space at the point $x \in \mathcal{M}$.

Remark 3.3 (Validity of the Single Support Phase Model)   *The single support phase model is valid when $F_1^v > 0$ and $|\frac{F_1^h}{F_1^v}| < \mu_s$, where*

$$
\begin{aligned}
F_1 &:= \begin{bmatrix} F_1^h \\ F_1^v \end{bmatrix} \\
&= m_{\text{tot}} \frac{\partial f_1}{\partial q} \ddot{q} + m_{\text{tot}} \frac{\partial}{\partial q} \left(\frac{\partial f_1}{\partial q} \dot{q} \right) \dot{q} + m_{\text{tot}} \begin{bmatrix} 0 \\ g_0 \end{bmatrix} \\
&= m_{\text{tot}} \frac{\partial f_1}{\partial q} D^{-1} (Bu - C\dot{q} - G) + m_{\text{tot}} \frac{\partial}{\partial q} \left(\frac{\partial f_1}{\partial q} \dot{q} \right) \dot{q} + m_{\text{tot}} \begin{bmatrix} 0 \\ g_0 \end{bmatrix}
\end{aligned}
\tag{3.8}
$$

and μ_s denotes the ground reaction force at the end of leg-1 and the static friction coefficient, respectively [52]. Also q_5 is the cyclic variable during single support, that is, $\frac{\partial \mathcal{K}_s}{\partial q_5} = 0$, in which $\mathcal{K}_s(q_s, \dot{q}_s) := \frac{1}{2} \dot{q}_s' D(q_b) \dot{q}_s$ denotes the total kinetic energy of the mechanical system in the single support phase.

Remark 3.4 (Angular Momentum Balance in Single Support)   *From Proposition B.9 of Ref. [18, p. 427], the last row of equation (3.7) can be expressed as[4]*

$$
\dot{\sigma}_s = -G_5(q) = -m_{\text{tot}} g_0 f_1^h(q),
\tag{3.9}
$$

in which $\sigma_s := D_5(q_b)\dot{q}$ represents the angular momentum of the mechanical system about the end of leg-1, D_5 is the last row of the matrix D, G_5 is the last row of the gravity vector G, and f_1^h denotes the horizontal components of f_1.

3.2.4 Dynamics of the Double Support Phase

During double support, the mechanical system has three DOF. We decompose q as $q = (q_d', q_i')'$, where q_d and q_i denote dependent and independent configuration variables, respectively. Next, we define

$$
\tilde{Q} := \left\{ q = (q_d', q_i')' \in Q_s \mid p_2(q) = (L_s, 0)', \text{rank} \frac{\partial p_2}{\partial q_d}(q) = 2 \right\},
\tag{3.10}
$$

where $p_2(q) := f_1(q) - f_2(q)$ represents the Cartesian coordinates of the end of leg-2 with respect to the end of leg-1 and L_s is a constant denoting the step length. If $\tilde{Q}$ is a nonempty set, by the Implicit Function Theorem, there exists a unique function Ψ such that $q_d = \Psi(q_i)$ for any $q = (q_d', q_i')' \in \tilde{Q}$. Without loss of generality, we assume that $q_d := (q_1, q_2)'$ and $q_i := (q_3, q_4, q_5)'$ (see Fig. 3.1). For the later purposes, define Q_i

---

[4] Since q_5 is measured so that it decreases in the clockwise direction, we do not make use of Proposition B.11 of Ref. [18, p. 430].

as a simply connected and open subset of $(0, \frac{\pi}{2}) \times (0, \frac{3\pi}{2}) \times (-\frac{\pi}{2}, \frac{\pi}{2})$ and $T : Q_i \rightarrow \mathbb{R}^5$ by

$$T(q_i) := \begin{bmatrix} \Psi(q_i) \\ q_i \end{bmatrix}.$$

The following theorem presents a reduced-order model with three DOF and four control inputs, called the *constrained dynamics*,[5] for the evolution of the mechanical system during double support.

Theorem 3.1 (Constrained Dynamics of Double Support) *Let $\tilde{Q}$ be a nonempty set. Then, the constrained dynamics for the double support phase is given by*

$$D_\psi(q_i)\ddot{q}_i + C_\psi(q_i, \dot{q}_i)\dot{q}_i + G_\psi(q_i) = B_\psi(q_i)u, \tag{3.11}$$

where

$$D_\psi(q_i) := \frac{\partial T}{\partial q_i}' D \circ T \frac{\partial T}{\partial q_i}$$

$$C_\psi(q_i, \dot{q}_i) := \frac{\partial T}{\partial q_i}' C\left(T, \frac{\partial T}{\partial q_i}\dot{q}_i\right)\frac{\partial T}{\partial q_i}\dot{q}_i + \frac{\partial T}{\partial q_i}' D \circ T \frac{\partial}{\partial q_i}\left(\frac{\partial T}{\partial q_i}\dot{q}_i\right)$$

$$G_\psi(q_i) := \frac{\partial T}{\partial q_i}' G \circ T$$

$$B_\psi(q_i) := \frac{\partial T}{\partial q_i}' B$$

and "$\circ$" represents function composition.

Proof. During the double support phase, applying the principal of virtual work in equation (3.1) yields

$$A(q_b)\ddot{q} + \bar{C}(q_b, \dot{q})\dot{q} = Bu - \frac{\partial f_1}{\partial q}'(q) F_1 - \frac{\partial f_2}{\partial q}'(q) F_2$$

$$m_{\text{tot}}\ddot{p}_{\text{cm}} + m_{\text{tot}} \begin{bmatrix} 0 \\ g0 \end{bmatrix} = F_1 + F_2, \tag{3.12}$$

where $F_1 := (F_1^h, F_1^v)' \in \mathbb{R}^2$ and $F_2 := (F_2^h, F_2^v)' \in \mathbb{R}^2$ denote the ground reaction forces at the end of leg-1 and leg-2, respectively. By assuming that the end of leg-1 is on the origin of the world frame, $p_{\text{cm}}(q) = f_1(q)$, and as a consequence, the last two

---

[5] The terminology of a *constrained dynamics* is taken from Ref. [70, p. 157].

rows of equation (3.12) yield

$$F_1 = m_{\text{tot}} \frac{\partial f_1}{\partial q} \ddot{q} + m_{\text{tot}} \frac{\partial}{\partial q} \left(\frac{\partial f_1}{\partial q} \dot{q} \right) \dot{q} + m_{\text{tot}} \begin{bmatrix} 0 \\ g_0 \end{bmatrix}. \tag{3.13}$$

Substituting equation (3.13) into the first five rows of matrix equation (3.12) results in

$$D(q_b) \ddot{q} + C(q_b, \dot{q}) \dot{q} + G(q) = B u + \frac{\partial p_2'}{\partial q}(q) F_2. \tag{3.14}$$

In the case of walking on the flat ground during the double support phase, $p_2(q) = (L_s, 0)'$, and hence, the second time derivative of p_2 is zero,

$$\ddot{p}_2(q, \dot{q}, \ddot{q}) = \frac{\partial p_2}{\partial q}(q) \ddot{q} + \frac{\partial}{\partial q} \left(\frac{\partial p_2}{\partial q}(q) \dot{q} \right) \dot{q} = 0_{2 \times 1}. \tag{3.15}$$

By definition of $\tilde{Q}$, $q \in \tilde{Q}$ implies that rank $\frac{\partial p_2}{\partial q}(q) = 2$. Thus, solving equations (3.14) and (3.15) simultaneously for $\ddot{q}$ and F_2 yields

$$F_2 = \left(\frac{\partial p_2}{\partial q} D^{-1} \frac{\partial p_2'}{\partial q} \right)^{-1} \left(\frac{\partial p_2}{\partial q} D^{-1}(C\dot{q} + G - Bu) - \frac{\partial}{\partial q} \left(\frac{\partial p_2}{\partial q} \dot{q} \right) \dot{q} \right) \tag{3.16}$$

and

$$D(q_b) \ddot{q} + C_d(q, \dot{q}) \dot{q} + G_d(q) = B_d(q) u, \tag{3.17}$$

where

$$C_d(q, \dot{q}) := \Pi C + \frac{\partial p_2'}{\partial q} \left(\frac{\partial p_2}{\partial q} D^{-1} \frac{\partial p_2'}{\partial q} \right)^{-1} \frac{\partial}{\partial q} \left(\frac{\partial p_2}{\partial q} \dot{q} \right)$$

$$G_d(q) := \Pi G$$

$$B_d(q) := \Pi B$$

$$\Pi(q) := I_{5 \times 5} - \frac{\partial p_2'}{\partial q} \left(\frac{\partial p_2}{\partial q} D^{-1} \frac{\partial p_2'}{\partial q} \right)^{-1} \frac{\partial p_2}{\partial q} D^{-1}.$$

We remark that matrix Π is a projection matrix because $\frac{\partial p_2}{\partial q} D^{-1} \Pi = 0_{2 \times 5}$ and $\Pi^2 = \Pi$ (idempotent). Due to the fact that for any $(q', \dot{q}')' \in T\tilde{Q}$,

$$\frac{\partial p_2}{\partial q}(q) \dot{q} = 0_{2 \times 1},$$

it follows that

$$\frac{\partial p_2}{\partial q} \circ T(q_i) \frac{\partial T}{\partial q_i}(q_i) = 0_{2\times 3}$$

and consequently,

$$\frac{\partial T'}{\partial q_i}(q_i) \Pi \circ T(q_i) = \frac{\partial T'}{\partial q_i}(q_i)$$

for every $q_i \in \mathcal{Q}_i$. Multiplying equation (3.17) by $\frac{\partial T'}{\partial q_i}(q_i)$ from the left and substituting the relations $\dot{q} = \frac{\partial T}{\partial q_i}\dot{q}_i$ and $\ddot{q} = \frac{\partial T}{\partial q_i}\ddot{q}_i + \frac{\partial}{\partial q_i}(\frac{\partial T}{\partial q_i}\dot{q}_i)\dot{q}_i$ completes the proof. ■

For reasons that will be discussed in Sections 3.3 and 3.4 (see Lemma 3.3 and Remark 3.10), we assume that u_1 and u_2 are two predetermined continuously differentiable functions of independent configuration variables, that is, $u_1 = u_1(q_i)$ and $u_2 = u_2(q_i)$. Then, the dynamics of the double support phase in equation (3.11) can be rewritten in the following form:

$$D_\psi \ddot{q}_i + C_\psi \dot{q}_i + G_\psi - \frac{\partial \Psi'}{\partial q_i}\begin{bmatrix} u_1(q_i) \\ u_2(q_i) \end{bmatrix} = \beta_\psi \begin{bmatrix} u_3 \\ u_4 \end{bmatrix}, \tag{3.18}$$

where $\beta_\psi := [I_{2\times 2}\, 0_{2\times 1}]'$. The functions $u_1(q_i)$ and $u_2(q_i)$ will be determined in Section 3.5. Furthermore, by introducing $x_d := (q_i', \dot{q}_i')'$ as the state vector for the double support phase, a state equation for equation (3.18) is

$$\dot{x}_d = f_d(x_d, u_1(x_d), u_2(x_d)) + g_d(x_d)\begin{bmatrix} u_3 \\ u_4 \end{bmatrix}.$$

The state space is also taken as

$$\mathcal{X}_d := T\tilde{\mathcal{Q}}_i := \{x_d := (q_i', \dot{q}_i')' | q_i \in \tilde{\mathcal{Q}}_i, \dot{q}_i \in \mathbb{R}^3\},$$

where $\tilde{\mathcal{Q}}_i := \{q_i \in \mathcal{Q}_i\, |\, T(q_i) \in \tilde{\mathcal{Q}}\}$.

Remark 3.5 (Validity of the Double Support Phase Model)   *The double support phase model is valid when $F_1^v > 0$, $|\frac{F_1^h}{F_1^v}| < \mu_s$, $F_2^v > 0$, and $|\frac{F_2^h}{F_2^v}| < \mu_s$.*

3.2.5 Impact Model

In this section, an impact model is obtained for describing the state of the mechanical system at the beginning of the double support phase (i.e., after impact) in terms of the state at the end of the single support phase (i.e., before impact). We shall assume that the impact is inelastic and instantaneous. Also, it is assumed that the stance leg does not leave the ground after impact. Let $I_{R1} := (I_{R1}^h, I_{R1}^v)' \in \mathbb{R}^2$ and $I_{R2} := (I_{R2}^h, I_{R2}^v)' \in \mathbb{R}^2$ represent the intensity of the impulsive ground reaction forces at the end of leg-1 and leg-2, respectively. For development of the impact map, we make use of the flight phase model. Let $\dot{q}_f^-$ and $\dot{q}_f^+$ be the generalized velocity of the mechanical system just before and after impact, respectively. From the impact model of Ref. [91], integration of equation (3.12) over the infinitesimal time interval of the impact yields

$$\begin{bmatrix} A & 0_{5\times2} \\ 0_{2\times5} & m_{\text{tot}} I_{2\times2} \end{bmatrix} (\dot{q}_f^+ - \dot{q}_f^-) = \begin{bmatrix} -\frac{\partial f_1}{\partial q}' \\ I_{2\times2} \end{bmatrix} I_{R1} + \begin{bmatrix} -\frac{\partial f_2}{\partial q}' \\ I_{2\times2} \end{bmatrix} I_{R2}. \qquad (3.19)$$

Since the robot is in single support before impact, $\dot{p}_{\text{cm}}^- = \frac{\partial f_1}{\partial q}(q)\dot{q}^-$. Moreover, the fact that the stance leg does not leave the ground after impact implies that $\dot{p}_{\text{cm}}^+ = \frac{\partial f_1}{\partial q}(q)\dot{q}^+$. Consequently, the first five rows of matrix equation (3.19) in combination with the last two rows imply that $I_{R1} = \Sigma_{12}(q) I_{R2}$, where

$$\Sigma_{12}(q) := -\left(I_{2\times2} + m_{\text{tot}} \frac{\partial f_1}{\partial q} A^{-1} \frac{\partial f_1}{\partial q}' \right)^{-1} \left(I_{2\times2} + m_{\text{tot}} \frac{\partial f_1}{\partial q} A^{-1} \frac{\partial f_2}{\partial q}' \right).$$

Since the impact is assumed inelastic, $\frac{\partial p_2}{\partial q}(q)\dot{q}^+ = 0_{2\times1}$ and thus,

$$\frac{\partial p_2}{\partial q} \left(\dot{q}^- - A^{-1} \left(\frac{\partial f_1}{\partial q}' \Sigma_{12} + \frac{\partial f_2}{\partial q}' \right) I_{R2} \right) = 0_{2\times1}. \qquad (3.20)$$

Definition 3.1 (Nonsingular Impact) *The impact model is nonsingular if*

$$\det \left(\frac{\partial p_2}{\partial q} A^{-1} \left(\frac{\partial f_1'}{\partial q} \Sigma_{12} + \frac{\partial f_2'}{\partial q} \right) \right) \neq 0.$$

Let the impact model be nonsingular. Then, equation (3.20) implies that

$$\begin{aligned}
I_{R2} &= \left(\frac{\partial p_2}{\partial q} A^{-1} \left(\frac{\partial f_1}{\partial q}' \Sigma_{12} + \frac{\partial f_2}{\partial q}' \right) \right)^{-1} \frac{\partial p_2}{\partial q} \dot{q}^- \\
&=: \Sigma_2(q) \dot{q}^- \\
I_{R1} &= \Sigma_{12}(q) \Sigma_2(q) \dot{q}^- =: \Sigma_1(q) \dot{q}^-.
\end{aligned} \qquad (3.21)$$

Finally, $\dot{q}^+$ is given by $\dot{q}^+ = \Delta^d_{\dot{q},s}(q)\dot{q}^-$, where

$$\Delta^d_{\dot{q},s}(q) := I_{5\times5} - A^{-1}\left(\frac{\partial f_1}{\partial q}'\Sigma_1 + \frac{\partial f_2}{\partial q}'\Sigma_2\right).$$

By defining $\pi_i := [0_{3\times2}\ I_{3\times3}]$, the transition map from the single support phase to the double support phase can be expressed as $x^+_d = \Delta^d_s(x^-_s)$, where

$$\Delta^d_s(x^-_s) = \Delta^d_s(q^-_s, \dot{q}^-_s) := \begin{bmatrix} \Delta^d_{q_i,s}(q^-_s) \\ \Delta^d_{\dot{q}_i,s}(q^-_s)\dot{q}^-_s \end{bmatrix}.$$

Moreover, $\Delta^d_{q_i,s}(q^-_s) := \pi_i q^-_s$, $\Delta^d_{\dot{q}_i,s}(q^-_s) := \pi_i \Delta^d_{\dot{q},s}(q^-_s)$ and the superscripts "$-$" and "$+$" denote the state of the mechanical system just before and after the discrete transitions.

Remark 3.6 (Validity of the Impact Model) *Following the double impact conditions presented in Refs. [67] and [92] (see equations (32)–(35)), the impact model is valid when $I^v_{R1} > 0$, $|\frac{I^h_{R1}}{I^v_{R1}}| < \mu_s$, $|\frac{I^h_{R2}}{I^v_{R2}}| < \mu_s$, and $\dot{p}^{2+}_1 \leq 0$.*

3.2.6 Transition from the Double Support Phase to the Single Support Phase

For simplifying the analysis of the hybrid model of walking in Section 3.4, it is assumed that the transition from the double support phase to the single support phase occurs at a predetermined point in the configuration space of the double support phase. During this transition, the position and velocity remain continuous. Hence, the transition map can be expressed as $x^+_s = \Delta^s_d(x^-_d)$, in which

$$\Delta^s_d(x^-_d) = \Delta^s_d(q^-_i, \dot{q}^-_i) := \begin{bmatrix} \Delta^s_{q,d}(q^-_i) \\ \Delta^s_{\dot{q},d}(q^-_i)\dot{q}^-_i \end{bmatrix}.$$

In addition, $\Delta^s_{q,d}(q^-_i) := RT(q^-_i)$, $\Delta^s_{\dot{q},d}(q^-_i) := R\frac{\partial T}{\partial q_i}(q^-_i)$ and R is a relabling matrix to swap the role of the legs, with the property $RR = I_{5\times5}$. The validity of this transition condition is confirmed by designing the control law in the single support phase so that it leads to $\ddot{y}_2 > 0$ at the beginning of the single support phase. For this purpose, as in Ref. [55], it is assumed that discontinuities of the control inputs in transitions are allowed.

3.2.7 Hybrid Model of Walking

The overall model of walking can be expressed as a nonlinear hybrid system consisting of two state manifolds that correspond to the single and double support phases as follows:

$$\Sigma_s : \begin{cases} \dot{x}_s = f_s(x_s) + g_s(x_s)\,u, & x_s^- \notin \mathcal{S}_s^d \\ x_d^+ = \Delta_s^d(x_s^-), & x_s^- \in \mathcal{S}_s^d \\ \mathcal{S}_s^d := \{x_s \in \mathcal{X}_s \mid H_s^d(x_s) = 0\} \end{cases}$$

$$\Sigma_d : \begin{cases} \dot{x}_d = f_d(x_d, u_1, u_2) + g_d(x_d) \begin{bmatrix} u_3 \\ u_4 \end{bmatrix}, & x_d^- \notin \mathcal{S}_d^s \\ x_s^+ = \Delta_d^s(x_d^-), & x_d^- \in \mathcal{S}_d^s \\ \mathcal{S}_d^s := \{x_d \in \mathcal{X}_d \mid H_d^s(x_d) = 0\}. \end{cases}$$

$$(3.22)$$

In this model, transition from the single support phase to the double support phase occurs when the height of the swing leg end becomes zero. Thus, $H_s^d(x_s) := y_2(q_s)$ (see Fig. 3.1). Following the assumption of Section 3.2.6, we define the switching hypersurface $\mathcal{S}_d^s$ as the zero level set of the smooth function $H_d^s : \mathcal{X}_d \to \mathbb{R}$ by $H_d^s(x_d) := x_H \circ T(q_i) - x_{H,d}^-$, where $x_H(q)$ is the horizontal position of the hip joint and $x_{H,d}^-$ is a constant threshold to be determined.

3.3 CONTROL LAWS FOR THE SINGLE AND DOUBLE SUPPORT PHASES

In order to reduce the dimension of the hybrid model of walking to simplify the stabilization problem for the desired periodic orbit in each of the continuous phases, a finite-time attractive and invariant submanifold is created by a continuous control law. Specifically, the control laws in the single and double support phases are chosen as time-invariant feedback based on zeroing holonomic output functions with the uniform vector relative degree 2. This control strategy will result in a two-dimensional zero dynamics manifold on each of the state manifolds, that is, holonomic quantities that are to be controlled are dependent on a holonomic quantity that is a strictly monotonic function of time on a typical walking gait.

3.3.1 Single Support Phase Control Law

As in Ref. [52], consider the following holonomic output function for the dynamics of the mechanical system in the single support phase

$$y_s := h_s(q_s) := q_b - h_{d,s} \circ \theta_s(q_s), \tag{3.23}$$

where[6] $\theta_s(q_s) := \frac{q_1}{2} + q_2 - q_5 =: C_o q_s$ is the angle of the virtual leg with respect to the world frame and $h_{d,s} : \mathbb{R} \to \mathbb{R}^4$ is at least a twice continuously differentiable function that specifies the desired evolution of the body angles in terms of θ_s. In Section 3.5, the function $h_{d,s}$ will be chosen such that the holonomic output function in equation (3.23) vanishes on the single support phase of a desired periodic solution of the open-loop hybrid model of walking in equation (3.22).

It is assumed that there exists an open set $\tilde{\mathcal{Q}}_s \subset \mathcal{Q}_s$ such that for any $q_s \in \tilde{\mathcal{Q}}_s$ the decoupling matrix

$$L_{g_s} L_{f_s} h_s(q_s) = \frac{\partial h_s}{\partial q_s}(q_s) D^{-1}(q_s) B$$

is invertible. For the later uses, let $\mathcal{O} = \mathcal{O}_s \cup \mathcal{O}_d$ denote a desired feasible period-one solution of the open-loop hybrid model of walking in equation (3.22) that is transversal to $\mathcal{S}_s^d$ and $\mathcal{S}_d^s$, where $\mathcal{O}_s := \mathcal{O} \cap \mathcal{X}_s$ and $\mathcal{O}_d := \mathcal{O} \cap \mathcal{X}_d$. Furthermore, suppose that $\mathcal{O}_s \subset T\tilde{\mathcal{Q}}_s$. By Lemma 1 of Ref. [52], since the decoupling matrix $L_{g_s} L_{f_s} h_s(q_s)$ is invertible on $\tilde{\mathcal{Q}}_s$ and the holonomic output function $h_s(q_s)$ vanishes on $\mathcal{O}_s$, the set

$$Z_s := \{x_s \in T\tilde{\mathcal{Q}}_s | h_s(x_s) = 0_{4\times1}, L_{f_s} h_s(x_s) = 0_{4\times1}\}$$

is an embedded two-dimensional submanifold of $T\mathcal{Q}_s$. Moreover, suppose that $\mathcal{S}_s^d \cap Z_s$ is an embedded one-dimensional submanifold of $T\mathcal{Q}_s$.[7] Then, on the manifold $\mathcal{S}_s^d \cap Z_s$, the configuration variables are determined [52]. In particular, let $\pi_q : (q, \dot{q}) \mapsto q$ be a canonical projection map. Then, $\pi_q(\mathcal{S}_s^d \cap Z_s) = \{q_s^-\}$, and for the later purposes, let $\theta_s^- := \theta_s(q_s^-)$.

Next, let $v_s : \mathbb{R}^4 \times \mathbb{R}^4 \to \mathbb{R}^4$ be a continuous function such that the origin for the closed-loop system $\ddot{y}_s = v_s(y_s, \dot{y}_s)$ is globally finite-time stable.[8] Then, the continuous time-invariant feedback law

$$
\begin{aligned}
u_s(x_s) &:= -(L_{g_s} L_{f_s} h_s(x_s))^{-1} \left(L_{f_s}^2 h_s(x_s) - v_s(h_s(x_s), L_{f_s} h_s(x_s)) \right) \\
&= -\underbrace{\left(\frac{\partial h_s}{\partial q_s} D^{-1} B \right)^{-1}}_{L_{g_s} L_{f_s} h_s} \left(\underbrace{\frac{\partial}{\partial q_s} \left(\frac{\partial h_s}{\partial q_s} \dot{q}_s \right) \dot{q}_s - \frac{\partial h_s}{\partial q_s} D^{-1}(C\dot{q}_s + G)}_{L_{f_s}^2 h_s} - v_s \right) \quad (3.24)
\end{aligned}
$$

---

[6] In this chapter, it is assumed that the femur and tibia links are of equal length.

[7] See hypothesis HH5 of Ref. [18, p. 126].

[8] References [46, 93] describe a method for designing the continuous function v_s. In particular, by applying the continuous feedback law $v = \frac{1}{\epsilon^2} \psi(y, \epsilon\dot{y})$, in which

$$\psi(y, \epsilon\dot{y}) := -\text{sign}(\epsilon\dot{y}) |\epsilon\dot{y}|^\alpha - \text{sign}(\phi(y, \epsilon\dot{y})) |\phi(y, \epsilon\dot{y})|^{\frac{\alpha}{2-\alpha}},$$

$0 < \alpha < 1, \epsilon > 0$ and

$$\phi(y, \epsilon\dot{y}) := y + \frac{1}{2-\alpha} \text{sign}(\epsilon\dot{y}) |\epsilon\dot{y}|^{2-\alpha},$$

the origin for the scalar double integrator $\ddot{y} = v$ is globally finite-time stable.

renders Z_s locally finite-time attractive (see Definition 2.8, Section 2.3) and forward invariant[9] under the closed-loop dynamics of the single support phase [52]. By definition, Z_s is the *single support phase zero dynamics manifold* and $\dot{z}_s = f_{zero,s}(z_s)$ is the *single support phase zero dynamics*, where[10]

$$f_{zero,s}(z_s) := f_s^*|_{Z_s}(z_s)$$
$$f_s^*(x_s) := f_s(x_s) + g_s(x_s)\,u_s^*(x_s)$$
$$u_s^*(x_s) := -(L_{g_s}L_{f_s}h_s(x_s))^{-1}L_{f_s}^2 h_s(x_s).$$

Following Ref. [52], (θ_s, σ_s) is a valid set of local coordinates for Z_s. Furthermore, the single support phase zero dynamics can be expressed by Ref. [52],

$$\dot{\theta}_s = \kappa_1(\theta_s)\,\sigma_s$$
$$\dot{\sigma}_s = \kappa_2(\theta_s), \tag{3.25}$$

in which

$$\kappa_1(\theta_s) := \left.\frac{\partial \theta_s}{\partial q_s}\lambda_s\right|_Z = \left.C_o\lambda_s\right|_Z$$
$$\kappa_2(\theta_s) := \left.-G_5\right|_Z$$
$$\lambda_s(q_s) := \begin{bmatrix}\frac{\partial h_s}{\partial q_s}(q_s)\\ D_5(q_s)\end{bmatrix}^{-1}\begin{bmatrix}0_{4\times 1}\\ 1\end{bmatrix}.$$

It is also shown that the zero dynamics of the single support phase has the Lagrangian $\mathcal{L}_{zero,s} := K_{zero,s} - V_{zero,s}$, where

$$K_{zero,s}(\sigma_s) := \frac{1}{2}(\sigma_s)^2$$
$$V_{zero,s}(\theta_s) := -\int_{\theta_s^+}^{\theta_s}\frac{\kappa_2(\xi)}{\kappa_1(\xi)}d\xi$$

and θ_s^+ is a constant value that will be determined later. Moreover, $\mathcal{S}_s^d \cap Z_s$ can be expressed as

$$\mathcal{S}_s^d \cap Z_s = \{(q', \dot{q}')'|q = q_s^-, \dot{q} = \lambda_s(q_s^-)\sigma_s^-, \sigma_s^- \in \mathbb{R}\}. \tag{3.26}$$

---

[9] A set $\mathcal{Z}$ is said to be *forward invariant* under the dynamics $\dot{x} = f(x)$ if for every $x_0 \in \mathcal{Z}$, there exists $t_1 > 0$ such that $\varphi^f(t; x_0) \in \mathcal{Z}$ for $t \in [0, t_1)$, where $\varphi^f(t; x_0)$ represents the maximal solution of the differential equation $\dot{x} = f(x)$ with the initial condition x_0. Furthermore, from Proposition B.1 of Ref. [18, p. 384], if $\mathcal{Z}$ is forward invariant, then for all $x \in \mathcal{Z}$, $f(x) \in T_x\mathcal{Z}$.

[10] In our notation, $f|_{\mathcal{M}}$ represents the restriction of the function f to the set $\mathcal{M}$.

Remark 3.7 (Invertibility of the Decoupling Matrix on $\mathcal{O}_s$) *From Proposition 6.1 of Ref. [18, p. 158] and equation (3.25), if on the orbit $\mathcal{O}_s$, the time evolution of θ_s is an increasing function of time (i.e., $\dot{\theta}_s > 0$), invertibility of the decoupling matrix $L_{g_s} L_{f_s} h_s(q_s)$ on $\mathcal{O}_s$ is equivalent to the angular momentum about the stance leg end (i.e., σ_s) being nonzero during the single support phase, because it can be shown that $1/\kappa_1(\theta_s)$ is the determinant of the decoupling matrix in the coordinates (q_b, θ_s).*

3.3.2 Double Support Phase Control Law

Analogous to the development for the single support phase, a holonomic output function $h_d(q_i)$ with dimension two is chosen for the constrained dynamics of the double support phase. The output function is chosen as a vector with relative degree $(2, 2)'$ with respect to the control inputs $(u_3, u_4)'$ on an open subset of the configuration space $\tilde{\mathcal{Q}}_i$. Solution of the output zeroing problem by the control inputs $(u_3, u_4)'$ results in a two-dimensional zero dynamics manifold. However, the control inputs u_1 and u_2 are not employed in the output zeroing problem. We will employ them to ensure validity of the double support phase model and minimization of the norm of the control input on $\mathcal{O}_d$ in Section 3.5. To make this notion precise, we define the following holonomic output function:

$$y_d := h_d(q_i) := \varphi(q_i) - h_{d,d} \circ \bar{x}_H(q_i), \tag{3.27}$$

where $\varphi(q_i)$ represents the quantities that are to be controlled. In particular, it consists of the vertical displacement of the hip joint and trunk angle,

$$\varphi(q_i) := \begin{bmatrix} \bar{y}_H(q_i) \\ e_3' q_i \end{bmatrix},$$

where $\bar{y}_H(q_i) = y_H \circ T(q_i)$ is the vertical displacement of the hip joint and $e_3' q_i = q_5$ is the trunk angle.[11] Moreover, $\bar{x}_H(q_i) = x_H \circ T(q_i)$ is the horizontal displacement of the hip joint and the function $h_{d,d}(\bar{x}_H)$ is at least a C^2 function that specifies the desired evolution of $\varphi(q_i)$ in terms of $\bar{x}_H$. The function $h_{d,d}(\bar{x}_H)$ will be constructed such that the holonomic output function in equation (3.27) vanishes on $\mathcal{O}_d$. Forcing y_d to be zero will result in the evolution of the vertical displacement of the hip joint and trunk angle to be constrained to the horizontal displacement of the hip joint.

To introduce a valid coordinate transformation on $T\tilde{\mathcal{Q}}_i$, we first present the following lemma by which a coordinate transformation will be used to obtain the zero dynamics corresponding to the output function (3.27) during the double support phase.

---

[11] Throughout this book, $e_i \in \mathbb{R}^n$ is defined by $e_i := [0 \; \cdots \; \underbrace{1}_{i\text{th}} \; \cdots 0]'$.

Lemma 3.1 (Coordinate Transformation on $\tilde{\mathcal{Q}}_i$)   *The mapping $\Phi_d : \tilde{\mathcal{Q}}_i \to \mathbb{R}^3$ by*

$$\Phi_d(q_i) := \begin{bmatrix} h_d(q_i) \\ \bar{x}_H(q_i) \end{bmatrix}$$

is a diffeomorphism to its image.

Proof. It is sufficient to show that the Jacobian matrix $\frac{\partial}{\partial q_i} \Phi_d(q_i)$ has full rank in $\tilde{\mathcal{Q}}_i$. Since the rank of a matrix does not change by adding a multiple of a row to another row,

$$\operatorname{rank} \frac{\partial}{\partial q_i} \Phi_d(q_i) = \operatorname{rank} \begin{bmatrix} \frac{\partial \varphi}{\partial q_i}(q_i) \\ \frac{\partial \bar{x}_H}{\partial q_i}(q_i) \end{bmatrix}. \tag{3.28}$$

Next, define the mapping $\Xi : \mathcal{Q}_s \to \mathbb{R}^5$ by

$$\Xi(q) := \begin{bmatrix} x_2(q) \\ y_2(q) \\ x_H(q) \\ y_H(q) \\ e'_5 q \end{bmatrix}.$$

For the biped robot described previously, it can be shown that

$$\det \frac{\partial \Xi}{\partial q}(q) = l_t^2 l_f^2 \sin(q_1) \sin(q_3),$$

where l_t and l_f represent the length of the tibia and femur links, respectively. Since $\mathcal{Q}_s$ is a simply connected and open subset of $[(0, \frac{\pi}{2}) \times (0, \frac{3\pi}{2})]^2 \times (-\frac{\pi}{2}, \frac{\pi}{2})$, for every $q \in \mathcal{Q}_s$, $\det \frac{\partial \Xi}{\partial q}(q) \neq 0$. Hence, Ξ is diffeomorphism to its image. Next, define $\bar{\Xi} : \tilde{\mathcal{Q}}_i \to \mathbb{R}^5$ as the restriction of Ξ to $\tilde{\mathcal{Q}}_i$, that is, $\bar{\Xi}(q_i) := \Xi \circ T(q_i)$. The facts that for any $q \in \mathcal{Q}_s$, $\operatorname{rank} \frac{\partial \Xi}{\partial q}(q) = 5$, and for any $q_i \in \tilde{\mathcal{Q}}_i$, $T(q_i) \in \mathcal{Q}_s$ imply that for any $q_i \in \tilde{\mathcal{Q}}_i$,

$$\operatorname{rank} \frac{\partial \bar{\Xi}}{\partial q_i}(q_i) = \operatorname{rank} \left(\frac{\partial \Xi}{\partial q} \circ T(q_i) \frac{\partial T}{\partial q_i}(q_i) \right)$$

$$= \operatorname{rank} \frac{\partial T}{\partial q_i}(q_i) = 3.$$

Furthermore,

$$
\bar{\Xi}(q_i) = \begin{bmatrix} L_s \\ 0 \\ \bar{x}_H(q_i) \\ \bar{y}_H(q_i) \\ e'_3 q_i \end{bmatrix},
$$

which, in turn, in combination with equation (3.28) implies that rank $\frac{\partial \bar{\Xi}}{\partial q_i}(q_i) =$ rank $\frac{\partial \Phi_d}{\partial q_i}(q_i) = 3$ for every $q_i \in \tilde{Q}_i$. ■

Now assume that there exists an open set $\check{Q}_i \subset \tilde{Q}_i$ such that for every $q_i \in \check{Q}_i$, the decoupling matrix

$$
L_{g_d} L_{f_d} h_d(q_i) = \frac{\partial h_d}{\partial q_i}(q_i) D_\psi^{-1}(q_i) \beta_\psi
$$

is invertible. Furthermore, suppose that $\mathcal{O}_d \subset T\check{Q}_i$. Since the decoupling matrix $L_{g_d} L_{f_d} h_d(q_i)$ is invertible on $\check{Q}_i$ and h_d vanishes on $\mathcal{O}_d$, the set

$$
Z_d := \{x_d \in T\check{Q}_i | h_d(x_d) = 0_{2\times 1}, L_{f_d} h_d(x_d) = 0_{2\times 1}\}
$$

is an embedded two-dimensional submanifold of $T\tilde{Q}_i$. We remark that Z_d is independent of $u_1(q_i)$ and $u_2(q_i)$ because h_d is a holonomic output function.

Lemma 3.2 *Let $\mathcal{S}_d^s \cap Z_d \neq \phi$. Then, $\mathcal{S}_d^s \cap Z_d$ is an embedded one-dimensional submanifold of $T\tilde{Q}_i$.*

Proof. $\mathcal{S}_d^s \cap Z_d$ can be expressed as $\mathcal{S}_d^s \cap Z_d = \{x_d \in T\check{Q}_i | F_d(x_d) = 0_{5\times 1}\}$, where

$$
F_d(x_d) := \begin{bmatrix} h_d(x_d) \\ \bar{x}_H(x_d) - x_{H,d}^- \\ L_{f_d} h_d(x_d) \end{bmatrix}.
$$

Since $\mathcal{S}_d^s \cap Z_d \neq \phi$ and by Lemma 3.1, the mapping $\Phi_d : \tilde{Q}_i \to \mathbb{R}^3$ is a diffeomorphism to its image, there exists a unique point $\bar{q}_{id} \in \tilde{Q}_i$ such that

$$
\Phi_d(\bar{q}_{id}) = \begin{bmatrix} h_d(\bar{q}_{id}) \\ \bar{x}_H(\bar{q}_{id}) \end{bmatrix} = \begin{bmatrix} 0_{2\times 1} \\ x_{H,d}^- \end{bmatrix}.
$$

In other words, on $\mathcal{S}_d^s \cap Z_d$, the configuration variables are determined. In particular, let $\pi_{q_i} : (q_i, \dot{q}_i) \mapsto q_i$ be the canonical projection map. Then, $\pi_{q_i}(\mathcal{S}_d^s \cap Z_d) = \{\bar{q}_{id}\}$.

In addition,

$$
\operatorname{rank} \frac{\partial F_d}{\partial x_d}(x_d) = \operatorname{rank}
\begin{bmatrix}
\frac{\partial h_d}{\partial q_i}(q_i) & 0_{2 \times 3} \\[4pt]
\frac{\partial \bar{x}_H}{\partial q_i}(q_i) & 0_{1 \times 3} \\[4pt]
\frac{\partial}{\partial q_i}\left(\frac{\partial h_d}{\partial q_i}(q_i)\dot{q}_i \right) & \frac{\partial h_d}{\partial q_i}(q_i)
\end{bmatrix},
$$

which in combination with Lemma 3.1 implies that for every $x_d \in S_d^s \cap Z_d$, $\operatorname{rank} \frac{\partial F_d}{\partial x_d}(x_d) = 5$. ∎

Analogous to the derivation for the single support phase, the feedback law for the double support phase is chosen to be a continuous time-invariant feedback law having the following form:

$$
\begin{bmatrix} u_{3d}(x_d) \\ u_{4d}(x_d) \end{bmatrix}
= - \left(L_{g_d} L_{f_d} h_d(x_d) \right)^{-1} \left(L_{f_d}^2 h_d(x_d, u_{1d}(x_d), u_{2d}(x_d)) - v_d(h_d(x_d), L_{f_d} h_d(x_d)) \right)
$$

$$
= - \underbrace{\left(\frac{\partial h_d}{\partial q_i} D_\psi^{-1} \beta_\psi \right)^{-1}}_{L_{g_d} L_{f_d} h_d} \Bigg(\underbrace{\frac{\partial}{\partial q_i}\left(\frac{\partial h_d}{\partial q_i}\dot{q}_i \right)\dot{q}_i - \frac{\partial h_d}{\partial q_i} D_\psi^{-1} \left(C_\psi \dot{q}_i + G_\psi - \frac{\partial \Psi'}{\partial q_i} \begin{bmatrix} u_{1d} \\ u_{2d} \end{bmatrix} \right)}_{L_{f_d}^2 h_d}
$$

$$
- v_d \Bigg), \tag{3.29}
$$

where $v_d : \mathbb{R}^2 \times \mathbb{R}^2 \to \mathbb{R}^2$ is a continuous function such that the origin for the closed-loop system $\ddot{y}_d = v_d(y_d, \dot{y}_d)$ is globally finite-time stable. The feedback law in equation (3.29) renders Z_d locally finite-time attractive and forward invariant under the closed-loop dynamics of the double support phase. By definition, Z_d is the *double support phase zero dynamics manifold* and $\dot{z}_d = f_{\text{zero},d}(z_d, u_{1d}(z_d), u_{2d}(z_d))$ is the *double support phase zero dynamics*, where

$$
f_{\text{zero},d}(z_d, u_{1d}(z_d), u_{2d}(z_d)) := f_d^* |_{Z_d}(z_d, u_{1d}(z_d), u_{2d}(z_d))
$$

$$
f_d^*(x_d, u_{1d}(x_d), u_{2d}(x_d)) := f_d(x_d, u_{1d}(x_d), u_{2d}(x_d)) + g_d(x_d) \begin{bmatrix} u_{3d}^*(x_d) \\ u_{4d}^*(x_d) \end{bmatrix}
$$

$$
\begin{bmatrix} u_{3d}^*(x_d) \\ u_{4d}^*(x_d) \end{bmatrix} := -(L_{g_d} L_{f_d} h_d(x_d))^{-1} L_{f_d}^2 h_d(x_d, u_{1d}(x_d), u_{2d}(x_d)).
$$

$$
\tag{3.30}
$$

From Lemma 3.1, $\Phi_d(q_i) = [h'_d(q_i), \bar{x}_H(q_i)]'$ is a valid coordinate transformation on $\check{Q}_i$ and thus,

$$
\begin{bmatrix} \eta_1 \\ \eta_2 \\ \vartheta_1 \\ \vartheta_2 \end{bmatrix} = \begin{bmatrix} h_d(q_i) \\ L_{f_d} h_d(q_i, \dot{q}_i) \\ \bar{x}_H(q_i) \\ L_{f_d} \bar{x}_H(q_i, \dot{q}_i) \end{bmatrix}
$$

is a valid coordinate transformation on $T\check{Q}_i$. Consequently, on the manifold Z_d, q_i and $\dot{q}_i$ can be given by

$$
q_i = \Phi_d^{-1}\left(\begin{bmatrix} 0_{2\times 1} \\ \vartheta_1 \end{bmatrix} \right)
$$

$$
\dot{q}_i = \frac{\partial \Phi_d}{\partial q_i}(q_i)^{-1} \begin{bmatrix} 0_{2\times 1} \\ 1 \end{bmatrix} \vartheta_2 =: \lambda_d(q_i)\, \vartheta_2.
$$

Now we are able to present the main result of this section that is expressed as the following lemma. This lemma proposes a closed form for the zero dynamics of the double support phase in the local coordinates $(\bar{x}_H, \bar{v}_{xH})$, where $\bar{v}_{xH}$ denotes the horizontal velocity of the hip joint.

Lemma 3.3 (Double Support Phase Zero Dynamics)   *Assume that $u_1 = u_{1d}(q_i)$ and $u_2 = u_{2d}(q_i)$. Then, the double support phase zero dynamics can be expressed as*

$$
\begin{aligned}
\dot{\bar{x}}_H &= \bar{v}_{xH} \\
\dot{\bar{v}}_{xH} &= \omega_1(\bar{x}_H) + \omega_2(\bar{x}_H)\, \bar{v}_{xH}^2,
\end{aligned}
\tag{3.31}
$$

in which

$$
\omega_1(\bar{x}_H) := \frac{\partial \bar{x}_H}{\partial q_i} D_\psi^{-1} \Gamma \left(\frac{\partial \Psi'}{\partial q_i} \begin{bmatrix} u_{1d} \\ u_{2d} \end{bmatrix} - G_\psi \right)
$$

$$
\begin{aligned}
\omega_2(\bar{x}_H) := &-\frac{\partial \bar{x}_H}{\partial q_i} D_\psi^{-1} \Gamma \bar{C}_\psi \lambda_d \\
&-\frac{\partial \bar{x}_H}{\partial q_i} D_\psi^{-1} \beta_\psi \left(\frac{\partial h_d}{\partial q_i} D_\psi^{-1} \beta_\psi \right)^{-1} \frac{\partial}{\partial q_i} \left(\frac{\partial h_d}{\partial q_i} \lambda_d \right) \lambda_d \\
&+ \lambda'_d \frac{\partial^2 \bar{x}_H}{\partial q_i^2} \lambda_d,
\end{aligned}
$$

$\Gamma(q_i) := I_{3\times 3} - \beta_\psi (\frac{\partial h_d}{\partial q_i} D_\psi^{-1} \beta_\psi)^{-1} \frac{\partial h_d}{\partial q_i} D_\psi^{-1}$ *and* $\bar{C}_\psi(q_i) := C_\psi(q_i, \lambda_d(q_i))$.

The proof is given in Appendix A.1. Our aim is to provide closed-form expression for the solutions of the double support phase zero dynamics. To achieve this goal, note that if on the zero dynamics manifold, $\bar{v}_{xH} \neq 0$, equation (3.31) can be rewritten as follows:

$$\frac{d\bar{v}_{xH}}{d\bar{x}_H} = \omega_2(\bar{x}_H)\,\bar{v}_{xH} + \frac{\omega_1(\bar{x}_H)}{\bar{v}_{xH}},$$

which is a type of Bernoulli's equation. Substituting $\bar{z}_{xH} := (\bar{v}_{xH})^2$ reduces the Bernoulli's equation to a first order nonhomogeneous linear equation with the following form:

$$\frac{d\bar{z}_{xH}}{d\bar{x}_H} - 2\omega_2(\bar{x}_H)\,\bar{z}_{xH} = 2\omega_1(\bar{x}_H), \tag{3.32}$$

for which the solutions can be expressed in closed form. To show this, let $x_{H,d}^+$ be a constant scalar such that $x_{H,d}^+ < x_{H,d}^-$. Then, equation (3.32) over the interval $[x_{H,d}^+, x_{H,d}^-]$ with the initial condition $\bar{z}_{xH}(x_{H,d}^+) = \bar{z}_{xH}^+ := (\bar{v}_{xH}^+)^2$ has the following solution:

$$\bar{z}_{xH}(\bar{x}_H) = \Omega_2(\bar{x}_H)(-\mathcal{W}_{\text{zero},d}(\bar{x}_H) + \bar{z}_{xH}^+), \tag{3.33}$$

where

$$\Omega_2(\bar{x}_H) := \exp\left(2\int_{x_{H,d}^+}^{\bar{x}_H} \omega_2(\xi)d\xi\right)$$

$$\mathcal{W}_{\text{zero},d}(\bar{x}_H) := -2\int_{x_{H,d}^+}^{\bar{x}_H} \frac{\omega_1(\xi)}{\Omega_2(\xi)}d\xi.$$

Note that since $\bar{z}_{xH} = (\bar{v}_{xH})^2 > 0$ and $\Omega_2(\bar{x}_H) > 0$, this solution is valid as long as $\bar{z}_{xH}^+ > \mathcal{W}_{\text{zero},d}^{\max}$, where

$$\mathcal{W}_{\text{zero},d}^{\max} := \max_{x_{H,d}^+ \leq \bar{x}_H \leq x_{H,d}^-} \mathcal{W}_{\text{zero},d}(\bar{x}_H).$$

Moreover, $\mathcal{S}_d^s \cap Z_d$ can be expressed by

$$\mathcal{S}_d^s \cap Z_d = \{(q_i', \dot{q}_i')' | q_i = q_{id}^-, \dot{q}_i = \lambda_d(q_{id}^-)\,\bar{v}_{xH}^-, \ \bar{v}_{xH}^- \in \mathbb{R}\}.$$

3.4   HYBRID ZERO DYNAMICS (HZD)

The concept of HZD was introduced in Ref. [52]. In order to reduce the dimension of the hybrid model of walking, by assumption of hybrid invariance, the zero dynamics manifolds of the single and double support phases can be assembled into a hybrid restricted dynamics called HZD. The hybrid restricted dynamics will result in a low-dimensional test to investigate the stability properties of a periodic orbit of the open-loop hybrid model of walking that is also an integral curve of HZD. This section presents the HZD for the walking model. To achieve this result, let $\Delta_s^d(\mathcal{S}_s^d \cap Z_s) \subset Z_d$ and $\Delta_d^s(\mathcal{S}_d^s \cap Z_d) \subset Z_s$. Then, HZD for the hybrid system in equation (3.22) can be defined as follows:

$$
\Sigma_{\text{zero}} : \begin{cases}
\dot{z}_s = f_{\text{zero},s}(z_s) & z_s^- \notin \mathcal{S}_s^d \cap Z_s \\
z_d^+ = \Delta_{\text{zero},s}^d(z_s^-) & z_s^- \in \mathcal{S}_s^d \cap Z_s \\
\dot{z}_d = f_{\text{zero},d}(z_d) & z_d^- \notin \mathcal{S}_d^s \cap Z_d \\
z_s^+ = \Delta_{\text{zero},d}^s(z_d^-) & z_d^- \in \mathcal{S}_d^s \cap Z_d,
\end{cases}
\tag{3.34}
$$

where $\Delta_{\text{zero},s}^d$ and $\Delta_{\text{zero},d}^s$ are restrictions of the switching maps Δ_s^d and Δ_d^s to the manifolds Z_s and Z_d, respectively.

3.4.1 Analysis of HZD in the Single Support Phase

Let $\Delta_d^s(\mathcal{S}_d^s \cap Z_d) \subset Z_s$. In the local coordinates (θ_s, σ_s) for the manifold Z_s, the values of the quantities θ_s and σ_s at the beginning of the single support phase can be expressed as

$$
\begin{aligned}
\theta_s^+ &:= \theta_s(q_s^+) = \theta_s \circ \Delta_{q,d}^s(q_{id}^-) \\
\sigma_s^+ &= D_5(q_s^+)\dot{q}_s^+ \\
&= D_5 \circ \Delta_{q,d}^s(q_{id}^-)\, \Delta_{\dot{q},d}^s(q_{id}^-)\, \lambda_d(q_{id}^-)\, \bar{v}_{xH}^- \\
&=: \delta_d^s(q_{id}^-)\, \bar{v}_{xH}^-.
\end{aligned}
$$

Consequently, the restricted transition map $\Delta_{\text{zero},d}^s : \mathcal{S}_d^s \cap Z_d \to Z_s$ can be given by

$$
\Delta_{\text{zero},d}^s(x_{H,d}^-, \bar{v}_{xH}^-) := \begin{bmatrix} \theta_s^+ \\ \delta_d^s(q_{id}^-)\, \bar{v}_{xH}^- \end{bmatrix},
\tag{3.35}
$$

where $q_{id}^- = \Phi_d^{-1}([0_{1\times 2}, x_{H,d}^-]')$. Following the results in Ref. [52], when the robot takes a step, the angular momentum about the stance leg end is nonzero. Thus, $\zeta_s := \frac{1}{2}(\sigma_s)^2$ is a valid coordinate transformation. Furthermore, since the single support phase zero dynamics is Lagrangian, $E_{\text{zero},s} := K_{\text{zero},s} + V_{\text{zero},s}$ is stationary on Z_s and consequently,

$$
\frac{1}{2}(\sigma_s^-)^2 - \frac{1}{2}(\sigma_s^+)^2 = \zeta_s^- - \zeta_s^+ = -V_{\text{zero},s}(\theta_s^-).
$$

By introducing $\bar{z}_{xH}^- := (\bar{v}_{xH}^-)^2$, the *restricted generalized Poincaré map of the single support phase* can be defined as $\rho_s : \mathcal{S}_d^s \cap Z_d \to \mathcal{S}_s^d \cap Z_s$ by

$$\rho_s(\bar{z}_{xH}^-) := \zeta_s^+ - V_{\text{zero},s}(\theta_s^-)$$

$$= \frac{1}{2}(\delta_d^s)^2 \, \bar{z}_{xH}^- - V_{\text{zero},s}(\theta_s^-).$$

Due to the fact that $\zeta_s^- = \frac{1}{2}(\sigma_s^-)^2 > 0$, the domain of definition of ρ_s can also be given by

$$\mathcal{D}_{\rho_s} := \left\{ \bar{z}_{xH}^- > 0 \,\Big|\, \frac{1}{2}(\delta_d^s)^2 \, \bar{z}_{xH}^- > V_{\text{zero},s}^{\max} \right\},$$

where

$$V_{\text{zero},s}^{\max} := \max_{\theta_s^+ \leq \theta_s \leq \theta_s^-} V_{\text{zero},s}(\theta_s).$$

Remark 3.8 (Upper Bound in $\mathcal{D}_{\rho_s}$) *As in Ref. [52], there exists an upper bound in the domain of definition $\mathcal{D}_{\rho_s}$. This upper bound is the largest value of $\bar{z}_{xH}^-$ such that the ground reaction force at the stance leg end is admissible (see Remark 3.3).*

The following lemma determines the set of all points in $\mathcal{S}_d^s \cap Z_d$ for which the transition from the double support phase to the single support phase can occur on HZD. Using this lemma, the domain of definition of ρ_s (i.e., $\mathcal{D}_{\rho_s}$) can be modified.

Lemma 3.4 (Transition Condition from DS to SS on HZD)[12] *Let $\Delta_d^s(\mathcal{S}_d^s \cap Z_d) \subset Z_s$. Then, there exist functions $\bar{\omega}_1, \bar{\omega}_2 : \mathbb{R} \to \mathbb{R}$ such that the transition condition from the double support phase to the single support phase on HZD can be expressed as*

$$\bar{\omega}_1(x_{H,d}^-) + \bar{\omega}_2(x_{H,d}^-)\, \bar{z}_{xH}^- > 0.$$

The proof is given in Appendix A.2. From Lemma 3.4, the domain of definition $\mathcal{D}_{\rho_s}$ is also modified as follows:

$$\mathcal{D}_{\rho_s} := \left\{ \bar{z}_{xH}^- > 0 \,\Big|\, \frac{1}{2}(\delta_d^s)^2 \, \bar{z}_{xH}^- > V_{\text{zero},s}^{\max}, \ \bar{\omega}_1 + \bar{\omega}_2 \, \bar{z}_{xH}^- > 0 \right\}.$$

---

[12] DS and SS represent the double support and single support phases, respectively.

3.4.2 Analysis of HZD in the Double Support Phase

Let $\Delta_s^d(S_s^d \cap Z_s) \subset Z_d$. In the local coordinates $(\bar{x}_H, \bar{v}_{xH})$ for the manifold Z_d, the initial values of $\bar{x}_H$ and $\bar{v}_{xH}$ can be expressed as

$$
\begin{aligned}
x_{H,d}^+ &:= \bar{x}_H(q_i^+) = \bar{x}_H \circ \Delta_{q_i,s}^d(q_s^-) \\
\bar{v}_{xH}^+ &= \frac{\partial \bar{x}_H}{\partial q_i}(q_i^+)\dot{q}_i^+ \\
&= \frac{\partial \bar{x}_H}{\partial q_i} \circ \Delta_{q_i,s}^d(q_s^-)\,\Delta_{\dot{q}_i,s}^d(q_s^-)\,\lambda_s(q_s^-)\,\sigma_s^- \\
&=: \delta_s^d(q_s^-)\,\sigma_s^- .
\end{aligned}
$$

Hence, the restricted transition map $\Delta_{\mathrm{zero},s}^d : S_s^d \cap Z_s \to Z_d$ can be given by

$$
\Delta_{\mathrm{zero},s}^d(\theta_s^-, \sigma_s^-) := \begin{bmatrix} x_{H,d}^+ \\ \delta_s^d(q_s^-)\,\sigma_s^- \end{bmatrix}, \tag{3.36}
$$

where $q_s^- = \Phi_s^{-1}([0_{1\times4}, \theta_s^-]')$ and

$$
\Phi_s(q_s) := \begin{bmatrix} h_s(q_s) \\ \theta_s(q_s) \end{bmatrix}.
$$

From equation (3.33),

$$
\bar{z}_{xH}^- = \Omega_2(x_{H,d}^+)(\bar{z}_{xH}^+ - W_{\mathrm{zero},d}(x_{H,d}^+)) \tag{3.37}
$$

and thus, the *restricted generalized Poincaré map of the double support phase* can be expressed as $\rho_d : S_s^d \cap Z_s \to S_d^s \cap Z_d$ by

$$
\begin{aligned}
\rho_d(\zeta_s^-) &:= \Omega_2(x_{H,d}^+)((\bar{v}_{xH}^+)^2 - W_{\mathrm{zero},d}(x_{H,d}^+)) \\
&= \Omega_2(x_{H,d}^+)(2(\delta_s^d)^2\,\zeta_s^- - W_{\mathrm{zero},d}(x_{H,d}^+)).
\end{aligned}
$$

The domain of definition of ρ_d is also given by

$$
\mathcal{D}_{\rho_d} := \left\{ \zeta_s^- > 0 \mid 2(\delta_d^s)^2\,\zeta_s^- > W_{\mathrm{zero},d}^{\max} \right\}.
$$

Remark 3.9 (Upper Bound in $\mathcal{D}_{\rho_d}$)   *There exists an upper bound in the domain of definition $\mathcal{D}_{\rho_d}$ due to admissibility of the ground reaction forces at the end of the legs during the double support phase.*

3.4.3 Restricted Poincaré Return Map

This section deals with a procedure for constructing a restricted Poincaré return map for the hybrid model of walking (see Section 2.3). Moreover, the fundamental results of this section that are developed to test the stability behavior of a periodic orbit of HZD are summarized. First, we shall define the restricted Poincaré return map. Let $\Delta_s^d(S_s^d \cap Z_s) \subset Z_d$ and $\Delta_d^s(S_d^s \cap Z_d) \subset Z_s$. Then, in the local coordinates (θ_s, ζ_s) for the manifold Z_s, the *restricted Poincaré return map* is defined as $\rho : S_s^d \cap Z_s :\rightarrow S_s^d \cap Z_s$ by

$$
\begin{aligned}
\rho(\zeta_s^-) &:= \rho_s \circ \rho_d(\zeta_s^-) \\
&= (\delta_d^s)^2 (\delta_s^d)^2 \, \Omega_2(x_{H,d}^-) \, \zeta_s^- - \frac{1}{2}(\delta_d^s)^2 \, \Omega_2(x_{H,d}^-) \, W_{\text{zero},d}(x_{H,d}^-) - V_{\text{zero},s}(\theta_s^-).
\end{aligned}
$$

Moreover, the domain of definition of ρ can be given by

$$
\mathcal{D}_\rho := \{\zeta_s^- \in \mathcal{D}_{\rho_d} | \rho_d(\zeta_s^-) \in \mathcal{D}_{\rho_s}\}.
$$

The following lemma is an important result that enables ρ to be considered as the Poincaré return map for a system with impulse effects. Consequently, the results developed for the existence and stability analysis of the periodic orbits in systems with impulse effects, see Theorem 2.1, Section 2.2, can be applied to the hybrid model of walking.

Lemma 3.5 (HZD as a System with Impulse Effects)   *Let $\Delta_s^d(S_s^d \cap Z_s) \subset Z_d$ and $\Delta_d^s(S_d^s \cap Z_d) \subset Z_s$. Assume that walking occurs from left to right. Then, in the coordinates (θ_s, σ_s) for the manifold Z_s, ρ is also a Poincaré return map for the system with impulse effects*

$$
\Sigma_{\text{zero},s} : \begin{cases} \dot{z}_s = f_{\text{zero},s}(z_s), & z_s^- \notin S_s^d \cap Z_s \\ z_s^+ = \Delta_{\text{zero},s}(z_s^-), & z_s^- \in S_s^d \cap Z_s, \end{cases} \tag{3.38}
$$

where $\Delta_{\text{zero},s}(\theta_s^-, \sigma_s^-) := [\theta_s^+, \varpi(\sigma_s^-)]'$ and

$$
\varpi(\sigma_s^-) := \delta_d^s \sqrt{\Omega_2(x_{H,d}^-)\left((\delta_s^d)^2(\sigma_s^-)^2 - W_{\text{zero},d}(x_{H,d}^-)\right)}. \tag{3.39}
$$

Proof. From the procedure for constructing $\Sigma_{\text{zero},s}$, $\Delta_{\text{zero},s} = \Delta_{\text{zero},d}^s \circ \rho_d$. Moreover, Theorem 2.1 of Chapter 2 immediately implies that ρ is a Poincaré return map for $\Sigma_{\text{zero},s}$. Since walking is from left to right (i.e., $\bar{v}_{xH} > 0$), equations (3.36) and (3.35) in combination with equation (3.37) yield equation (3.39). ∎

Definition 3.2 (Continuously Differentiable HZD)   *HZD is said to be continuously differentiable if* $f_{\text{zero},s} : Z_s \rightarrow TZ_s$, $f_{\text{zero},d} : Z_d \rightarrow TZ_d$, $\Delta^d_{\text{zero},s} : \mathcal{S}^d_s \cap Z_s \rightarrow Z_d$ *and* $\Delta^s_{\text{zero},d} : \mathcal{S}^s_d \cap Z_d \rightarrow Z_s$ *are* C^1.

To investigate the stability behavior of the periodic orbits of HZD, we prove the following theorem that is the main result of this section. This theorem establishes an equivalence between the stability of the periodic orbits of HZD and the equilibrium points of the discrete-time system $\zeta_s^-[k+1] = \rho(\zeta_s^-[k])$ with the state space $\mathcal{S}^d_s \cap Z_s$.

Theorem 3.2 (Exponentially Stable Periodic Orbits of HZD)   *Assume that* $\mathcal{S}^d_{s} \cap Z_s$ *and* $\mathcal{S}^s_d \cap Z_d$ *are embedded one-dimensional submanifolds of* $T\mathcal{Q}_s$ *and* $T\tilde{\mathcal{Q}}_i$, *respectively. Moreover, suppose that* $\Delta^d_s(\mathcal{S}^d_s \cap Z_s) \subset Z_d$, $\Delta^s_d(\mathcal{S}^s_d \cap Z_d) \subset Z_s$, *and HZD is* C^1. *By defining*

$$\zeta_s^* := -\frac{\frac{1}{2}(\delta^s_d)^2 \, \Omega_2(x^-_{H,d}) \, W_{\text{zero},d}(x^-_{H,d}) + V_{\text{zero},s}(\theta^-_s)}{1-\mu},$$

in which

$$\mu := (\delta^s_d)^2 (\delta^d_s)^2 \, \Omega_2(x^-_{H,d}), \tag{3.40}$$

the following statements are true.

1. *If* $\zeta_s^* \in \mathcal{D}_\rho$, *then* ζ_s^* *is the fixed point of* ρ.
2. *If* $\zeta_s^* \in \mathcal{D}_\rho$, *then* ζ_s^* *is a locally exponentially stable equilibrium point of* $\zeta_s^-[k+1] = \rho(\zeta_s^-[k])$ *if and only if* $\mu < 1$.
3. *HZD has a nontrivial periodic orbit transversal to* $\mathcal{S}^d_s \cap Z_s$ *and* $\mathcal{S}^s_d \cap Z_d$ *if and only if* $\mu \neq 1$ *and* $\zeta_s^* \in \mathcal{D}_\rho$.
4. *HZD has an exponentially stable periodic orbit transversal to* $\mathcal{S}^d_s \cap Z_s$ *and* $\mathcal{S}^s_d \cap Z_d$ *if and only if* $\zeta_s^* \in \mathcal{D}_\rho$ *and* $\mu < 1$.

Proof. By considering the fact that μ is nonnegative, all of the statements are immediate consequences of Theorems 2.2 and 2.4 of Chapter 2 and Lemma 3.5. ∎

Remark 3.10 (Effect of $u_{1d}(q_i)$ and $u_{2d}(q_i)$ on Stability of Periodic Orbits)   *Since during the double support phase, $u_1 = u_{1d}(q_i)$ and $u_2 = u_{2d}(q_i)$, from Lemma 3.3, $\omega_2(\bar{x}_H)$ and thereby $\Omega_2(x^-_{H,d})$ and μ are independent of these control inputs. Thus, the functions $u_{1d}(q_i)$ and $u_{2d}(q_i)$ do not affect the stability of the fixed point. However, the existence of a limit cycle and the value of ζ_s^* are affected by the choice of $u_{1d}(q_i)$ and $u_{2d}(q_i)$.*

Remark 3.11 (Using u_1 and u_2 to Zero the Output Function)   *By using u_1 and u_2 to zero the holonomic output function $h_d(q_i)$, the decoupling matrix can be expressed as $L_{g_d} L_{f_d} h_d(q_i) = \frac{\partial h_d}{\partial q_i} D_\psi^{-1} \frac{\partial \Psi}{\partial q_i}'$. It can be shown that the (2×2) upper submatrix of*

$\frac{\partial \Psi}{\partial q_i}'$ is full rank on $\tilde{Q}_i$ and its determinant is equal to $\frac{\sin q_1}{\sin q_3}$. Thus, from Lemma 3.1, $1 \leq \operatorname{rank} L_{g_d} L_{f_d} h_d(q_i) \leq 2$, for every $q_i \in \tilde{Q}_i$. However, small numerical values of q_1 and q_3 (as can be observed on a typical gait) may result in significant errors while computing u_1 and u_2, due to the term $\frac{\sin q_1}{\sin q_3}$ in $(L_{g_d} L_{f_d} h_d(q_i))^{-1}$.

3.5 DESIGN OF AN HZD CONTAINING A PRESPECIFIED PERIODIC SOLUTION

The objective of this section is to design an HZD containing a desired feasible period-one solution of the open-loop hybrid model. For this purpose, the sample-based virtual constraints method introduced in Ref. [94] is used. However, since in the double support phase, u_1 and u_2 are not employed for the output zeroing problem and the open-loop control input corresponding to a trajectory is not unique, the sample-based virtual constraints method is not sufficient to achieve the objective. This section presents a design method that ensures that a desired feasible periodic solution is an integral curve of HZD and the control input associated with this solution in the double support phase has minimum norm. The required conditions will be specified as we proceed.

Let $\mathcal{O} := \mathcal{O}_s \cup \mathcal{O}_d$ be a period-one solution of the open-loop hybrid model in equation (3.22). Suppose that $\mathbf{q}_s(t)$, $0 \leq t < T_s$ and $\mathbf{q}_d(t)$, $T_s \leq t < T_s + T_d =: T$ represent the time evolutions of the configuration variables on $\mathcal{O}_s$ and $\mathcal{O}_d$, respectively. Moreover, T_s and T_d are the time durations of the single and double support phases on $\mathcal{O}$. For the later purposes, let $\mathbf{q}_{id}(t)$ denote the set of independent configuration variables of $\mathbf{q}_d(t)$.

3.5.1 Design of the Output Functions

Assume that the following hypotheses of periodic orbit are satisfied:[13]

(HPO1) $\mathbf{q}_s(t)$ and $\mathbf{q}_d(t)$ are at least three-times continuously differentiable on $[0, T_s)$ and $[T_s, T)$, respectively.

(HPO2) $\mathcal{O}_s$ is transversal to $\mathcal{S}_s^d$ and $\mathcal{O}_d$ is transversal to $\mathcal{S}_d^s$.

(HPO3) $\Theta_s(t)$, the time evolution of θ_s in the single support phase, is a strictly increasing function of time (i.e., $\inf_{0 \leq t < T_s} \dot{\Theta}_s(t) > 0$).

(HPO4) $\mathbf{x}_{Hd}(t)$, the time evolution of x_H in the double support phase, is a strictly increasing function of time (i.e., $\inf_{T_s \leq t < T} \dot{\mathbf{x}}_{Hd}(t) > 0$).

(HPO5) The angular momentum about the stance leg end during the single support phase is nonzero.[14]

---

[13] Hypotheses HPO1, HPO2, HPO3, and HPO5 are taken from Ref. [18, p. 162].

[14] Hypotheses HPO3 and HPO5, together with Remark 3.7, imply the invertibility of the decoupling matrix $L_{g_s} L_{f_s} h_s(q_s)$ on the orbit $\mathcal{O}_s$.

Next define

$$h_{d,s}(\theta_s) := \mathbf{q}_{bs}(t)\Big|_{t=\Theta_s^{-1}(\theta_s)} \tag{3.41}$$

and

$$h_{d,d}(\bar{x}_H) := \begin{bmatrix} \mathbf{y}_{Hd}(t) \\ \mathbf{q}_{5d}(t) \end{bmatrix}\Bigg|_{t=\mathbf{x}_{Hd}^{-1}(\bar{x}_H)}, \tag{3.42}$$

where $\mathbf{q}_{bs}(t)$ is the time evolution of the body configuration variables on $\mathcal{O}_s$, and $\mathbf{y}_{Hd}(t)$ and $\mathbf{q}_{5d}(t)$ are the time evolutions of the vertical displacement of the hip joint and torso angle on $\mathcal{O}_d$, respectively. Furthermore, suppose that $\mathcal{O}$ satisfies the following additional hypothesis.

(HPO6) For every $t \in [T_s, T)$, the decoupling matrix $L_{g_d}L_{f_d}h_d(\mathbf{q}_{id}(t))$ is invertible.

Hypotheses HPO3, HPO5, and HPO6 in combination with Remark 3.7 imply that there exist open sets $\tilde{\mathcal{Q}}_s \subset \mathcal{Q}_s$ and $\check{\mathcal{Q}}_i \subset \tilde{\mathcal{Q}}_i$ such that the decoupling matrices $L_{g_s}L_{f_s}h_s(q_s)$ and $L_{g_g}L_{f_d}h_d(q_i)$ are invertible on them, and thus the single and double support phase zero dynamics (Z_s and Z_d) exist. Moreover, by HPO2, Theorem 6.2 of Ref. [18, p. 163], and Lemma 3.2, $\mathcal{S}_s^d \cap Z_s$ and $\mathcal{S}_d^s \cap Z_d$ are embedded one-dimensional submanifolds of $T\mathcal{Q}_s$ and $T\tilde{\mathcal{Q}}_i$, respectively. The following lemma presents the main result of this section that utilizes the previously mentioned construction procedure for holonomic output functions in the single and double support phases to establish that the zero dynamics manifolds Z_s and Z_d are *hybrid invariant* for the hybrid model of walking.

Lemma 3.6 (Hybrid Invariance)   *Let $\mathcal{O}$ be a periodic orbit of the hybrid model in equation (3.22) satisfying hypotheses HPO1–HPO6. Then, for the output functions in equations (3.23) and (3.27) in combination with equations (3.41) and (3.42), the corresponding zero dynamics manifolds (Z_s and Z_d) are hybrid invariant, that is, $\Delta_s^d(\mathcal{S}_s^d \cap Z_s) \subset Z_d$ and $\Delta_d^s(\mathcal{S}_d^s \cap Z_d) \subset Z_s$. Moreover, HZD exists.*

Proof. Let (θ_s, σ_s) and $(\bar{x}_H, \bar{v}_{xH})$ be the local coordinates for Z_s and Z_d, respectively. By HPO2, $\sigma_s^* \neq 0$ and $\bar{v}_{xH}^* \neq 0$, where σ_s^* and $\bar{v}_{xH}^*$ are the values of the quantities σ_s and $\bar{v}_{xH}$ in the points $x_s^* := \mathcal{O}_s \cap \mathcal{S}_s^d$ and $x_d^* := \mathcal{O}_d \cap \mathcal{S}_d^s$, respectively. Since $\mathcal{S}_s^d \cap Z_s$ is an embedded one-dimensional submanifold of $T\mathcal{Q}_s$, equation (3.26) implies that on $\mathcal{S}_s^d \cap Z_s$, $q = q_s^-$ and $\dot{q} = \lambda_s(q_s^-)\sigma_s^-$, where $\sigma_s^- \in \mathbb{R}$. Furthermore, from the construction procedure for $h_{d,d}$,

$$h_d \circ \Delta_{q_i,s}^d(q_s^-) = 0_{2\times 1}$$

$$L_{f_d} h_d \circ \Delta_s^d(q_s^-, \lambda_s(q_s^-)\sigma_s^*) = \frac{\partial h_d}{\partial q_i} \circ \Delta_{q_i,s}^d(q_s^-) \Delta_{\dot{q}_i,s}^d(q_s^-) \lambda_s(q_s^-)\sigma_s^* \quad (3.43)$$

$$= 0_{2\times 1}.$$

Since $L_{f_d} h_d \circ \Delta_s^d(q_s^-, \lambda_s(q_s^-)\sigma_s^-)$ is linear with respect to σ_s^- and $\sigma_s^* \neq 0$, equation (3.43) implies that for every $\sigma_s^- \in \mathbb{R}$, $L_{f_d} h_d \circ \Delta_s^d(q_s^-, \lambda_s(q_s^-)\sigma_s^-) = 0_{2\times 1}$ and hence, $\Delta_s^d(\mathcal{S}_s^d \cap Z_s) \subset Z_d$. In a similar manner, it can be shown that $\Delta_d^s(\mathcal{S}_d^s \cap Z_d) \subset Z_s$. The existence of HZD is immediate. ∎

3.5.2 Design of u_{1d} and u_{2d}

In the single support phase, if $\mathbf{q}_s(t)$ is known, the corresponding open-loop control input $\mathbf{u}_s(t)$ is determined uniquely. In particular, let $H_0 := [I_{4\times 4} \, 0_{4\times 1}]$, then,

$$\mathbf{u}_s(t) = H_0(D\ddot{\mathbf{q}}_s(t) + C\dot{\mathbf{q}}_s(t) + G). \quad (3.44)$$

However, in the double support phase, if $\mathbf{q}_d(t)$ is known, the corresponding open-loop control input is not unique because the mechanical system is overactuated. The following lemma presents the set of C^1 open-loop control inputs that correspond to $\mathcal{O}_d$.

Lemma 3.7 (C^1 Open-Loop Control Inputs Corresponding to $\mathcal{O}_d$) *Let $\mathcal{O}$ be a periodic orbit of the open-loop hybrid system in equation (3.22) satisfying HPO1. Furthermore, assume that the step length is nonzero (i.e., $L_s \neq 0$) and rank $\frac{\partial p_2}{\partial q} = 2$ on $\mathcal{O}_d$. Then, C^1 open-loop control input vector corresponding to $\mathcal{O}_d$ is not unique and belongs to the set $\mathcal{U}_d(\mathcal{O}_d)$, where*[15]

$$\mathcal{U}_d(\mathcal{O}_d) := \left\{ u_d : [T_s, T) \to \mathbb{R}^4 \mid u_d(t) = u_d^0(t) - \frac{\partial x_2'}{\partial q_b} \Upsilon^h(t), \ \Upsilon^h \in C^1([T_s, T), \mathbb{R}) \right\}$$

and for every $t \in [T_s, T)$,

$$u_d^0(t) := \left(H_0 - \frac{1}{L_s} \frac{\partial y_2'}{\partial q_b} e_5' \right) (D\ddot{\mathbf{q}}_d(t) + C\dot{\mathbf{q}}_d(t) + G). \quad (3.45)$$

Moreover, there exists a unique $F_2(t)$ for every $u_d(t) \in \mathcal{U}_d(\mathcal{O}_d)$ such that equation (3.14) is satisfied.

---

[15] In our notation, $C^k(I, \mathbb{R})$ denotes the set of all C^k functions $f : I \to \mathbb{R}$.

The proof is given in Appendix A.3. Define

$$\Lambda_u := \{u \in \mathbb{R}^4 \mid |u_i| < u_{\max}, i = 1, \ldots, 4\} \tag{3.46}$$

and

$$\Lambda_f := \{F := (F^h, F^v)' \in \mathbb{R}^2 \mid F^v > 0, |F^h| < \mu_s |F^v|\} \tag{3.47}$$

as the admissible regions for the control input and ground reaction forces, respectively, where $u_{\max}$ is a positive scalar.

Definition 3.3 (Feasible Trajectory of the Hybrid Model of Walking) *The trajectory $\mathcal{O}$ of the open-loop hybrid model of walking is feasible if*

1. *the constraints on the joint angles and angular velocities are satisfied on $\mathcal{O}$;*
2. *the open-loop control input and ground reaction force at the stance leg end corresponding to $q_s(t)$ are admissible;*
3. *there exists at least one admissible control input $u_d(t) \in \mathcal{U}_d(\mathcal{O}_d)$ such that the ground reaction forces at the leg ends corresponding to $q_d(t)$ and $u_d(t)$ are admissible;*
4. *the impact model on $\mathcal{O}$ is nonsingular and the impulsive ground reaction forces are admissible (see Remark 3.6). Moreover, $\ddot{y}_2(0) > 0$, where $y_2(t)$ is the time evolution of the vertical displacement of the swing leg end with respect to the world frame on $\mathcal{O}$.*

For the later purposes, assume that the periodic orbit $\mathcal{O}$ satisfies the following additional hypothesis.

(HPO7) The step length of the periodic orbit is positive (i.e., $L_s > 0$), rank $\frac{\partial p_2}{\partial q_d} = 2$ on $\mathcal{O}_d$, and $\mathcal{O}$ is feasible.[16]

Now let $\mathcal{O}$ be a periodic orbit of the hybrid system in equation (3.22) satisfying HPO1–HPO7. Since all of the open-loop control inputs corresponding to $\mathcal{O}_d$ are not admissible, it is difficult to design a time-invariant controller for the double support phase such that $\mathcal{O}_d$ is an integral curve of the closed-loop system. To overcome this difficulty, we make use of u_{1d} and u_{2d} to ensure admissibility of the double support phase controller. Assume that $u_{1d} = u_{1d}(\bar{x}_H)$ and $u_{2d} = u_{2d}(\bar{x}_H)$. On the manifold $Z_d, \bar{x}_H$ lies in $[x_{H,d}^+, x_{H,d}^-]$. Thus, for any $\bar{x}_H \in [x_{H,d}^+, x_{H,d}^-]$, let u_1 and u_2 be the values of the feedback laws $u_{1d}(\bar{x}_H)$ and $u_{2d}(\bar{x}_H)$, respectively. From equation (3.29), u_{3d}

---

[16] As discussed previously, rank $\frac{\partial p_2}{\partial q_d} = 2$ on $\mathcal{O}_d$, together with the Implicit Function Theorem, implies the existence of a unique function Ψ such that $q_d = \Psi(q_i)$ for every $q = (q_d', q_i')' \in \tilde{Q}_i$.

and u_{4d} on $\mathcal{O}_d$ are affine functions with respect to $(u_1, u_2)'$,

$$\begin{bmatrix} u_{3d}(t, u_1, u_2) \\ u_{4d}(t, u_1, u_2) \end{bmatrix} = \bar{U}_0(t) + \bar{U}_1(t) \begin{bmatrix} u_1 \\ u_2 \end{bmatrix}, \tag{3.48}$$

where from HPO4, the quantity t can be expressed in terms of $\bar{x}_H$, in particular, $t = \mathbf{x}_{Hd}^{-1}(\bar{x}_H)$. Equation (3.48) in combination with equations (3.16) and (3.13) yield the following affine relations for the ground reaction forces on $\mathcal{O}_d$:

$$F_1(t, u_1, u_2) = F_{01}(t) + F_{11}(t) \begin{bmatrix} u_1 \\ u_2 \end{bmatrix}$$

$$F_2(t, u_1, u_2) = F_{02}(t) + F_{12}(t) \begin{bmatrix} u_1 \\ u_2 \end{bmatrix}.$$

Next, for every $\bar{x}_H \in [x_{H,d}^+, x_{H,d}^-]$, evaluate $t = \mathbf{x}_{Hd}^{-1}(\bar{x}_H)$, $\bar{U}_0(t)$, $\bar{U}_1(t)$, $F_{01}(t)$, $F_{11}(t)$, $F_{02}(t)$, and $F_{22}(t)$ on $\mathcal{O}_d$ and define the following nonlinear optimization problem for determining u_1 and u_2:

$$\begin{aligned} &\min_{u_1, u_2} \frac{1}{2} \|u_d(t, u_1, u_2)\|_2^2 \\ &\text{s.t.} \quad u_d(t, u_1, u_2) \in \mathcal{U}_d(\mathcal{O}_d) \\ &\quad\quad (u_1, u_2)' \in \Lambda_a(\mathcal{O}_d, \bar{x}_H), \end{aligned} \tag{3.49}$$

where for every $\bar{x}_H \in [x_{H,d}^+, x_{H,d}^-]$, $\Lambda_a(\mathcal{O}_d, \bar{x}_H)$ is defined to be the set of all points $(u_1, u_2)' \in \mathbb{R}^2$ for which the control input vector and the ground reaction forces at the leg ends are admissible on the trajectory $\mathcal{O}_d$,

$$\Lambda_a(\mathcal{O}_d, \bar{x}_H) := \{(u_1, u_2)' \in \mathbb{R}^2 \,|\, u_d(t, u_1, u_2) \in \Lambda_u, F_1(t, u_1, u_2), F_2(t, u_1, u_2) \in \Lambda_f\}.$$

By the constraint $u_d \in \mathcal{U}_d(\mathcal{O}_d)$ in the optimization problem (3.49), $\mathcal{O}_d$ is an integral curve of the closed-loop system in the double support phase. Furthermore, the constraint $(u_1, u_2)' \in \Lambda_a(\mathcal{O}_d, \bar{x}_H)$ for every $\bar{x}_H \in [x_{H,d}^+, x_{H,d}^-]$ imposes the admissibility of u_d on $\mathcal{O}_d$. Note that from equation (3.48), u_d on $\mathcal{O}_d$ can be given by

$$\begin{aligned} u_d(t, u_1, u_2) &= \begin{bmatrix} 0_{2\times 1} \\ \bar{U}_0(t) \end{bmatrix} + \begin{bmatrix} I_{2\times 2} \\ \bar{U}_1(t) \end{bmatrix} \begin{bmatrix} u_1 \\ u_2 \end{bmatrix} \\ &=: U_0(t) + U_1(t) \begin{bmatrix} u_1 \\ u_2 \end{bmatrix}. \end{aligned} \tag{3.50}$$

The following lemma implies that for every $\bar{x}_H \in [x_{H,d}^+, x_{H,d}^-]$, the constraint $u_d(t, u_1, u_2) \in \mathcal{U}_d(\mathcal{O}_d)$ can be expressed as an affine equality constraint with respect to $(u_1, u_2)'$.

Lemma 3.8 *Let $\mathcal{O}$ be a periodic orbit of the open-loop hybrid model in equation (3.22) satisfying HPO1–HPO7. Then, there exist functions $V_0(t) \in \mathbb{R}$ and $0_{1\times 2} \neq V_1(t) \in \mathbb{R}^{1\times 2}$ such that on the periodic orbit $\mathcal{O}$, the constraint $u_d(t, u_1, u_2) \in \mathcal{U}_d(\mathcal{O}_d)$ can be expressed as*

$$
V_0(t) + V_1(t) \begin{bmatrix} u_1 \\ u_2 \end{bmatrix} = 0.
$$

Proof. By definition of the set $\mathcal{U}_d(\mathcal{O}_d)$ and equation (3.50), the constraint $u_d(t, u_1, u_2) \in \mathcal{U}_d(\mathcal{O}_d)$ is equivalent to

$$
U_0(t) + U_1(t) \begin{bmatrix} u_1 \\ u_2 \end{bmatrix} = \mathbf{u}_d^0(t) - \frac{\partial x_2'}{\partial q_b}(t) \Upsilon^h(t) \tag{3.51}
$$

for some $\Upsilon^h(t) \in C^1([T_s, T), \mathbb{R})$. Since $U_1'(t)U_1(t) = I_{2\times 2} + \bar{U}_1'(t)\bar{U}_1(t)$ is positive definite, equation (3.51) results in

$$
\begin{bmatrix} u_1 \\ u_2 \end{bmatrix} = \left(U_1'(t)U_1(t)\right)^{-1} U_1'(t) \left(\mathbf{u}_d^0(t) - U_0(t) - \frac{\partial x_2'}{\partial q_b}(t)\Upsilon^h(t)\right). \tag{3.52}
$$

Next, define

$$
V_2(t) := \left(U_1'(t)U_1(t)\right)^{-1} U_1'(t) \frac{\partial x_2'}{\partial q_b}(t)
$$
$$
\mathcal{P}(t) := \{y := (y_1, y_2) \in \mathbb{R}^{1\times 2} | yV_2(t) = 0\}.
$$

Choosing an arbitrary nonzero $V_1(t) \in \mathcal{P}(t)$ and $V_0(t)$ in the following form:

$$
V_0(t) := -V_1(t)\left(U_1'(t)U_1(t)\right)^{-1} U_1'(t)(\mathbf{u}_d^0(t) - U_0(t)) \tag{3.53}
$$

completes the proof.   ■

By HPO7, the trajectory $\mathcal{O}$ is feasible and as a consequence, the solution space of the optimization problem (3.49) is nonempty. Since the set $\Lambda_a(\mathcal{O}_d, \bar{x}_H)$ is open for every $\bar{x}_H \in [x_{H,d}^+, x_{H,d}^-]$, if the optimal solution of (3.49) exists, it is also the solution of the following optimization problem:

$$\min_{u_1, u_2} \frac{1}{2} \begin{bmatrix} u_1 u_2 \end{bmatrix} U_1'(t) U_1(t) \begin{bmatrix} u_1 \\ u_2 \end{bmatrix} + U_0'(t) U_1(t) \begin{bmatrix} u_1 \\ u_2 \end{bmatrix}$$

$$\text{s.t.} \quad V_0(t) + V_1(t) \begin{bmatrix} u_1 \\ u_2 \end{bmatrix} = 0. \tag{3.54}$$

Because $Q := U_1' U_1 = I_{2 \times 2} + \bar{U}_1' \bar{U}_1$ is positive definite and the cost function is quadratic, by applying Lagrange multipliers, the global minimum of the constrained optimization problem can be obtained online as follows:

$$\begin{bmatrix} u_1^*(\bar{x}_H) \\ u_2^*(\bar{x}_H) \end{bmatrix} = -Q^{-1} \left(U_1' U_0 + (V_1 Q^{-1} V_1')^{-1} (V_0 - V_1 Q^{-1} U_1' U_0) V_1' \right). \tag{3.55}$$

Remark 3.12 *It can be easily shown that the optimal solution (3.55) is independent of the choice $0_{1 \times 2} \neq V_1(t) \in \mathcal{P}(t)$.*

Next assume that the periodic orbit $\mathcal{O}$ fulfills hypotheses HPO1–HPO7 and the additional following hypothesis.

(HPO8) For every $\bar{x}_H \in [x_{H,d}^+, x_{H,d}^-]$, $U_1'(t) \frac{\partial x_2'}{\partial q_b}(t) \neq 0_{2 \times 1}$ and $(u_1^*(\bar{x}_H), u_2^*(\bar{x}_H))' \in \Lambda_a(\mathcal{O}_d, \bar{x}_H)$.

Then, choose $u_{1d}(\bar{x}_H) = u_1^*(\bar{x}_H)$ and $u_{2d}(\bar{x}_H) = u_2^*(\bar{x}_H)$. By this choice, the closed-loop control input on an open neighborhood of $\mathcal{O}_d$ is also admissible. In addition, $\mathcal{O}_d$ is an integral curve of the closed-loop system in the double support phase. However, on $\mathcal{O}_d$ the control input vector is not necessarily identical to $\mathbf{u}_d(t)$ from Definition 3.3.

If the condition $(u_1^*(\bar{x}_H), u_2^*(\bar{x}_H))' \in \Lambda_a(\mathcal{O}_d, \bar{x}_H)$ is not satisfied for some $\bar{x}_H \in [x_{H,d}^+, x_{H,d}^-]$, we can define the following optimization problem for determining u_1 and u_2:

$$\min_{u_1, u_2} \frac{1}{2} \begin{bmatrix} u_1 & u_2 \end{bmatrix} U_1'(t) U_1(t) \begin{bmatrix} u_1 \\ u_2 \end{bmatrix} + U_0'(t) U_1(t) \begin{bmatrix} u_1 \\ u_2 \end{bmatrix}$$

$$\text{s.t.} \quad V_0(t) + V_1(t) \begin{bmatrix} u_1 \\ u_2 \end{bmatrix} = 0$$

$$-u_{\max} + \varepsilon \leq u_i(t, u_1, u_2) \leq u_{\max} - \varepsilon, \quad i = 1, \ldots, 4$$

$$F_i^h(t, u_1, u_2) - \mu_s F_i^v(t, u_1, u_2) \leq -\varepsilon, \quad i = 1, 2$$

$$-F_i^h(t, u_1, u_2) - \mu_s F_i^v(t, u_1, u_2) \leq -\varepsilon, \quad i = 1, 2, \tag{3.56}$$

where ε is a positive scalar. We remark that the set of admissible ground reaction forces can also be expressed as

$$\Lambda_f = \{F := (F^h, F^v)' \in \mathbb{R}^2 | F^h - \mu_s F^v < 0, -F^h - \mu_s F^v < 0\}.$$

The solution of this latter optimization problem can be determined for a finite number of points in the interval $[x_{H,d}^+, x_{H,d}^-]$ that are interpolated, for example, by cubic splines.

Remark 3.13 (Cubic Splines to Compute $h_{d,s}(\theta_s)$ and $h_{d,d}(\bar{x}_H)$) *Following Propositions 6.2 and 6.3 of Ref. [18, pp. 163–164], since $\mathcal{O}$ is obtained from a motion planning algorithm (see Section 3.7), we can produce the desired functions $h_{d,s}(\theta_s)$ and $h_{d,d}(\bar{x}_H)$ by sampling the orbits $\mathcal{O}_s$ and $\mathcal{O}_d$ and applying cubic spline interpolations between the samples.*

Remark 3.14 (Cubic Splines to Compute $u_{1d}(\bar{x}_H)$ and $u_{2d}(\bar{x}_H)$) *In general, it is difficult to obtain a closed-form expression for $t = x_{Hd}^{-1}(\bar{x}_H)$. Moreover, $q_d(t)$ is known at a finite number of samples. Thus, cubic spline interpolation can be used to compute $u_{1d}(\bar{x}_H)$ and $u_{2d}(\bar{x}_H)$ on the interval $[x_{H,d}^+, x_{H,d}^-]$.*

Remark 3.15 (Validity of the Transition Model from DS to SS) *By HPO7, the vertical acceleration of the end of leg-2 is positive at the beginning of the single support phase on the periodic orbit (i.e., $\ddot{y}_2(0) > 0$). Since $\Delta_d^s : \mathcal{S}_d^s \to \mathcal{X}_s$ and $u_s : \mathcal{X}_s \to \mathbb{R}^4$ are continuous, there exists an open set $\tilde{\mathcal{S}}_d^s \subset \mathcal{S}_d^s$ such that for every $x_d^- \in \tilde{\mathcal{S}}_d^s$, $\ddot{y}_2 > 0$ at the beginning of the single support phase. Hence, the transition from the double support phase to the single support phase is valid on an open neighborhood of $\mathcal{O}$.*

3.6 STABILIZATION OF THE PERIODIC ORBIT

This section addresses the problem of stabilization for a desired periodic orbit $\mathcal{O}$ satisfying hypotheses HPO1–HPO7 or a set of weaker hypotheses as described in Section 2.4. If $\mu \geq 1$ in Theorem 3.2, the periodic orbit is not asymptotically stable. For stabilizing $\mathcal{O}$, the holonomic outputs in equations (3.23) and (3.27) will be modified. The modification procedure consists of (i) adding *augmentation functions* that are finitely parameterized functions [18, p. 164], such as Bézier polynomials to the holonomic output functions, and (ii) updating the parameters of the augmentation functions on a stride-to-stride basis. The idea of updating the parameters of HZD and output functions was introduced in Refs. [57–59] for stabilization, thereby improving the convergence rate and regulation of the average walking rate. In this chapter, we make use of an *event-based controller* to update the parameters of the augmentation functions at the end of each single support phase. The purpose of updating the parameters is to achieve hybrid invariance and stabilization.

Now define the modified outputs in the following forms:

$$h_s(q_s; \alpha) := h_s(q_s) - \text{fcn}_1(s_s(q_s); \alpha)$$
$$h_d(q_i; \beta) := h_d(q_i) - \text{fcn}_2(s_d(q_i); \beta), \tag{3.57}$$

where $\alpha := [\alpha_0 \, \alpha_1 \, ... \, \alpha_{M_s-1} \, \alpha_{M_s}] \in \mathcal{A}$ and $\beta := [\beta_0 \, \beta_1 \, ... \, \beta_{M_d-1} \, \beta_{M_d}] \in \mathcal{B}$ are the parameter matrices for the single and double support phases, respectively. Furthermore, $\mathcal{A} \subset \mathbb{R}^{4 \times (M_s+1)}$ and $\mathcal{B} \subset \mathbb{R}^{2 \times (M_d+1)}$ are also open sets. The functions $\text{fcn}_1 : [0, 1] \times \mathcal{A} \to \mathbb{R}^4$ and $\text{fcn}_2 : [0, 1] \times \mathcal{B} \to \mathbb{R}^2$ are defined as Bézier polynomials of degree M_s and M_d, that is,[17]

$$\text{fcn}_1(s_s; \alpha) := \sum_{k=0}^{M_s} \frac{M_s!}{k!(M_s - k)!} \alpha_k s_s^k (1 - s_s)^{M_s-k}$$

$$\text{fcn}_2(s_d; \beta) := \sum_{k=0}^{M_d} \frac{M_d!}{k!(M_d - k)!} \beta_k s_d^k (1 - s_d)^{M_d-k},$$

where $\alpha_i \in \mathbb{R}^4$, $i = 0, 1, \ldots, M_s$ and $\beta_j \in \mathbb{R}^2$, $j = 0, 1, \ldots, M_d$ denote the ith and jth columns of α and β, respectively. Also $s_s(q_s)$ and $s_d(q_i)$ are given by

$$s_s(q_s) := \frac{\theta_s(q_s) - \theta_s^+}{\theta_s^- - \theta_s^+}$$

$$s_d(q_i) := \frac{\bar{x}_H(q_i) - x_{H,d}^+}{x_{H,d}^- - x_{H,d}^+}.$$

The control laws in the single and double support phases (i.e., within-stride controllers) are modified by replacing $h_s(q_s)$ by $h_s(q_s; \alpha)$ in equation (3.24) and replacing $h_d(q_i)$ by $h_d(q_i; \beta)$ in equation (3.29), respectively. By this method, the single support phase control law is parameterized by α and denoted $u_s(x_s; \alpha)$. Also, the double support phase control law is denoted $(u_{3d}(x_d; \beta), u_{4d}(x_d; \beta))'$. It is worth mentioning that $u_{1d}(x_d)$ and $u_{2d}(x_d)$ do not depend on β and are identical to those developed in Section 3.5. Next, let $Z_{s,\alpha}$ and $Z_{d,\beta}$ represent the zero dynamics manifolds corresponding to $h_s(q_s; \alpha)$ and $h_d(q_i; \beta)$, respectively. Define $\alpha^* := 0_{4 \times (M_s+1)}$ and $\beta^* := 0_{2 \times (M_d+1)}$. Then, $Z_{s,\alpha^*} = Z_s$ and $Z_{d,\beta^*} = Z_d$.

Remark 3.16 (Fundamental Properties of Bézier Polynomials) *For any coefficient matrix* $\alpha = \text{col}\{\alpha_i\}_{i=0}^M := [\alpha_0, \ldots, \alpha_M]$, *the Bézier polynomial*

$$\mathcal{B}(s, \alpha) := \sum_{i=0}^{M} \frac{M!}{i!(M - i)!} \alpha_i \, s^i \, (1 - s)^{M-i}$$

---

[17] The idea of using Bézier polynomials as augmentation functions is taken from Ref. [94].

has the following properties [18, p. 139]:

$$(i) \quad \mathcal{B}(0, \alpha) = \alpha_0$$

$$(ii) \quad \mathcal{B}(1, \alpha) = \alpha_M$$

$$(iii) \quad \frac{\partial}{\partial s}\mathcal{B}(0, \alpha) = M(\alpha_1 - \alpha_0)$$

$$(iv) \quad \frac{\partial}{\partial s}\mathcal{B}(1, \alpha) = M(\alpha_M - \alpha_{M-1}).$$

In addition,

$$(v) \quad \frac{\partial^2}{\partial s^2}\mathcal{B}(0, \alpha) = M(M - 1)(\alpha_2 - 2\alpha_1 + \alpha_0)$$

$$(vi) \quad \frac{\partial^2}{\partial s^2}\mathcal{B}(1, \alpha) = M(M - 1)(\alpha_M - 2\alpha_{M-1} + \alpha_{M-2}).$$

Lemma 3.9 (Hybrid Invariance for Parameterized Manifolds)   *Let $\mathcal{O}$ be a periodic orbit of the hybrid model (3.22) satisfying hypotheses HPO1–HPO7. Assume that the desired functions $h_{d,s}(\theta_s)$ and $h_{d,d}(\bar{x}_H)$ are defined as those of equations (3.41) and (3.42), respectively. Moreover, suppose that $M_s \geq 3$, $M_d \geq 3$, $\alpha_0 = \alpha_1 = \alpha_{M_s-1} = \alpha_{M_s} = 0_{4\times1}$, and $\beta_0 = \beta_1 = \beta_{M_d-1} = \beta_{M_d} = 0_{2\times1}$. Then, the following statements are true:*

1. $\mathcal{S}_s^d \cap Z_{s,\alpha} = \mathcal{S}_s^d \cap Z_s$
2. $\mathcal{S}_d^s \cap Z_{d,\beta} = \mathcal{S}_d^s \cap Z_d$
3. $\Delta_d^s(\mathcal{S}_d^s \cap Z_{d,\beta}) \subset Z_{s,\alpha}$
4. $\Delta_s^d(\mathcal{S}_s^d \cap Z_{s,\alpha}) \subset Z_{d,\beta}.$

Proof. By properties of Bézier polynomials, see Remark 3.16, the proof is straight-forward.   ∎

HZD exists for $\alpha = \alpha^* = 0_{4\times(M_s+1)}$ and $\beta = \beta^* = 0_{2\times(M_d+1)}$. Furthermore, Lemma 3.9 and the continuity of the modified outputs with respect to α and β imply that there exists $\varepsilon > 0$ such that for any α and β with the property $\|\alpha\| < \varepsilon$ and $\|\beta\| < \varepsilon$, the corresponding HZD exists. For any $M_s \geq 3$ and $M_d \geq 3$, the matrices $\alpha \in \mathcal{A}$ and $\beta \in \mathcal{B}$ are said to be *regular*[18] if $\alpha_0 = \alpha_1 = \alpha_{M_s-1} = \alpha_{M_s} = 0_{4\times1}$, $\beta_0 = \beta_1 = \beta_{M_d-1} = \beta_{M_d} = 0_{2\times1}$, and the HZD corresponding to $h_s(q_s; \alpha)$ and $h_d(q_i; \beta)$ exists. Figure 3.2 illustrates the geometry of the HZD for some regular α and β.

---

[18] The terminology of a *regular parameter* follows Definition 6.1 of [18, p. 140].

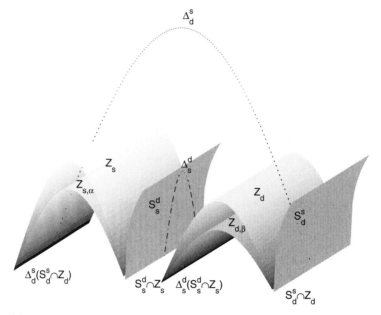

Figure 3.2   The geometry of the HZD for some regular α and β. From Lemma 3.9, $S_s^d \cap Z_{s,\alpha} = S_s^d \cap Z_s$, $S_d^s \cap Z_{d,\beta} = S_d^s \cap Z_d$, $\Delta_d^s(S_d^s \cap Z_{d,\beta}) \subset Z_{s,\alpha}$ and $\Delta_s^d(S_s^d \cap Z_{s,\alpha}) \subset Z_{d,\beta}$.

For the regular matrices α and β, the parametric restricted Poincaré return map $\rho_{\alpha,\beta} : S_s^d \cap Z_s \to S_s^d \cap Z_s$ is defined by $\rho_{\alpha,\beta}(\zeta_s^-) := \rho_{s,\alpha} \circ \rho_{d,\beta}(\zeta_s^-)$, where $\rho_{s,\alpha} : S_d^s \cap Z_d \to S_s^d \cap Z_s$ and $\rho_{d,\beta} : S_s^d \cap Z_s \to S_d^s \cap Z_d$ are the parametric restricted generalized Poincaré maps for the single and double support phases, respectively. Let $\rho(\zeta_s^-; \alpha, \beta) := \rho_{\alpha,\beta}(\zeta_s^-)$. Then, the following discrete-time system can be defined to study the stabilization problem:

$$\zeta_s^-[k+1] = \rho(\zeta_s^-[k]; \alpha[k], \beta[k]), \tag{3.58}$$

where $S_s^d \cap Z_s$ is the one-dimensional state space of equation (3.58) and $\alpha_{ij}[k]$, $i = 1, ..., 4$, $j = 2, ..., M_s - 2$ and $\beta_{ij}[k]$, $i = 1, 2$, $j = 2, ..., M_d - 2$ are the control inputs. Linearization of the discrete-time system in equation (3.58) about $(\zeta_s^-, \alpha, \beta) = (\zeta_s^*, \alpha^*, \beta^*)$ results in

$$\delta\zeta_s^-[k+1] = a\delta\zeta_s^-[k] + \sum_{i=1}^{4}\sum_{j=2}^{M_s-2} b_{\alpha_{ij}} \delta\alpha_{ij}[k] + \sum_{i=1}^{2}\sum_{j=2}^{M_d-2} b_{\beta_{ij}} \delta\beta_{ij}[k], \tag{3.59}$$

where

$$a := \frac{\partial \rho}{\partial \zeta_s^-}(\zeta_s^*; \alpha^*, \beta^*) = \mu$$

$$b_{\alpha_{ij}} := \frac{\partial \rho}{\partial \alpha_{ij}}(\zeta_s^*; \alpha^*, \beta^*)$$

$$b_{\beta_{ij}} := \frac{\partial \rho}{\partial \beta_{ij}}(\zeta_s^*; \alpha^*, \beta^*).$$

Moreover, $\delta\zeta_s^-[k] := \zeta_s^-[k] - \zeta_s^*$, $\delta\alpha_{ij}[k] := \alpha_{ij}[k] - \alpha_{ij}^* = \alpha_{ij}[k]$ and $\delta\beta_{ij}[k] := \beta_{ij}[k] - \beta_{ij}^* = \beta_{ij}[k]$.

Theorem 3.3 (Static Event-Based Update Laws)   *Let $\mathcal{O}$ be a periodic orbit of the hybrid model (3.22) satisfying hypotheses HPO1–HPO7. Moreover, suppose that M_s, $M_d \geq 3$, $\alpha_0 = \alpha_1 = \alpha_{M_s-1} = \alpha_{M_s} = 0_{4\times1}$, and $\beta_0 = \beta_1 = \beta_{M_d-1} = \beta_{M_d} = 0_{2\times1}$. Assume that $b \neq 0_{1\times p}$, where $b := (b_\alpha, b_\beta)$, $b_\alpha := (b_{\alpha_{12}}, \ldots, b_{\alpha_{4M_s-2}})$, $b_\beta := (b_{\beta_{12}}, \ldots, b_{\beta_{2M_d-2}})$, and $p := 4(M_s - 3) + 2(M_d - 3)$. Then, there exist scalars $K_{\alpha_{ij}}, i = 1, \ldots, 4, j = 2, \ldots, M_s - 2$, and $K_{\beta_{ij}}, i = 1, 2, j = 2, \ldots, M_d - 2$ such that by using the static event-based update laws*

$$\alpha_{ij}(\zeta_s^-) = -K_{\alpha_{ij}}(\zeta_s^- - \zeta_s^*)$$
$$\beta_{ij}(\zeta_s^-) = -K_{\beta_{ij}}(\zeta_s^- - \zeta_s^*),$$ (3.60)

ζ_s^* is a locally exponentially stable equilibrium point for the closed-loop discrete-time system $\zeta_s^-[k + 1] = \rho_{cl}(\zeta_s^-[k])$, where $\rho_{cl}(\zeta_s^-) := \rho(\zeta_s^-; \alpha(\zeta_s^-), \beta(\zeta_s^-))$.

Proof. $b \neq 0_{1\times p}$ implies the controllability of (a, b). Controllability of (a, b) implies the existence of $K_\alpha := (K_{\alpha_{12}}, \ldots, K_{\alpha_{4M_s-2}})'$ and $K_\beta := (K_{\beta_{12}}, \ldots, K_{\beta_{2M_d-2}})'$ such that $|\mu_{cl}| < 1$, where $\mu_{cl} := a - b_\alpha K_\alpha - b_\beta K_\beta$. Since $\mu_{cl} = \frac{\partial \rho_{cl}}{\partial \zeta_s^-}(\zeta_s^-)|_{\zeta_s^- = \zeta_s^*}$, $|\mu_{cl}| < 1$ follows that ζ_s^* is a locally exponentially stable equilibrium point for the closed-loop discrete-time system that completes the proof.   ∎

Remark 3.17 (Asymptotic Stability by Static Event-Based Update Laws)   *Theorem 2.5 of Chapter 2 in combination with Theorem 3.3 guarantee that $\mathcal{O}$ is an asymptotically stable periodic orbit for the closed-loop hybrid model of walking.*

3.7 MOTION PLANNING ALGORITHM

The objective of this section is to present an algorithm for designing a period-one orbit $\mathcal{O}$ of the open-loop hybrid model of walking in equation (3.22) satisfying hypotheses HPO1–HPO7. Like many papers in the literature of the bipedal gait design (e.g., [61, 73, 84]), the algorithm developed in this chapter is based on a finite-dimensional nonlinear optimization problem with equality and inequality constraints.

3.7.1   Motion Planning Algorithm for the Single Support Phase

The motion planning algorithm for the single support phase is based on the Spong normal form [1, 95]. To make this notion precise, let $\Omega(q_s, \dot{q}_s) := C(q_s, \dot{q}_s)\dot{q}_s + G(q_s)$ and partition the dynamical equation (3.7) as follows:

$$D_{bb}(q_b)\ddot{q}_b + D_{b5}(q_b)\ddot{q}_5 + \Omega_b(q_s, \dot{q}_s) = u$$
$$D_{5b}(q_b)\ddot{q}_b + D_{55}(q_b)\ddot{q}_5 + \Omega_5(q_s, \dot{q}_s) = 0,$$

in which $D_{bb} \in \mathbb{R}^{4 \times 4}$ and $D_{b5} \in \mathbb{R}^{4 \times 1}$ denote the upper left and right submatrices of D, and $D_{5b} \in \mathbb{R}^{1 \times 4}$ and $D_{55} \in \mathbb{R}$ represent the lower left and right submatrices of D, respectively. Furthermore, $\Omega_b \in \mathbb{R}^{4 \times 1}$ and $\Omega_5 \in \mathbb{R}$ are the first four rows and last row of the vector Ω. Applying the static feedback law

$$u = \bar{D}_{bb}(q_b)\, v_b + \bar{\Omega}_b(q_s, \dot{q}_s),$$

where

$$\bar{D}_{bb}(q_b) := D_{bb} - D_{b5}D_{55}^{-1}D_{5b}$$
$$\bar{\Omega}_b(q_s, \dot{q}_s) := \Omega_b - D_{b5}D_{55}^{-1}\Omega_5,$$

yields the following partially feedback linearized result that is known as the *Spong normal form*:

$$
\begin{aligned}
\ddot{q}_b &= v_b \\
\ddot{q}_5 &= -D_{55}^{-1}(q_b)\, D_{5b}(q_b)\, v_b - \bar{\Omega}_5(q_s, \dot{q}_s),
\end{aligned}
\tag{3.61}
$$

in which $\bar{\Omega}_5(q_s, \dot{q}_s) := D_{55}^{-1}\Omega_5$. The motion planning algorithm during the single support phase is an extension of that developed in Ref. [73]. From the first four rows of the partially feedback linearized equation (3.61), body angles can be controlled independently. Hence, we choose the following polynomial evolution of time for the body angles during single support:

$$\mathbf{q}_{bs}(t) = \sum_{i=0}^{m_s} a_i t^i, \quad 0 \le t \le T_s,
\tag{3.62}$$

where m_s is an integer with the property $m_s \ge 4$. It is obvious that the coefficients $a_i \in \mathbb{R}^4$ for $i = 0, 1, \ldots, m_s$ can be obtained uniquely if $\mathbf{q}_{bs}(t)$ fulfills the following boundary conditions:

$$\mathbf{q}_{bs}(0) = q_{bs}^i$$
$$\dot{\mathbf{q}}_{bs}(0) = \dot{q}_{bs}^i$$
$$\mathbf{q}_{bs}(T_s) = q_{bs}^f \qquad (3.63)$$
$$\dot{\mathbf{q}}_{bs}(T_s) = \dot{q}_{bs}^f$$
$$\mathbf{q}_{bs}(t_s^j) = q_{bs}^j, \quad j = 1, \ldots, m_s - 3,$$

where the superscripts "i" and "f" will designate the initial and final conditions, respectively. Moreover, in equation (3.63), $t_s^j := \frac{j}{m_s - 2} T_s$, $j = 1, \ldots, m_s - 3$ denote a set of intermediate times during the single support phase at which the body angles are identical to q_{bs}^j. For the later purposes, define the intermediate body angles vector as

$$q_{bs}^{int} := (q_{bs}^{'1}, \ldots, q_{bs}^{'m_s-3})'.$$

Next, to determine the time evolution of the torso angle during single support, we restrict our attention to the Spong normal form. The last row of matrix equation (3.61) implies that the evolution of the torso angle (i.e., $\mathbf{q}_{5s}(t)$) can be described by the following differential equation:

$$\ddot{\mathbf{q}}_{5s}(t) = -D_{55}^{-1}(\mathbf{q}_{bs}(t))\, D_{5b}(\mathbf{q}_{bs}(t))\, \ddot{\mathbf{q}}_{bs}(t) - \bar{\Omega}_5(\mathbf{q}_s(t), \dot{\mathbf{q}}_s(t)). \qquad (3.64)$$

Hence, by assuming that $\mathbf{q}_{5s}(t)$ satisfies the following boundary conditions

$$\mathbf{q}_{5s}(T_s) = q_{5s}^f$$
$$\dot{\mathbf{q}}_{5s}(T_s) = \dot{q}_{5s}^f,$$

differential equation (3.64) can be integrated numerically on the time interval $[0, T_s]$. Also, from equations (3.44) and (3.8), $\mathbf{u}_s(t)$ and thereby $\mathbf{F}_{1s}(t)$ can be determined uniquely. Next, let us define $q_s^f := (q_{bs}^{'f}, q_{5s}^f)'$ and $\dot{q}_s^f := (\dot{q}_{bs}^{'f}, \dot{q}_{5s}^f)'$ as the position and velocity vectors of the mechanical system at the end of the single support phase. Due to the fact that the swing leg contacts the ground at the end of single support, q_s^f satisfies the following equality constraint:

$$y_2(q_s^f) = 0. \qquad (3.65)$$

3.7.2 Motion Planning Algorithm for the Double Support Phase

Assume that $\mathbf{q}_d(t)$ (i.e., the time evolution of the configuration variables during the double support phase) is such that $\text{rank}\frac{\partial p_2}{\partial q}(\mathbf{q}_d(t)) = 2$ for every $t \in [T_s, T]$. Using

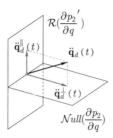

Figure 3.3   Geometrical description of the motion planning algorithm during double support. In equation (3.66), it is assumed that rank($\frac{\partial p_2}{\partial q}(\mathbf{q}_d(t))$) = 2 and $\ddot{\mathbf{q}}_d(t)$ can be expressed as $\ddot{\mathbf{q}}_d(t) = \ddot{\mathbf{q}}_d^{\parallel}(t) + \ddot{\mathbf{q}}_d^{\perp}(t)$. (See the color version of this figure in color plates section.)

this assumption and equation (3.15), $\mathbf{q}_d(t)$ fulfills the following differential equation:

$$\ddot{\mathbf{q}}_d(t) = -\left(\frac{\partial p_2}{\partial q}\right)^+ \frac{\partial}{\partial q}\left(\frac{\partial p_2}{\partial q}\dot{\mathbf{q}}_d(t)\right)\dot{\mathbf{q}}_d(t) + \left(\frac{\partial p_2}{\partial q}\right)^- \lambda(t), \qquad (3.66)$$

where $(\frac{\partial p_2}{\partial q})^+$ and $(\frac{\partial p_2}{\partial q})^-$ are the pseudo inverse and projection matrices due to $\frac{\partial p_2}{\partial q}$, respectively, that is,

$$\left(\frac{\partial p_2}{\partial q}\right)^+ := \frac{\partial p_2'}{\partial q}\left(\frac{\partial p_2}{\partial q}\frac{\partial p_2'}{\partial q}\right)^{-1}$$

$$\left(\frac{\partial p_2}{\partial q}\right)^- := I_{5\times5} - \frac{\partial p_2'}{\partial q}\left(\frac{\partial p_2}{\partial q}\frac{\partial p_2'}{\partial q}\right)^{-1}\frac{\partial p_2}{\partial q},$$

and $\lambda : [T_s, T] \rightarrow \mathbb{R}^5$ is an arbitrary continuously differentiable function. Figure 3.3 illustrates a geometric description for differential kinematic inversion problem (3.66). In this equation, $\ddot{\mathbf{q}}_d(t)$ can be expressed as

$$\ddot{\mathbf{q}}_d(t) = \ddot{\mathbf{q}}_d^{\parallel}(t) + \ddot{\mathbf{q}}_d^{\perp}(t),$$

where

$$\ddot{\mathbf{q}}_d^{\parallel}(t) := -\left(\frac{\partial p_2}{\partial q}\right)^+ \frac{\partial}{\partial q}\left(\frac{\partial p_2}{\partial q}\dot{\mathbf{q}}_d(t)\right)\dot{\mathbf{q}}_d(t)$$

$$\ddot{\mathbf{q}}_d^{\perp}(t) := \left(\frac{\partial p_2}{\partial q}\right)^- \lambda(t).$$

From the properties of the pseudo inverse and projection matrices, for every $t \in [T_s, T]$, $\ddot{\mathbf{q}}_d^{\parallel}(t)$ is perpendicular to $\ddot{\mathbf{q}}_d^{\perp}(t)$, that is, $\ddot{\mathbf{q}}_d^{\parallel}(t) \perp \ddot{\mathbf{q}}_d^{\perp}(t)$. Furthermore, $\ddot{\mathbf{q}}_d^{\parallel}(t) \in \mathcal{R}(\frac{\partial p_2'}{\partial q})$ and $\ddot{\mathbf{q}}_d^{\perp}(t) \in \mathcal{N}ull(\frac{\partial p_2}{\partial q})$, where $\mathcal{R}$ and $\mathcal{N}ull$ represent the range and null spaces of a matrix, respectively. Next, let us define

$$\lambda(t) := \sum_{i=0}^{m_d} \lambda_i (t - T_s)^i, \quad T_s \le t \le T$$

for some $m_d \in \mathbb{Z}^+$ and $\lambda_i \in \mathbb{R}^5$, $i = 0, 1, \ldots, m_d$. λ_i is a vector of five components but its projection by $\frac{\partial p_2}{\partial q}$ has only three independent components.[19] Next, assume that $\mathbf{q}_d(t)$ satisfies the boundary conditions

$$\mathbf{q}_d(T) = q_d^f$$
$$\dot{\mathbf{q}}_d(T) = \dot{q}_d^f,$$

where q_d^f and $\dot{q}_d^f$ denote the position and velocity vectors of the mechanical system at the end of double support. Then, the equation of motion (3.66) can be integrated numerically on the interval $[T_s, T]$. The fact that the end of leg-2 is stationary during the double support phase also implies that q_d^f and $\dot{q}_d^f$ satisfy the following equality constraints:

$$p_2(q_d^f) = p_2(q_s^f)$$
$$\frac{\partial p_2}{\partial q}(q_d^f)\dot{q}_d^f = 0_{2\times 1}. \tag{3.67}$$

Moreover, suppose that the step length of the biped robot is nonzero (i.e., $L_s = x_2(q_s^f) = x_2(q_d^f) \neq 0$) and choose the following polynomial evolution of time for $\Upsilon^h(t)$ defined in Lemma 3.7:

$$\Upsilon^h(t) := \sum_{i=0}^{m_d} \Upsilon_i (t - T_s)^i, \quad T_s \le t \le T,$$

where $\Upsilon_i \in \mathbb{R}$, $i = 0, 1, \ldots, m_d$. In this case, from the proof of Lemma 3.7, the time evolutions of $\mathbf{F}_{2d}(t)$ and $\mathbf{u}_d(t)$ can be given by

$$\mathbf{F}_{2d}(t) = \begin{bmatrix} 0 \\ \frac{1}{L_s} \end{bmatrix} e_5'(D\ddot{\mathbf{q}}_d(t) + C\dot{\mathbf{q}}_d(t) + G) + \begin{bmatrix} \Upsilon^h(t) \\ 0 \end{bmatrix}$$

$$\mathbf{u}_d(t) = \mathbf{u}_d^0(t) - \frac{\partial x_2'}{\partial q_b}\Upsilon^h(t).$$

Finally, equation (3.13) determines $\mathbf{F}_{1d}(t)$ uniquely.

---

[19] It is worth noting that instead of solving the kinematic inversion problem of $p_2(q) = (L_s, 0)'$, during the double support phase, we make use of *differential* kinematic inversion in equation (3.66) that, in turn, simplifies the kinematic inversion problem but increases the number of variables during the optimization process of the motion planning algorithm.

3.7.3 Constructing a Period-One Orbit for the Open-Loop Hybrid Model of Walking

By considering Definition 3.3, a solution of the open-loop hybrid model of walking given in equation (3.22) is constructed by piecing together the trajectories of the single and double support phases, according to the transition maps. Thus, the necessary and sufficient conditions by which the open-loop hybrid model of walking has a period-one solution can be expressed as the following equality constraints:

$$
\begin{aligned}
\mathbf{q}_d(T_s) &= q_s^f \\
\dot{\mathbf{q}}_d(T_s) &= \Delta_{\dot{q},s}^d(q_s^f)\,\dot{q}_s^f \\
\begin{bmatrix} q_{bs}^i \\ \mathbf{q}_{5s}(0) \end{bmatrix} &= R\,q_d^f \\
\begin{bmatrix} \dot{q}_{bs}^i \\ \dot{\mathbf{q}}_{5s}(0) \end{bmatrix} &= R\,\dot{q}_d^f.
\end{aligned}
\tag{3.68}
$$

We remark that from equation (3.68), $q_{bs}^i = H_0 R q_d^f$ and $\dot{q}_{bs}^i = H_0 R \dot{q}_d^f$. Thus, the evolution of the mechanical system during walking with non-instantaneous double support phase can be completely determined by the following vector of parameters:

$$
x := (q_s'^f,\, \dot{q}_s'^f,\, q_{bs}'^{\text{int}},\, q_d'^f,\, \dot{q}_d'^f,\, \bar{\lambda}',\, \bar{\Upsilon}',\, T_s,\, T_d)',
$$

where $\bar{\lambda} := (\lambda_0', \ldots, \lambda_{m_d}')'$ and $\bar{\Upsilon} := (\Upsilon_0, \ldots, \Upsilon_{m_d})'$. Next, to determine an admissible value for the vector of parameters x, we set up an optimization problem. The constraints of the optimization problem are composed of equality and inequality constraints.

3.7.3.1 Equality Constraints
The equality constraints are expressed as equations (3.65), (3.67), and (3.68).

3.7.3.2 Inequality Constraints
The inequality constraints can be expressed as hypotheses HPO2–HPO7. The constraints associated to the double impact model, studied here, are based on those presented in Refs. [67, 92].

3.7.3.3 Cost Function
A two-stage strategy is used to solve the motion planning algorithm. In the first stage, the cost function is chosen as 1 and by using the fmincon function of MATLAB's Optimization Toolbox, we search for a feasible periodic solution of the open-loop hybrid model of walking (3.22), which will be used in the next stage as an initial guess. To simplify the search procedure for a feasible periodic solution, the constraints can be added in a step-by-step manner. Following the results of Refs. [61, 73, 84], by using the fmincon function, the motion planning algorithm during the second stage is continued to minimize the following desired cost

function:

$$
\mathcal{J}(x) := \frac{1}{L_s} \left(\int_0^{T_s} \|\mathbf{u}_s(t)\|_2^2 \, dt + \int_{T_s}^{T} \|\mathbf{u}_d(t)\|_2^2 \, dt \right).
$$

3.8 NUMERICAL EXAMPLE FOR THE MOTION PLANNING ALGORITHM

The physical parameters of the walking robot are those of the planar biped robot, RABBIT (see Refs. [47, 96] for more details). On the trajectory $\mathcal{O}$ that is obtained by applying the motion planning algorithm,

$$
\mathbf{q}_{bs}(t) = \sum_{i=0}^{6} a_i t^i, \qquad 0 \le t < T_s
$$

$$
\Upsilon^h(t) = \sum_{i=0}^{6} \Upsilon_i (t - T_s)^i, \quad T_s \le t < T.
$$

Moreover, from equation (3.15), $\mathbf{q}_d(t)$, $T_s \le t < T$ is the solution of the following differential equation:

$$
\ddot{\mathbf{q}}_d(t) = - \left(\frac{\partial p_2}{\partial q} \right)^+ \frac{\partial}{\partial q} \left(\frac{\partial p_2}{\partial q} \dot{\mathbf{q}}_d(t) \right) \dot{\mathbf{q}}_d(t) + \left(\frac{\partial p_2}{\partial q} \right)^- \lambda(t)
$$

with the initial condition

$$
\mathbf{q}_d(T_s) = [0.8414, 2.8239, 0.2344, 2.6577, 1.6134]'(\mathrm{rad})
$$
$$
\dot{\mathbf{q}}_d(T_s) = [-0.8496, 0.4765, 0.7515, -0.3317, -0.4291]'(\mathrm{rad/s}),
$$

where $(\frac{\partial p_2}{\partial q})^+$ and $(\frac{\partial p_2}{\partial q})^-$ denote the pseudo inverse and projection matrices, respectively, and

$$
\lambda(t) = \sum_{i=0}^{6} \lambda_i (t - T_s)^i, \quad T_s \le t < T.
$$

The coefficients a_i, λ_i and Υ_i for $i = 0, \ldots, 6$ are given in Tables 3.1–3.3, respectively. We remark that for this orbit, the vectors of generalized coordinates and velocities immediately before the impact can be expressed as

$$
\mathbf{q}_s(T_s) = [0.8414, 2.8239, 0.2344, 2.6577, 1.6134]'(\mathrm{rad})
$$
$$
\dot{\mathbf{q}}_s(T_s) = [-0.8698, 0.4794, 0.8155, -0.3471, -0.4325]'(\mathrm{rad/s}).
$$

TABLE 3.1 Coefficients a_i, $i = 0, \ldots, 6$

a'_0	0.6635	2.4577	0.4308	3.0458
a'_1	0.7483	-0.3624	-0.5762	0.3019
$a'_2(10^1)$	-2.7279	0.1109	0.9938	0.6606
$a'_3(10^2)$	1.3010	0.1956	-0.1455	-0.9645
$a'_4(10^2)$	-2.5902	-0.4555	-0.2394	2.6304
$a'_5(10^2)$	2.3831	0.3001	0.5347	-2.6856
$a'_6(10^2)$	-0.8283	-0.0432	-0.2451	0.9470

TABLE 3.2 Coefficients λ_i, $i = 0, \ldots, 6$

λ'_0	0.6318	0.1778	0.1418	-0.1685	-0.2914
λ'_1	0.2220	-0.3819	-0.7544	0.2448	0.1614
λ'_2	0.9925	-0.4739	-0.7128	-0.3508	-0.3843
λ'_3	-0.1777	0.8382	0.3223	0.8882	0.5165
λ'_4	-0.4506	-0.1765	0.3838	-0.8357	-0.0285
λ'_5	0.8627	0.9618	0.7079	0.9734	0.8718
λ'_6	-0.1908	0.9414	0.1576	0.7550	0.8007

TABLE 3.3 Coefficients Υ_i, $i = 0, \ldots, 6$

Υ_0	-0.1980
Υ_1	0.2652
Υ_2	0.7318
Υ_3	-0.3610
Υ_4	0.2502
Υ_5	-0.9358
Υ_6	-0.2292

In addition, by using the impact model developed in Section 3.2.5,

$$\frac{\partial p_2}{\partial q} A^{-1} \left(\frac{\partial f'_1}{\partial q} \Sigma_{12} + \frac{\partial f'_2}{\partial q} \right) = \begin{bmatrix} -0.3882 & -0.1121 \\ -0.1121 & -0.0739 \end{bmatrix}$$

$$I_{R1} = \begin{bmatrix} 0.0085 \\ 0.0220 \end{bmatrix}$$

$$I_{R2} = \begin{bmatrix} 0.0832 \\ -0.1244 \end{bmatrix},$$

and consequently, the impact model is nonsingular. The feasibility conditions of the double impact model, presented in Refs. [67, 92] for a five-link planar bipedal robot with point feet, imply that the impact model is feasible, that is, $I^v_{R1} = 0.0221 > 0$,

$|\frac{I_{R1}^h}{I_{R1}^v}| = 0.3847 < \mu_s$, $|\frac{I_{R2}^h}{I_{R2}^v}| = 0.6667 < \mu_s$, and $\dot{p}_1^{2+} = 0$, where $\mu_s = \frac{2}{3}$. Following
the results of Refs. [67, 92] (see equations (3.32)–(3.35)), the condition $I_{R2}^v > 0$ is not
included in the feasibility conditions of the double impact mode. However, $I_{R1}^v > 0$
should be satisfied because the stance leg end is assumed to remain on the ground
during and after the impact. It is worth mentioning that the coordinates relabeling
to swap the roles of the legs occurs immediately after the double support phase (not
after the impact). In addition, an implicit condition is that before the impact the leg
is above the ground (see Fig. 3.4) and the velocity of the foot is directed downward.
Specially, $\frac{\partial p_2}{\partial q}(\mathbf{q}_s(T_s))\dot{\mathbf{q}}_s(T_s) = [-0.0184, -0.0001]'$ and sudden changes in angular
velocities during double impact are

$$\dot{\mathbf{q}}_d(T_s) - \dot{\mathbf{q}}_s(T_s) = [0.0202, -0.0029, -0.0640, 0.0154, 0.0034]'.$$

Sudden changes in the absolute angular velocities

$$\begin{bmatrix} \dot{\theta}_1 \\ \dot{\theta}_2 \\ \dot{\theta}_3 \\ \dot{\theta}_4 \\ \dot{\theta}_5 \end{bmatrix} = \begin{bmatrix} 1 & 1 & 0 & 0 & -1 \\ 0 & 1 & 0 & 0 & -1 \\ 0 & 0 & 1 & 1 & -1 \\ 0 & 0 & 0 & 1 & -1 \\ 0 & 0 & 0 & 0 & 1 \end{bmatrix} \begin{bmatrix} \dot{q}_1 \\ \dot{q}_2 \\ \dot{q}_3 \\ \dot{q}_4 \\ \dot{q}_5 \end{bmatrix}$$

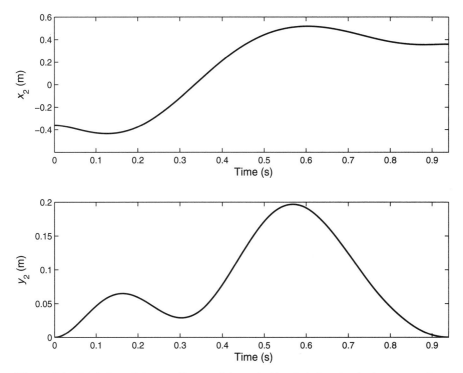

Figure 3.4 Evolution of the coordinates of the end of leg-2 during the single support phase
of the periodic orbit $\mathcal{O}$.

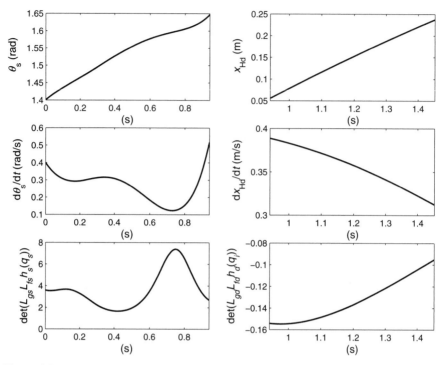

Figure 3.5 The validity of hypotheses HPO3–HPO6 for the optimal periodic motion. HPO1 is trivially satisfied. From Remark 3.7, if on the orbit $\mathcal{O}_s$ the time evolution of θ_s is an increasing function of time, invertibility of the decoupling matrix $L_{g_s}L_{f_s}h_s(q_s)$ on $\mathcal{O}_s$ is equivalent to the angular momentum about the stance leg end being nonzero during the single support phase (HPO5).

during impact can also be given by $[0.0139, -0.0063, -0.0521, 0.0120, 0.0034]'$, which are in the acceptable range reported in Ref. [97] (using both the Integration and Newtonian methods) for a five-link bipedal model with point feet and parameters close to those of RABBIT. The desired periodic motion $\mathcal{O}$ also has a period of $T = T_s + T_d = 0.9443 + 0.5576 = 1.5018$(s), a step length of $L_s = 0.3602$(m), and an average walking speed of $0.2398(\frac{m}{s})$. We remark that in solving the motion planning algorithm by using the `fmincon` function of MATLAB's Optimization Toolbox, the average walking speed of the robot was not fixed; in fact, when the average walking speed was chosen outside the interval $[0.21, 0.34]$, the `fmincon` function could not converge to a feasible periodic solution satisfying hypotheses HPO3–HPO7 (see Figs. 3.4 and 3.5). From equation (3.40), $\mu = 1.2522$ and, consequently, the periodic orbit is not stable without applying event-based update laws. Table 3.4 also represents the desired gait statistics.

Figures 3.6 and 3.7 show the angular position and velocity of the knee, hip, and torso joints during two consecutive steps of the optimal motion, respectively. In plotting the

TABLE 3.4   Desired Gait Statistics

ζ_s^*	$\bar{z}_{xH}^*$	$\Omega_2(x_{H,d}^-)$	$V_{zero,s}(\theta_s^-)$	$\mathcal{W}_{zero,d}(x_{H,d}^-)$
40.8587	0.0971	1.1799	−14.0342	0.0671
$V_{zero,s}^{max}$	$\mathcal{W}_{zero,d}^{max}$	δ_s^d	δ_d^s	μ
28.4940	0.0671	−0.0397	−25.9528	1.2522

results, the stance and swing legs are generalized to the double support phase. In the double support phase (DS), the stance leg is defined to be the leg that was the stance leg in the previous single support phase (SS). The definition of the swing leg in the double support phase is analogous. The control inputs during two consecutive steps of the optimal motion are also depicted in Fig. 3.8. From Fig. 3.8, $\|u\|_{\mathcal{L}_\infty} < u_{max}$ and at the transitions between the continuous phases, the control inputs have discontinuity. Figure 3.9 shows the horizontal and vertical components of the ground reaction forces at the end of the legs during two consecutive steps.

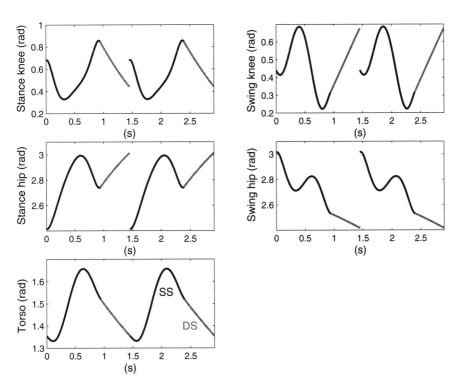

Figure 3.6   Angular positions of the knee, hip, and torso joints during two consecutive steps of the optimal motion. The discontinuities are due to the coordinate relabling for swapping the role of the legs. (See the color version of this figure in color plates section.)

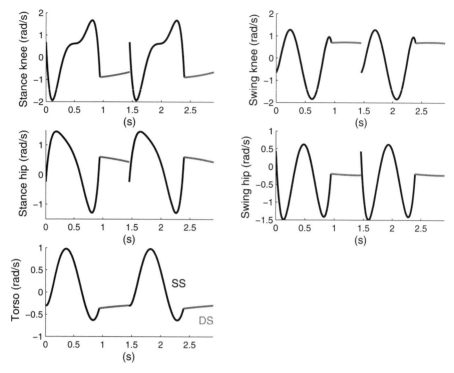

Figure 3.7   Angular velocities of the knee, hip and torso joints during two consecutive steps of the optimal motion. The discontinuities are due to the coordinate relabling for swapping the role of the legs. (See the color version of this figure in color plates section.)

3.9   SIMULATION RESULTS OF THE CLOSED-LOOP HYBRID SYSTEM

This section presents a numerical example for the proposed control strategy to asymptotically stabilize the desired period-one orbit $\mathcal{O}$ for the hybrid model of walking in equation (3.22). Figure 3.10 depicts the plot of the functions $h_{d,s}(\theta_s)$, $V_{\text{zero},s}(\theta_s)$, $h_{d,d}(\bar{x}_H)$, $u_{1d}(\bar{x}_H)$, $u_{2d}(\bar{x}_H)$, $\Omega_2(\bar{x}_H)$, and $\mathcal{W}_{\text{zero},d}(\bar{x}_H)$.

3.9.1   Effect of Double Support Phase on Angular Momentum Transfer and Stabilization

Let σ_1 and σ_2 be the angular momenta of the robot about the end of leg-1 and leg-2, respectively. From the angular momentum balance theorem[20] with the clockwise

---

[20] From the angular momentum balance theorem, the time derivative of the angular momentum about a fixed point is equal to the sum of moments applied by the external forces about that point.

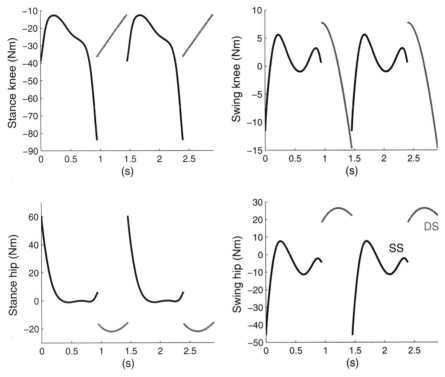

Figure 3.8 Open-loop control inputs corresponding to the knee and hip joints during two consecutive steps of the optimal motion. Two types of discontinuity due to the transitions between the continuous phases are shown in the graphs. (See the color version of this figure in color plates section.)

convention, the time derivatives of σ_1 and σ_2 in double support can be expressed as

$$
\begin{aligned}
\dot{\sigma}_1 &= m_{\text{tot}} g_0 x_{\text{cm}} - L_s F_2^v \\
\dot{\sigma}_2 &= - m_{\text{tot}} g_0 (L_s - x_{\text{cm}}) + L_s F_1^v,
\end{aligned}
\tag{3.69}
$$

where $F_i^v, i = 1, 2$ denotes the vertical component of the ground reaction force at the end of leg-i. Next, assume that the impact occurs at time $t = 0$ and the subscripts "s" and "d" represent the single and double support phases, respectively. Then the variation of σ_2 during double support can be expressed as

$$
\begin{aligned}
\sigma_{2,d}^- - \sigma_{2,d}^+ &:= \sigma_2(t_d^-) - \sigma_2(0^+) \\
&= \int_{0^+}^{t_d^-} -m_{\text{tot}} g_0 (L_s - x_{\text{cm}}) + L_s F_1^v \, dt,
\end{aligned}
\tag{3.70}
$$

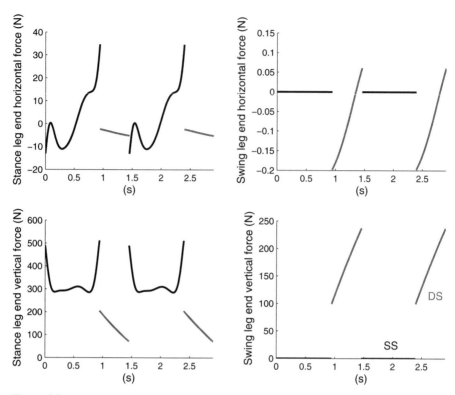

Figure 3.9   Horizontal and vertical components of the ground reaction forces experienced at the end of the legs during two consecutive steps of the optimal motion. (See the color version of this figure in color plates section.)

where $\sigma_{2,d}^- := \sigma_2(t_d^-)$, $\sigma_{2,d}^+ := \sigma_2(0^+)$, and $t = 0^+$ and $t = t_d^-$ denote the time instances just after the impact and just before the takeoff, respectively. In addition, t_d represents the time duration of the double support phase. At impact, σ_2 is not affected by the impulsive reaction force I_{R2} because I_{R2} acts at the end of leg-2. Hence,

$$\sigma_{2,d}^+ - \sigma_{2,s}^- = I_{R1}^v L_s, \qquad (3.71)$$

in which $\sigma_{2,s}^- := \sigma_2(0^-)$, and $t = 0^-$ represents the time instance just before the impact. Furthermore, according to the principle of angular momentum transfer [18, pp. 421,430], $\sigma_{2,s}^-$ can be expressed in terms of $\sigma_{1,s}^-$, that is,

$$\sigma_{2,s}^- = \sigma_{1,s}^- + m_{tot} L_s \dot{y}_{cm,s}^-, \qquad (3.72)$$

where $\sigma_{1,s}^- := \sigma_1(0^-)$ and $\dot{y}_{cm,s}^-$ is the vertical component of the velocity of the COM immediately before impact. After relabeling, the roles of the legs are swapped and,

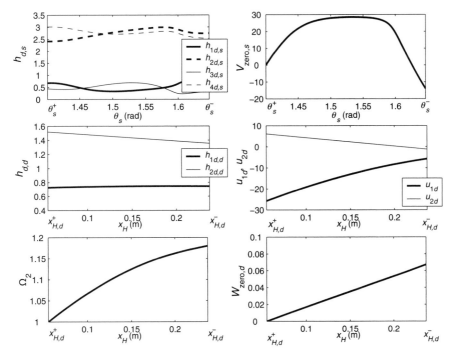

Figure 3.10 Graphs of the functions $h_{d,s}(\theta_s)$, $V_{\text{zero},s}(\theta_s)$, $h_{d,d}(\bar{x}_H)$, $u_{1d}(\bar{x}_H)$, $u_{2d}(\bar{x}_H)$, $\Omega_2(\bar{x}_H)$, and $\mathcal{W}_{\text{zero},d}(\bar{x}_H)$.

therefore,

$$\sigma_{1,s}^+ := \sigma_1(t_d^+) = \sigma_{2,d}^-, \tag{3.73}$$

in which $\sigma_{1,s}^+ := \sigma_1(t_d^+)$ is the value of σ_1 at the beginning of the next single support phase. Finally, equations (3.70)–(3.73) result in

$$\sigma_{1,s}^+ = \sigma_{1,s}^- + m_{\text{tot}} L_s \dot{y}_{\text{cm},s}^- + I_{R1}^v L_s$$

$$+ \int_{0^+}^{t_d^-} -m_{\text{tot}} g_0 (L_s - x_{\text{cm}}) + L_s F_1^v \, dt. \tag{3.74}$$

In addition, equation (3.26) implies that on the manifold $\mathcal{S}_s^d \cap Z_s$, $q = q_s^-$ and $\dot{q} = \lambda_s(q_s^-)\sigma_{1,s}^-$ which together with equation (3.21) and $p_{\text{cm}} = f_1(q)$ result in

$$\dot{y}_{\text{cm},s}^- = \frac{\partial f_1^v}{\partial q}(q_s^-)\dot{q} = \frac{\partial f_1^v}{\partial q}(q_s^-)\lambda_s(q_s^-)\sigma_{1,s}^- =: \lambda_{\text{cm},s}^v(q_s^-)\sigma_{1,s}^-$$

$$I_{R1}^v = \Sigma_1^v(q_s^-)\dot{q} = \Sigma_1^v(q_s^-)\lambda_s(q_s^-)\sigma_{1,s}^- =: \lambda_{I,1}^v(q_s^-)\sigma_{1,s}^-, \tag{3.75}$$

where f_1^v and Σ_1^v represent the second rows of f_1 and Σ_1, respectively. Consequently, equation (3.74) can be rewritten as follows:

$$\sigma_{1,s}^+ = \left(1 + m_{\mathrm{tot}} L_s \lambda_{\mathrm{cm},s}^v(q_s^-) + L_s \lambda_{I,1}^v(q_s^-)\right) \sigma_{1,s}^-$$
$$+ \int_{0+}^{t_d^-} -m_{\mathrm{tot}} g_0 (L_s - x_{\mathrm{cm}}) + L_s F_1^v \, dt. \tag{3.76}$$

By defining

$$\delta_{\mathrm{zero}} := 1 + m_{\mathrm{tot}} L_s \lambda_{\mathrm{cm},s}^v(q_s^-) + L_s \lambda_{I,1}^v(q_s^-)$$

$$\chi := \int_{0+}^{t_d^-} -m_{\mathrm{tot}} g_0 (L_s - x_{\mathrm{cm}}) + L_s F_1^v \, dt$$

and using equation (3.39), equation (3.76) can be rewritten as follows:

$$\sigma_{1,s}^+ = \varpi(\sigma_{1,s}^-) = \delta_{\mathrm{zero}}\, \sigma_{1,s}^- + \chi(\sigma_{1,s}^-). \tag{3.77}$$

It is worth mentioning that the effect of the double support phase on momentum transfer is given by the term χ that can also be expressed in terms of $\sigma_{1,s}^-$. To make this notion precise, we observe that

$$\chi = \int_{0+}^{t_d^-} -m_{\mathrm{tot}} g_0 (L_s - x_{\mathrm{cm}}) + L_s F_1^v \, dt$$
$$= \int_{x_{H,d}^+}^{x_{H,d}^-} \frac{-m_{\mathrm{tot}} g_0 (L_s - x_{\mathrm{cm}}(\bar{x}_H)) + L_s F_1^v(\bar{x}_H, \bar{v}_{xH})}{\bar{v}_{xH}} \, d\bar{x}_H,$$

where from equation (3.33) and $\bar{v}_{xH}^+ = \delta_s^d(q_s^-)\sigma_{1,s}^-$, $\bar{v}_{xH}$ can be expressed in terms of $\sigma_{1,s}^-$. Finally, we are able to represent effect of u_1 and u_2 on χ and stabilization.

Lemma 3.10 *Let $\mathcal{O} = \mathcal{O}_s \cup \mathcal{O}_d$ be a periodic orbit of the hybrid model (3.22) satisfying hypotheses HPO1–HPO7. Assume that $\sigma_{1,s}^{+*}$ and $\sigma_{1,s}^{-*}$ are the initial and final values of $\sigma_{1,s}$ on $\mathcal{O}_s$, respectively. Then,*

$$\frac{d\chi}{d\sigma_{1,s}^-}(\sigma_{1,s}^-)\big|_{\sigma_{1,s}^- = \sigma_{1,s}^{-*}}$$

is independent of the choice of the continuous functions $u_1 = u_{1d}(\bar{x}_H)$ and $u_2 = u_{2d}(\bar{x}_H)$.

Proof. As discussed previously, the single support phase zero dynamics is Lagrangian and therefore, $E_{\mathrm{zero}} = \frac{1}{2}(\sigma_{1,s})^2 + V_{\mathrm{zero},s}(\theta_s)$ is stationary. Let us consider the

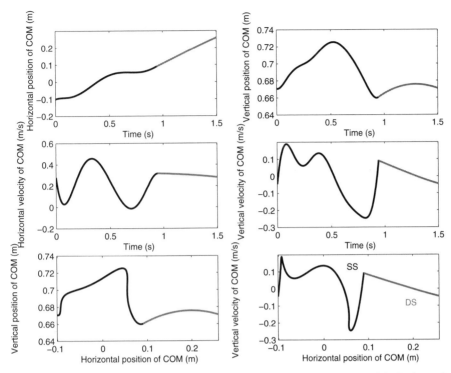

Figure 3.11 Plot of the vertical height and velocity of the COM versus time and the horizontal position of the COM on $\mathcal{O}$. At the impact, the velocity of the COM is not pointed downward. (See the color version of this figure in color plates section.)

restricted Poincaré return map in the coordinates $(\theta_s, \sigma_{1,s})$ by $\sigma_{1,s}^- \mapsto P(\sigma_{1,s}^-)$ and

$$
P(\sigma_{1,s}^-) := \sqrt{(\sigma_{1,s}^+)^2 - 2V_{\text{zero},s}(\theta_s^-)}
$$
$$
= \sqrt{(\varpi(\sigma_{1,s}^-))^2 - 2V_{\text{zero},s}(\theta_s^-)}.
$$

(3.78)

Then the derivative of P with respect to $\sigma_{1,s}^-$ evaluated at $\sigma_{1,s}^- = \sigma_{1,s}^{-*}$ can be expressed as

$$
\frac{dP}{d\sigma_{1,s}^-}(\sigma_{1,s}^-)\big|_{\sigma_{1,s}^- = \sigma_{1,s}^{-*}} = \frac{\varpi(\sigma_{1,s}^{-*})\frac{d\varpi}{d\sigma_{1,s}^-}(\sigma_{1,s}^{-*})}{\sqrt{(\varpi(\sigma_{1,s}^{-*}))^2 - 2V_{\text{zero},s}(\theta_s^-)}}
$$
$$
= \frac{\sigma_{1,s}^{+*}}{\sigma_{1,s}^{-*}}\left(\delta_{\text{zero}} + \frac{d\chi}{d\sigma_{1,s}^-}(\sigma_{1,s}^-)\big|_{\sigma_{1,s}^- = \sigma_{1,s}^{-*}}\right).
$$

(3.79)

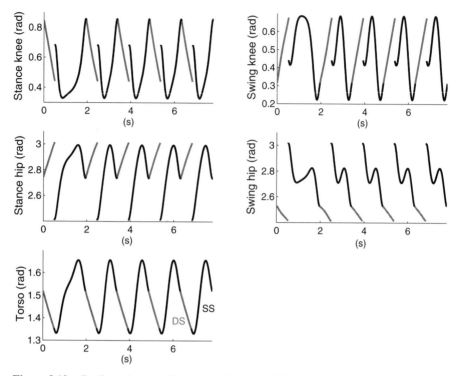

Figure 3.12 Configuration variables during five steps of the closed-loop simulation. Discontinuities in the graphs are due to coordinate relabling. (See the color version of this figure in color plates section.)

On the other hand, the restricted Poincaré return map in the coordinates (θ_s, ζ_s) for Z_s was expressed as $\rho(\zeta_s^-)$ in Section 3.4.3. In addition from Remark 4, $\mu = \frac{d\rho}{d\zeta_s^-}(\zeta_s^-)|_{\zeta_s^- = \zeta_s^{-*}}$ is independent of the continuous functions $u_1 = u_{1d}(q_i)$ and $u_2 = u_{2d}(q_i)$. Finally, the fact that δ_{zero} is dependent only on the orbit $\mathcal{O}_s$ and $P(\sigma_{1,s}^-) = \sqrt{2\rho(\frac{1}{2}(\sigma_{1,s}^-)^2)}$ completes the proof. ∎

Corollary 3.1 *The effects of the double support phase on angular momentum transfer and stabilization are given by χ and $\frac{d\chi}{d\sigma_{1,s}^-}(\sigma_{1,s}^-)|_{\sigma_{1,s}^- = \sigma_{1,s}^{-*}}$, respectively. In addition, from Lemma 3.10, $\frac{d\chi}{d\sigma_{1,s}^-}(\sigma_{1,s}^-)|_{\sigma_{1,s}^- = \sigma_{1,s}^{-*}}$ is independent of continuous functions $u_{1d}(q_i)$ and $u_{2d}(q_i)$.*

In this chapter, u_{1d} and u_{2d} are designed as the solution of a nonlinear optimization problem (3.56). By this means,

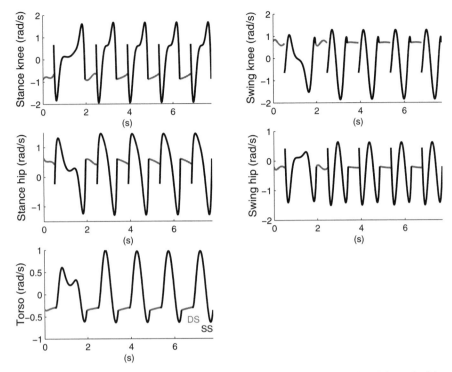

Figure 3.13   Velocity variables during five steps of the closed-loop simulation. Discontinuities in the graphs are due to coordinate relabling. (See the color version of this figure in color plates section.)

1. $\mathcal{O}_d$ is an integral curve of the double support phase dynamics (according to the equality constraint);
2. the control input and ground reaction forces corresponding to $\mathcal{O}_d$ are feasible (according to the inequality constraints);
3. the control input corresponding to $\mathcal{O}_d$ is minimum norm.

For the periodic orbit $\mathcal{O}$ obtained from the motion planning algorithm, the vertical component of the velocity of the COM just before the impact is positive (see Fig. 3.11), and hence, $\delta_{\text{zero}} \cong 1.1199 > 1$. In addition, $\frac{d\chi}{d\sigma_{1,s}^-}(\sigma_{1,s}^-)|_{\sigma_{1,s}^- = \sigma_{1,s}^{-*}} = -0.0018$. Hence,

$$\frac{\sigma_{1,s}^{+*}}{\sigma_{1,s}^{-*}}\left(\delta_{\text{zero}} + \frac{d\chi}{d\sigma_{1,s}^-}(\sigma_{1,s}^-)|_{\sigma_{1,s}^- = \sigma_{1,s}^{-*}}\right) = 1.2522 > 1.$$

Since the Jacobian of the restricted Poincaré return map without the double support phase is $\delta_{\text{zero}}^2 = 1.2541 > \mu$, the double support phase controller does not destabilize $\mathcal{O}$. However, it does not have sufficient strength to overcome the terms $\delta_{\text{zero}} > 1$ nor

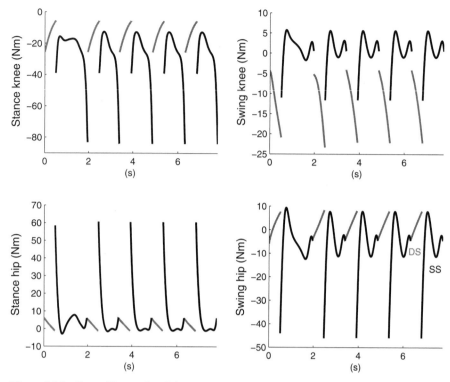

Figure 3.14   Control inputs (i.e., joint torques) during five steps of the closed-loop simulation. Discontinuities in the graphs are due to transition between the continuous phases. (See the color version of this figure in color plates section.)

stabilize $\mathcal{O}$ without applying the event-based update laws. We remark that in contrast to the approach proposed in Ref. [71], the proposed within-stride control law during the double support phase is continuous and utilizes input–output linearization to obtain a *nontrivial* HZD.[21] Since the periodic orbit is not stable without applying the second level of the control scheme, the output functions are modified as in equation (3.57). The degrees of the Bézier polynomials are chosen as $M_s = 4$ and $M_d = 4$. Then, the coefficients b_α and b_β can be calculated numerically as follows:

$$b_\alpha = [25.2776, 49.7273, -5.4290, -13.8925]$$
$$b_\beta = [298.2165, -25.2154],$$

---

[21] Since the approach of this chapter is based on Theorem 4.1 of Ref. [18, p. 89], the closed loop hybrid system has continuous vector fields during the single and double support phases to satisfy hypotheses HSH1 and HSH2 of Ref. [18, p. 83]. We also remark that while solving the motion planning algorithm by using the fmincon function, the inequality constraint $\dot{y}_{cm,s}^{-} < 0$ was not imposed on the motion planning algorithm.

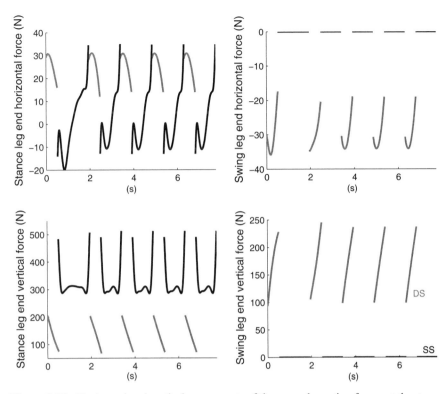

Figure 3.15   Horizontal and vertical components of the ground reaction forces at the stance and swing leg ends during five steps of the closed-loop simulation. The four graphs depict the admissibility of the ground reaction forces. (See the color version of this figure in color plates section.)

and consequently (a, b) is controllable. The gains of the static update laws K_α and K_β can be calculated via DLQR subject to the linearized system in equation (3.59).[22] Calculation for $Q = 1$ and $R = 100 \times I_{6\times6}$ by the dlqr function of MATLAB yields

$$K_\alpha = 10^{-3} \times [0.3371, 0.6631, -0.0724, -0.1853]'$$
$$K_\beta = 10^{-2} \times [0.3977, -0.0336]'.$$

In this case, the Jacobian of the closed-loop restricted Poincaré map ρ_{cl} evaluated at ζ_s^* can be calculated as $\mu_{cl} = a - b_\alpha K_\alpha - b_\beta K_\beta = 0.0013 < 1$. Thus, by using the event-based controller, the periodic orbit $\mathcal{O}$ is an asymptotically stable limit cycle for the closed-loop hybrid model of walking.

---

[22] In Ref. [61], the DLQR design method has been used in control of walking of an underactuated 3D biped robot.

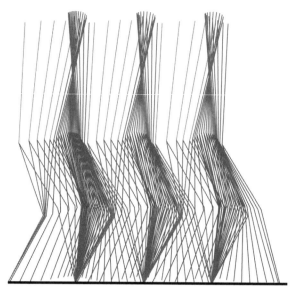

Figure 3.16   Stick diagram of the five-link, four-actuator biped robot taking three steps from left to right. (See the color version of this figure in color plates section.)

3.9.2   Effect of Event-Based Update Laws on Momentum Transfer and Stabilization

To investigate the effect of event-based update laws on angular momentum transfer and stabilization, let us assume that only the parameters of the double support phase controller (i.e., β) are updated and the parameters of the single support phase controller are held constant, that is, $\alpha = \alpha^*$. Then, χ can be expressed as $\chi(\sigma_{1,s}^-; \beta)$ and thereby,

$$\frac{dP}{d\sigma_{1,s}^-}(\sigma_{1,s}^-)\Big|_{\sigma_{1,s}^- = \sigma_{1,s}^{-*}} = \frac{\sigma_{1,s}^{+*}}{\sigma_{1,s}^{-*}} \left(\delta_{\text{zero}} + \frac{\partial \chi}{\partial \sigma_{1,s}^-}(\sigma_{1,s}^{-*}; \beta^*) - \frac{\partial \chi}{\partial \beta}(\sigma_{1,s}^{-*}; \beta^*)K_\beta \right).$$

According to this latter equation, $-\frac{\partial \chi}{\partial \beta}(\sigma_{1,s}^{-*}; \beta^*)K_\beta$ is an auxiliary term introduced by the event-based update laws to stabilize $\mathcal{O}$.

To confirm the analytical results obtained in this chapter, the simulation of the closed-loop hybrid model of walking is started at the end of single support phase. The initial condition of the configuration vector is assumed to be q_s^-. However, the initial condition for the velocity vector is chosen as the value of the velocity vector at the end of single support phase on the periodic orbit with an error of $+2°s^{-1}$ on each of its component. Figures 3.12–3.16 depict results of the simulation of the closed-loop system. Figures 3.12 and 3.13 represent the angular position and velocity of the knee, hip, and torso joints during five consecutive steps of the closed-loop system, respectively. The control inputs during five steps of walking are depicted in Fig. 3.14. Figure 3.15 also displays the horizontal and vertical components of the ground reaction forces at the end of the legs during five consecutive steps. Finally, the stick diagram of the five-link, four-actuator biped robot taking three steps from left to right is presented in Fig. 3.16.

CHAPTER 4

Asymptotic Stabilization of Periodic Orbits for Planar Monopedal Running

4.1 INTRODUCTION

This chapter presents an analytical approach for designing a two-level control law to asymptotically stabilize a desired period-one orbit during running by a planar monopedal robot. The monoped robot is a three-link, two-actuator planar mechanism in the sagittal plane with point foot. The desired periodic orbit is generated by the method developed in Ref. [73]. It is assumed that the model of monopedal running can be expressed by a hybrid system with two continuous phases, including stance phase (one leg on the ground) and flight phase (no leg on the ground), and discrete transitions between the continuous phases, including takeoff and landing (impact).

The configuration of the mechanical system is specified by the absolute orientation with respect to an inertial world frame and by the joint angles determining the shape of the robot. During the flight phase, the angular momentum of the mechanical system about its COM is conserved. To reduce the dimension of the full-order hybrid model of running, which, in turn, simplifies the stabilization problem of the desired orbit, as proposed in Ref. [55], we desire that the configuration of the mechanical system can be transferred from a specified initial pose (immediately after the takeoff) to a specified final pose (immediately before the landing) during flight phases. This problem is referred to as *landing in a fixed configuration* or *configuration determinism at landing* [18, p. 252]. However, the flight time and angular momentum about the COM may differ during consecutive steps. Consequently, the *reconfiguration problem* must be solved online. A number of control problems for reconfiguration of a planar multilink robot with zero angular momentum have been considered in the literature, for example, [74–78]. For the case that the angular momentum is not necessarily zero, a method based on the Averaging Theorem [79, Theorem 2.1] was presented in [80]

Hybrid Control and Motion Planning of Dynamical Legged Locomotion, Nasser Sadati, Guy A. Dumont,
Kaveh Akbari Hamed, and William A. Gruver.

such that for any value of the angular momentum, joint motions can reorient the multilink arbitrarily over an arbitrary time interval. However, when the angular momentum is not zero, this method cannot be employed online for solving the reconfiguration problem for monopedal running. For this reason, we present an online reconfiguration algorithm that solves this problem for given flight times and angular momenta [81, 82]. The algorithm proposed in this chapter is expressed using the methodology of reachability and optimal control for time-varying linear systems with input and state constraints.

Probably the most basic tool for analyzing the stability of periodic orbits of time-invariant dynamical systems described by ordinary differential equations is the Poincaré first return map that establishes an equivalence between the stability analysis of the periodic orbit for an nth-order continuous-time system and that of the corresponding equilibrium point for an $(n - 1)$th-order discrete-time system. Grizzle et al. [46] showed that the Poincaré return map can be applied to systems with impulse effects for analyzing the stability of periodic orbits. To reduce the dimension of the Poincaré return map during bipedal walking with one degree of underactuation, the strategy of using virtual constraints has been developed in Refs. [46–51]. For coordination of robot links, a set of holonomic output functions, referred to as virtual constraints, are defined and imposed to be zero by a feedback law [47]. For the case that the corresponding zero dynamics manifold is impact invariant, the concept of HZD was introduced in Ref. [52] that, in turn, results in a one-dimensional restricted Poincaré return map with a closed-form expression. To create impact invariance during bipedal walking with more than one degree of underactuation, a new approach based on parameterization of the outputs and updating their parameters in a stride-to-stride manner was presented in [60]. Following this strategy, asymptotically stable walking for an underactuated spatial biped robot is described in Ref. [61]. By using the virtual constraints approach, the configuration determinism at landing was proposed in [55] to obtain a closed-form expression for the one-dimensional restricted Poincaré return map of running by a five-link, four-actuator planar bipedal robot. Moreover, to ensure that the stance phase zero dynamics manifold is hybrid invariant under the closed-loop hybrid model of running, an additional constraint was imposed on the vector of generalized velocities at the end of flight phases. To satisfy the configuration determinism at landing and hybrid invariance, Ref. [55] utilized the approach of parameterized HZD. Specifically, on the basis of the Implicit Function Theorem and a *numerical* nonlinear optimization problem with an equality constraint, the parameters of the virtual constraints of the *flight phase* were updated in a step-by-step fashion during the discrete transition from stance to flight (i.e., takeoff). However, the stance phase controller was assumed to be fixed.

The main contribution of this chapter is to present an *analytical* approach for online generation of twice continuously differentiable (C^2) modified reference trajectories during flight phases of running to satisfy the configuration determinism at landing [81]. Moreover, by relaxing the constraint of Ref. [55] on the vector of generalized velocities at the end of the flight phases, we present a two-level control scheme based on the reconfiguration algorithm to asymptotically stabilize a desired periodic orbit.

In this scheme, within-stride controllers, including stance and flight phase controllers, are employed at the first level. The stance phase controller is chosen as a time-invariant and *parameterized* feedback law to generate a *family* of finite-time attractive, zero dynamics manifolds. An alternative approach based on continuous feedback law is employed here to track the modified reference trajectories generated by the reconfiguration algorithm during the flight phase. To generate a family of hybrid invariant manifolds, an event-based controller updates the parameters of the *stance phase controller* during the transition from flight to stance (i.e., impact) [81]. Consequently, the stability properties of the desired periodic orbit can be analyzed and modified by a one-dimensional discrete-time system defined on the basis of a restricted Poincaré return map (see Section 2.4).

4.2 MECHANICAL MODEL OF A MONOPEDAL RUNNER

4.2.1 The Monopedal Runner

A planar three-link monopedal robot with two ideal revolute joints and point foot (see Fig. 4.1) is considered throughout this chapter. The joints are controlled by internal actuators. Also, it is assumed that torques cannot be applied at the leg end. For the later uses, a coordinate frame is assumed to be attached to the ground called the *world frame*.

4.2.2 Dynamics of the Flight Phase

A convenient choice of the configuration variables consists of the body angles, the absolute orientation, and the absolute position of the monoped with respect to the

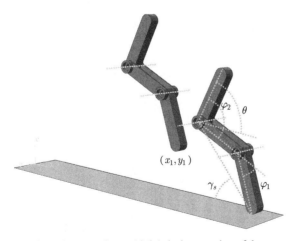

Figure 4.1   Flight (left) and stance phases (right) during running of the monopedal robot. The virtual leg is depicted by the dashed line connecting the stance leg end and the hip joint.

world frame. The body angles represented by $\varphi := (\varphi_1, \varphi_2)'$ describe the shape of the robot, where prime denotes matrix transpose. The absolute orientation of the robot is represented by θ, and the absolute position is represented by the Cartesian coordinates of its COM, $p_{cm} := (x_{cm}, y_{cm})'$. Consequently, the generalized coordinates during the flight phase are defined as $q_f := (\varphi', \theta, p'_{cm})' = (q', p'_{cm})'$, where $q := (\varphi', \theta)'$.

Following the notation of Ref. [18, Chapter 3], the dynamical model during the flight phase can be expressed as the following second-order equation:

$$A(\varphi)\ddot{q} + \bar{C}(\varphi, \dot{q})\dot{q} = Bu \tag{4.1}$$

$$m_{tot}\ddot{x}_{cm} = 0 \tag{4.2}$$

$$m_{tot}\ddot{y}_{cm} + m_{tot}g_0 = 0, \tag{4.3}$$

in which A is a (3×3) mass-inertia matrix, $\bar{C}$ is a (3×3) matrix containing the Coriolis and centrifugal terms, m_{tot} is the total mass of the robot, $u := (u_1, u_2)'$ is a vector of actuator torques, g_0 is the gravitational constant, and

$$B := \begin{bmatrix} I_{2\times2} \\ 0_{1\times2} \end{bmatrix}.$$

The configuration space for the flight phase, $\mathcal{Q}_f$, is also chosen to be a simply-connected and open subset of $(0, \frac{\pi}{2}) \times (0, \frac{3\pi}{2}) \times (-\frac{\pi}{2}, \frac{\pi}{2}) \times \mathbb{R}^2$. By introducing $x_f := (q'_f, \dot{q}'_f)'$ as the state vector, the evolution of the mechanical system during the flight phase can be expressed in the state space form $\dot{x}_f = f_f(x_f) + g_f(x_f)u$. Moreover, the state manifold for the flight phase is chosen as the tangent bundle of $\mathcal{Q}_f$, that is, $\mathcal{X}_f := T\mathcal{Q}_f$.

4.2.3 Dynamics of the Stance Phase

Using the principle of virtual work, a reduced-order model for describing the evolution of the mechanical system during the stance phase can be obtained as follows:

$$D(\varphi)\ddot{q} + C(\varphi, \dot{q})\dot{q} + G(q) = Bu, \tag{4.4}$$

where

$$D(\varphi) := A + m_{tot}\frac{\partial f_1}{\partial q}' \frac{\partial f_1}{\partial q}$$

$$C(\varphi, \dot{q}) := \bar{C} + m_{tot}\frac{\partial f_1}{\partial q}' \frac{\partial}{\partial q}\left(\frac{\partial f_1}{\partial q}\dot{q}\right)$$

$$G(q) := m_{tot}\frac{\partial f_1}{\partial q}' \begin{bmatrix} 0 \\ g_0 \end{bmatrix}$$

and f_1 is a smooth function of the configuration variables such that $(x_1, y_1)' = p_1(q_f) := p_{cm} - f_1(q) \in \mathbb{R}^2$ denotes the Cartesian coordinates of the leg end (see Fig. 4.1). By defining the state vector of the stance phase as $x_s := (q_s', \dot{q}_s')'$, where $q_s := q$ and $\dot{q}_s := \dot{q}$, equation (4.4) can be represented in state space form by $\dot{x}_s = f_s(x_s) + g_s(x_s)u$. The state manifold is chosen as $\mathcal{X}_s := T\mathcal{Q}_s$, in which $\mathcal{Q}_s$ is the configuration space of the stance phase and assumed to be a simply connected and open subset of $(0, \frac{\pi}{2}) \times (0, \frac{3\pi}{2}) \times (-\frac{\pi}{2}, \frac{\pi}{2})$.

4.2.4 Open-Loop Hybrid Model of Running

Following the modeling method presented in [55], the open-loop model of monopedal running can be expressed by a nonlinear hybrid system consisting of the following stance and flight phase state manifolds:

$$
\Sigma_s : \begin{cases} \dot{x}_s = f_s(x_s) + g_s(x_s)\,u, & x_s^- \notin \mathcal{S}_s^f \\ x_f^+ = \Delta_s^f(x_s^-), & x_s^- \in \mathcal{S}_s^f \\ \mathcal{S}_s^f := \{x_s \in \mathcal{X}_s \mid H_s^f(x_s) = 0\} \end{cases}
$$

$$(4.5)$$

$$
\Sigma_f : \begin{cases} \dot{x}_f = f_f(x_f) + g_f(x_f)\,u, & x_f^- \notin \mathcal{S}_f^s \\ x_s^+ = \Delta_f^s(x_f^-), & x_f^- \in \mathcal{S}_f^s \\ \mathcal{S}_f^s := \{x_f \in \mathcal{X}_f \mid H_f^s(x_f) = 0\}. \end{cases}
$$

In equation (4.5), the superscripts "$-$" and "$+$" denote the state of the hybrid system immediately before and after the switching between the state manifolds, respectively. We assume that the takeoff switching hypersurface $\mathcal{S}_s^f$ can be defined as the zero level set of the smooth function $H_s^f(x_s) := \gamma_s(q_s) - \gamma_s^-$, where $\gamma_s(q_s)$ is the angle of the virtual leg with respect to the world frame (see Fig. 4.1) and γ_s^- is a threshold value. Moreover, the impact switching hypersurface $\mathcal{S}_f^s$ is defined as the zero level set of $H_f^s(x_f) := y_1(q_f)$ (see Fig. 4.1). $\Delta_s^f : \mathcal{S}_s^f \rightarrow \mathcal{X}_f$ and $\Delta_f^s : \mathcal{S}_f^s \rightarrow \mathcal{X}_s$ also represent the takeoff and impact maps, respectively.

4.3 RECONFIGURATION ALGORITHM FOR THE FLIGHT PHASE

The conservation of angular momentum about the COM of the monopedal robot during the flight phase is expressed in the third row of matrix equation (4.1) that can be rewritten as follows[1]:

$$
\begin{aligned}
\dot{\theta} &= \frac{\sigma_{cm}}{A_{3,3}(\varphi)} - \sum_{i=1}^{2} \frac{A_{3,i}(\varphi)}{A_{3,3}(\varphi)} \dot{\varphi}_i \\
&= \frac{\sigma_{cm}}{A_{3,3}(\varphi)} - J(\varphi)\dot{\varphi},
\end{aligned} \tag{4.6}
$$

where σ_{cm} is a constant representing the angular momentum of the mechanical system about its COM and

$$
J(\varphi) := \frac{1}{A_{3,3}(\varphi)}[A_{3,1}(\varphi) \ A_{3,2}(\varphi)] \in \mathbb{R}^{1 \times 2}.
$$

Also $\mathcal{Q}_b$ is the body configuration space of the mechanical system and is assumed to be a simply connected and open subset of $(0, \frac{\pi}{2}) \times (0, \frac{3\pi}{2})$.

Remark 4.1 ($\dot{\varphi}$ as the Control Input for the Dynamical System (4.6)) *Since θ is a cyclic variable [1] for the mechanical system during the flight phase, the mass-inertia and Coriolis matrices in equation (4.1) are independent of θ. Hence, the right-hand side of equation (4.6) is expressed as a function of φ and $\dot{\varphi}$, and we can study the following dynamical system:*

$$
\begin{aligned}
\dot{\varphi} &= v \\
\dot{\theta} &= \frac{\sigma_{cm}}{A_{3,3}(\varphi)} - J(\varphi)v
\end{aligned} \tag{4.7}
$$

with the state space $\mathcal{Q} := \mathcal{Q}_b \times \mathbb{S}^1$ and the control v, where $\mathbb{S}^1 := [0, 2\pi)$ denotes the unit circle.

Fundamental Assumption We will assume that the takeoff and landing occur in fixed configurations. In particular, we present the following problem.

Problem 4.1 (Boundary Conditions on Configuration Variables) *Assume that a twice continuously differentiable (i.e., C^2) nominal trajectory $\varphi^* : [t_1^*, t_2^*] \to \mathcal{Q}_b$ (the evolution of body angles on a nominal trajectory) can transfer the configuration of the monoped robot during the flight phase from the initial condition $q_1^* := [\varphi'^*(t_1^*), \theta_1]'$ to the final condition $q_2^* := [\varphi'^*(t_2^*), \theta_2]'$ when the angular momentum about its COM is identically equal to σ_{cm}^*, that is,*

$$
\theta_2 = \theta_1 + \int_{t_1^*}^{t_2^*} \left(\frac{\sigma_{cm}^*}{A_{3,3}(\varphi^*(s))} - J(\varphi^*(s))\dot{\varphi}^*(s) \right) ds.
$$

---

[1] Because matrix $A(\varphi)$ is positive definite, $A_{3,3}(\varphi) > 0$ for any $\varphi \in \mathcal{Q}_b$.

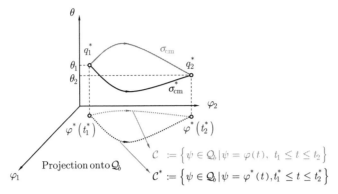

Figure 4.2 Geometric description of Problem 4.1 in the state space of the dynamical system given in equation (4.7). (See the color version of this figure in color plates section.)

Next, let the angular momentum about the COM be σ_{cm}, where $\sigma_{cm} \neq \sigma_{cm}^*$. The objective of this section is to present an online algorithm for generating the trajectory $\varphi : [t_1, t_2] \to \mathcal{Q}_b$ based on the nominal trajectory φ^* such that the configuration of the mechanical system can be transferred from the initial condition q_1^* to the final condition q_2^*, where $t_1 \neq t_1^*$ and $t_2 \neq t_2^*$. In other words, we look for a C^2 function φ such that (i) $\varphi(t_1) = \varphi^*(t_1^*)$, (ii) $\varphi(t_2) = \varphi^*(t_2^*)$, and (iii)

$$\theta_2 = \theta_1 + \int_{t_1}^{t_2} \left(\frac{\sigma_{cm}}{A_{3,3}(\varphi(t))} - J(\varphi(t)) \, \dot\varphi(t) \right) dt.$$

Figure 4.2 represents a geometric description for Problem 4.1 in the state space of system (4.7). In this figure, the nominal C^1 input $v^*(t) := \dot\varphi^*(t)$, $t_1^* \leq t \leq t_2^*$ transfers the state of the system from the initial point q_1^* to the final point q_2^* when the angular momentum about the COM is equal to σ_{cm}^*. The objective is to generate the C^1 input $v(t) := \dot\varphi(t)$, $t_1 \leq t \leq t_2$ transferring the state of the system from q_1^* to q_2^* when $\sigma_{cm} \neq \sigma_{cm}^*$. In addition,

$$C^* := \{\psi \in \mathcal{Q}_b | \psi = \varphi^*(t), t_1^* \leq t \leq t_2^*\}$$
$$C := \{\psi \in \mathcal{Q}_b | \psi = \varphi(t), \ t_1 \leq t \leq t_2\}$$

denote the projections of the nominal and generated trajectories onto the body configuration space $\mathcal{Q}_b$. The results of this section will be utilized in Section 4.4 to reduce the dimension of the full-order hybrid model of the monopedal robot and thereby simplify the stabilization problem in Section 4.5.

Integrating equation (4.6) over the time interval $[t_1, t_2]$ results in

$$\theta(t_2) = \theta_1 + \int_{t_1}^{t_2} \frac{\sigma_{cm}}{A_{3,3}(\varphi(t))} \, dt - \int_C J(\varphi) \, d\varphi. \qquad (4.8)$$

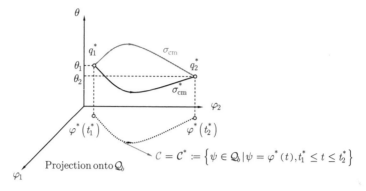

Figure 4.3 Geometric description of the proposed reconfiguration algorithm. In this algorithm, it is assumed that $\varphi(t) = \varphi^*(\tau(t))$, where $\tau : [t_1, t_2] \to [t_1^*, t_2^*]$ is the virtual time satisfying the constraints given in equation (4.9). Thus, $\mathcal{C} = \mathcal{C}^*$ and the geometric phases corresponding to the trajectories φ and φ^* are equal. (See the color version of this figure in color plates section.)

The second and third terms in the right-hand side of equation (4.8) are called the *dynamic* and *geometric phases*, respectively [80, 98]. The dynamic term is nonzero if and only if the angular momentum σ_{cm} is nonzero because $A_{3,3}(\varphi) > 0$ for every $\varphi \in \mathcal{Q}_b$. In addition, the geometric phase is expressed as a line integral along the path of joint angles (i.e., $\mathcal{C}$) and consequently, it depends only on the path $\mathcal{C}$ [80]. To simplify the analysis, we look for ways in which the geometric phase generated by the proposed reconfiguration algorithm is equal to that of the nominal trajectory φ^* (see Fig. 4.3). Toward that end, by assuming $\varphi(t) := \varphi^*(\tau(t))$, where $\tau : [t_1, t_2] \to [t_1^*, t_2^*]$ is the *virtual time* fulfilling the following constraints

$$
\begin{aligned}
&(i) \quad \tau(t_1) = t_1^* \\
&(ii) \quad \tau(t_2) = t_2^* \\
&(iii) \quad \inf_{t_1 \leq t \leq t_2} \dot{\tau}(t) > 0,
\end{aligned}
\tag{4.9}
$$

$\mathcal{C} = \mathcal{C}^*$ and equation (4.8) can be rewritten as follows:

$$
\theta(t_2) = \theta_1 + \int_{t_1^*}^{t_2^*} \frac{\sigma_{cm}}{A_{3,3}(\varphi^*(s))} \frac{ds}{\dot{\tau} \circ \tau^{-1}(s)} - \int_{\mathcal{C}^*} J(\varphi^*) \, d\varphi^*,
$$

and consequently,

$$
\theta(t_2) - \theta_2 = \int_{t_1^*}^{t_2^*} \frac{1}{A_{3,3}(\varphi^*(s))} \left(\frac{\sigma_{cm}}{\dot{\tau} \circ \tau^{-1}(s)} - \sigma_{cm}^* \right) ds.
$$

By defining $\mu(s) := \frac{1}{\dot{\tau} \circ \tau^{-1}(s)} > 0$ and $w(s) := \frac{1}{A_{3,3}(\varphi^*(s))} > 0$ for $s \in [t_1^*, t_2^*]$, and assuming $\sigma_{cm} \neq 0$, the condition $\theta(t_2) = \theta_2$ can be expressed as the following equality constraint:

$$\int_{t_1^*}^{t_2^*} w(s)\,\mu(s)\,ds = \frac{\sigma_{cm}^*}{\sigma_{cm}} \int_{t_1^*}^{t_2^*} w(s)\,ds. \tag{4.10}$$

Furthermore, from the definition of $\mu(s)$, $\dot{\tau}(t) = \frac{1}{\mu(\tau(t))}$, $t_1 \leq t \leq t_2$, and hence,

$$\int_{t_1^*}^{t_2^*} \mu(s)\,ds = t_2 - t_1. \tag{4.11}$$

By using the virtual time approach, we can present an alternative problem equivalent to Problem 4.1 in which the reconfiguration can be solved on the basis of reachability and optimal control formulations of a linear time-varying system with input constraints.

Problem 4.2 *Determination of $\mu(\tau) > 0$, $t_1^* \leq \tau \leq t_2^*$ such that the equality constraints in equations (4.10) and (4.11) are met is equivalent to determining the open-loop control input $\mu : [t_1^*, t_2^*] \to \mathbb{R}^{>0}$ that transfers the state of the following linear time-varying system in the virtual time domain:*

$$\Sigma : \begin{aligned} \dot{x}_1 &= w(\tau)\,\mu \\ \dot{x}_2 &= \mu \end{aligned} \tag{4.12}$$

from $(x_1(t_1^*), x_2(t_1^*))' = (0, 0)'$ to $(x_1(t_2^*), x_2(t_2^*))' = (x_1^f, x_2^f)'$, where $\dot{x}_i := \frac{d}{d\tau}x_i$ for $i = 1, 2$ and

$$\begin{aligned} x_1^f &:= \frac{\sigma_{cm}^*}{\sigma_{cm}} \int_{t_1^*}^{t_2^*} w(s)\,ds \\ x_2^f &:= t_2 - t_1. \end{aligned} \tag{4.13}$$

4.3.1 Determination of the Reachable Set

The purpose of this section is to determine the reachable set from the origin (at t_1^*) at time t_2^* for the system Σ. Since $\varphi(t) = \varphi^*(\tau(t))$, the following relations can be obtained for the first and second time derivatives of $\varphi(t)$:

$$\dot{\varphi}(t) = \frac{\partial \varphi^*}{\partial \tau}(\tau(t))\,\dot{\tau}(t)$$

$$\ddot{\varphi}(t) = \frac{\partial \varphi^*}{\partial \tau}(\tau(t))\,\ddot{\tau}(t) + \frac{\partial^2 \varphi^*}{\partial \tau^2}(\tau(t))\,\dot{\tau}^2(t),$$

and, hence, a discontinuity of μ (or equivalently, a discontinuity of $\dot{\tau}$) may result in an impulsive nature of $\ddot{\varphi}(t)$. In view of the actuator limitations, this latter fact implies that $\varphi(t)$ cannot be used as a reference trajectory for the joint angles. Thus, we present the following definition.

Definition 4.1 (Admissible Open-Loop Control Inputs for System Σ) *The set of admissible open-loop control inputs for system Σ is denoted by $\mathcal{U}_{m,M}$ and defined to be the set of all continuously differentiable functions $\tau \mapsto \mu(\tau) \in [m, M]$ defined on the interval $[t_1^*, t_2^*]$, where $0 < m < M$.*

We present a design method for obtaining an admissible open-loop control $\mu \in C^1([t_1^*, t_2^*], [m, M])$. For this purpose, we consider μ to be the output of a double integrator and study the following augmented system:

$$\Sigma_a : \begin{aligned} \dot{x}_1 &= w(\tau)\,x_3 \\ \dot{x}_2 &= x_3 \\ \dot{x}_3 &= x_4 \\ \dot{x}_4 &= v, \end{aligned}$$

which can be viewed as a cascade connection of two components. The first component is the system Σ in equation (4.12) with x_3 as input and the second component is the double integrator with a piecewise continuous function v as input. The admissibility of μ can be expressed as $m \leq x_3 \leq M$ that is a constraint on the state of the system Σ_a.

Definition 4.2 (Admissible Open-Loop Control Inputs for System Σ_a) *The set of admissible open-loop control inputs for system Σ_a is denoted by $\mathcal{V}_{L_1,L_2}$ and defined to be the set of all piecewise continuous functions $\tau \mapsto v(\tau) \in [L_1, L_2]$ defined on the interval $[t_1^*, t_2^*]$, where $L_1 < 0 < L_2$.*

Definition 4.3 (Reachable Set from the Origin) *For any $0 < m < M$, $L_1 < 0 < L_2$ and $(x_3^0, x_4^0)' \in \mathbb{R}^2$, define $\mathcal{A}_{m,M,L_1,L_2}(x_3^0, x_4^0)$ as the set of all points $(x_1^f, x_2^f)' \in \mathbb{R}^2$ for which there exists an open-loop control $v \in \mathcal{V}_{L_1,L_2}$ such that the state of the system Σ_a is transferred from the initial point $(0, 0, x_3^0, x_4^0)'$ at t_1^* to the final point $(x_1^f, x_2^f, x_3^f, x_4^f)'$ at t_2^* with the constraint $m \leq x_3(\tau) \leq M$, $t_1^* \leq \tau \leq t_2^*$, where x_3^f and x_4^f are free.*

It is clear that for every $x_3^0 \notin [m, M]$, $\mathcal{A}_{m,M,L_1,L_2}(x_3^0, x_4^0) = \phi$. To determine $\mathcal{A}_{m,M,L_1,L_2}$, we study two optimal control problems. From these problems, the optimal admissible open-loop control inputs, $v^{\max}(\tau)$, $v^{\min}(\tau) \in \mathcal{V}_{L_1,L_2}$, $t_1^* \leq \tau \leq t_2^*$, are determined such that the state of the augmented system Σ_a is to be transferred from the initial point $x^0 := (0, 0, x_3^0, x_4^0)'$ at t_1^* to the final point $(x_1(t_2^*), x_2(t_2^*), x_3(t_2^*), x_4(t_2^*))'$

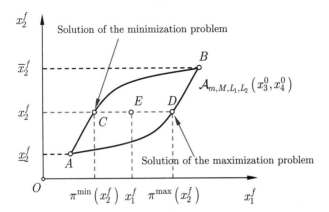

Figure 4.4   The reachable set $A_{m,M,L_1,L_2}(x_3^0, x_4^0)$. The solutions of the minimization and maximization problems for a given x_2^f are denoted by C and D, respectively. (See the color version of this figure in color plates section.)

at t_2^* with the property $m \leq x_3(\tau) \leq M$, $t_1^* \leq \tau \leq t_2^*$, while the performance measure

$$\mathcal{I}_a(v) := x_1(t_2^*)$$

is maximized (see point D in Fig. 4.4) and minimized (see point C in Fig. 4.4). Note that in these two optimal control problems, $x_2(t_2^*) = x_2^f$ is specified, whereas $x_3(t_2^*)$ and $x_4(t_2^*)$ are free.

The constraint $m \leq x_3(\tau) \leq M$ can be rewritten as the following inequality constraints:

$$S_1(x) := m - x_3 \leq 0$$
$$S_2(x) := x_3 - M \leq 0.$$

Next, we take successive virtual time derivatives of $S_1(x)$ and $S_2(x)$ until obtaining an expression that is explicitly dependent on v [99, p. 118]. This process will result in $\ddot{S}_1 = -v$ and $\ddot{S}_2 = v$. Now, define the following Hamiltonian function:

$$\begin{aligned}\mathcal{H}(x, p, \lambda, v, \tau) :=&\, p_1 w(\tau) x_3 + p_2 x_3 + p_3 x_4 + p_4 v + \lambda_1 \ddot{S}_1 + \lambda_2 \ddot{S}_2 \\ =&\, p_1 w(\tau) x_3 + p_2 x_3 + p_3 x_4 + (p_4 - \lambda_1 + \lambda_2) v,\end{aligned} \quad (4.14)$$

where $x := (x_1, x_2, x_3, x_4)'$, $p := (p_1, p_2, p_3, p_4)'$, and $\lambda := (\lambda_1, \lambda_2)'$ are the state, costate, and multiplier vectors, respectively. Furthermore, in equation (4.14),

$$\begin{aligned}\ddot{S}_i &= 0, \quad \text{on the constraint boundary (i.e., } S_i = 0) \\ \lambda_i &= 0, \quad \text{off the constraint boundary (i.e., } S_i < 0),\end{aligned}$$

for $i = 1, 2$, which can also be expressed as

$$
\begin{aligned}
v &= 0, &\text{on the constraint boundary (i.e., } S_i = 0) \\
\lambda_i &= 0, &\text{off the constraint boundary (i.e., } S_i < 0).
\end{aligned}
\tag{4.15}
$$

Necessary conditions for multipliers $\lambda_i(\tau)$, $i = 1, 2$ to minimize the performance measures[2] are

$$
\lambda_i(\tau) \geq 0, \quad \text{on the constraint boundary (i.e., } S_i = 0). \tag{4.16}
$$

The costates satisfy the following differential equations:

$$
\begin{aligned}
\dot{p}_1 &= 0 \\
\dot{p}_2 &= 0 \\
\dot{p}_3 &= -p_1 w(\tau) - p_2 \\
\dot{p}_4 &= -p_3,
\end{aligned}
$$

in which $\dot{p}_i := \frac{d}{d\tau} p_i$, $i = 1, \cdots, 4$. From here on, the superscripts "max" and "min" will denote the solutions of the maximization and minimization problems, respectively. We first study the maximization problem. Since the final values $x_3^{\max}(t_2^*)$, $x_3^{\max}(t_2^*)$, and $x_4^{\max}(t_2^*)$ are free, from Table 5.1 of [100, p. 200], $p_1^{\max}(t_2^*) = -1$ and $p_3^{\max}(t_2^*) = p_4^{\max}(t_2^*) = 0$. These boundary conditions in combination with the costate equations yield

$$
\begin{aligned}
p_3^{\max}(\tau; p_2^{\max}) &= -\int_\tau^{t_2^*} w(s)\, ds - p_2^{\max}(\tau - t_2^*) \\
p_4^{\max}(\tau; p_2^{\max}) &= -\int_\tau^{t_2^*}\int_s^{t_2^*} w(\eta)\, d\eta\, ds + \frac{p_2^{\max}}{2}(\tau - t_2^*)^2.
\end{aligned}
$$

From Pontryagin's Minimum Principle [101], $v^{\max}(\tau)$ is given by

$$
v^{\max}(\tau) = \begin{cases}
L_1 & p_4^{\max} - \lambda_1^{\max} + \lambda_2^{\max} > 0 \\
L_2 & p_4^{\max} - \lambda_1^{\max} + \lambda_2^{\max} < 0 \\
\text{undetermined} & p_4^{\max} - \lambda_1^{\max} + \lambda_2^{\max} = 0.
\end{cases}
\tag{4.17}
$$

Note that if $p_4^{\max}(\tau; p_2^{\max}) - \lambda_1^{\max}(\tau) + \lambda_2^{\max}(\tau)$ passes through the zero, a switching of the optimal control input $v^{\max}(\tau)$ occurs. Assume that $w(\tau)$ satisfies the following hypothesis:

---

[2] Note that the maximization of the performance measure $\mathcal{I}_a(v) := x_1(t_2^*)$ can be expressed as the minimization of $-\mathcal{I}_a(v)$.

(H1) $\dot{w}(\tau) := \frac{d}{d\tau} w(\tau)$ is not zero on the open set (t_1^*, t_2^*).

It will be shown that by hypothesis H1 the optimal control inputs $v^{\max}(\tau)$ and $v^{\min}(\tau)$ can switch at most once, and the singular condition does not occur. For this purpose, we present the following result.

Lemma 4.1 (Behavior of the Solutions for the Optimization Problems)   *Let $m < x_3^0 < M$ and $L_1 < 0 < L_2$. Assume that hypothesis H1 holds and x_4^0 is such that the optimal trajectories $x^{\max}(\tau)$ and $x^{\min}(\tau)$ of the system Σ_a exist. Then, the following statements are true.*

(a) The optimal trajectories do not enter onto the boundaries $S_1 = 0$ and $S_2 = 0$.

(b) The optimal control inputs $v^{\max}(\tau)$ and $v^{\min}(\tau)$ can switch at most once.

(c) The singular condition does not occur. In other words, the sets

$$\mathcal{T}_0^{\max} := \left\{ \tau \in [t_1^*, t_2^*] \,\middle|\, p_4^{\max}(\tau) - \lambda_1^{\max}(\tau) + \lambda_2^{\max}(\tau) = 0 \right\}$$

$$\mathcal{T}_0^{\min} := \left\{ \tau \in [t_1^*, t_2^*] \,\middle|\, p_4^{\min}(\tau) - \lambda_1^{\min}(\tau) + \lambda_2^{\min}(\tau) = 0 \right\}$$

are Lebesgue negligible.

Proof. To prove the statements (a), (b), and (c) of Lemma 4.1, the maximization problem will be studied. Similar reasonings can also be presented for the minimization problem.

If the optimal trajectory enters onto the constraint boundary $S_1 = 0$, S_2 will be negative and consequently from condition (4.15), $v^{\max} = 0$ and $\lambda_2^{\max} = 0$. Since $L_1 < 0 < L_2$, from the open-loop control input $v^{\max}(\tau)$ in equation (4.17), $v^{\max} = 0$ results in $\lambda_1^{\max} = p_4^{\max}$ that, in combination with the necessary conditions given in equation (4.16), yields $p_4^{\max} \geq 0$. Similarly, if the optimal trajectory enters onto the constraint boundary $S_2 = 0$, then $v^{\max} = 0$ and $\lambda_1^{\max} = 0$. Moreover, $\lambda_2^{\max} = -p_4^{\max}$ that implies that $p_4^{\max} \leq 0$.

Next, we study the roots of the nonlinear equation $p_4^{\max}(\tau; p_2^{\max}) = 0$ on the interval $[t_1^*, t_2^*]$. This equation can also be expressed as follows:

$$\mathcal{W}(\tau) = \frac{p_2^{\max}}{2} (\tau - t_2^*)^2, \tag{4.18}$$

where

$$\mathcal{W}(\tau) := \int_\tau^{t_2^*} \int_s^{t_2^*} w(\eta) \, d\eta \, ds.$$

For any $p_2^{max} \in \mathbb{R}$, t_2^* is the solution of equation (4.18). We claim that the equation $p_4^{max}(\tau; p_2^{max}) = 0$ has at most one root in the interval $[t_1^*, t_2^*)$. To show this, let $\bar{\tau} \in [t_1^*, t_2^*)$ exist such that $p_4^{max}(\bar{\tau}; p_2^{max}) = 0$. Then, p_2^{max} is unique and can be given by

$$p_2^{max} = \frac{2}{\left(\bar{\tau} - t_2^*\right)^2} \mathcal{W}(\bar{\tau}). \tag{4.19}$$

We remark that p_2^{max} is positive. If there exists $\tilde{\tau} \in [t_1^*, t_2^*)$ such that $\tilde{\tau} \neq \bar{\tau}$ and $p_4^{max}(\tilde{\tau}; p_2^{max}) = 0$, then equation (4.19) implies that

$$\frac{\mathcal{W}(\bar{\tau})}{\left(\bar{\tau} - t_2^*\right)^2} = \frac{\mathcal{W}(\tilde{\tau})}{\left(\tilde{\tau} - t_2^*\right)^2}.$$

Hence, it is sufficient to show that the function $\kappa : [t_1^*, t_2^*) \to \mathbb{R}$ by

$$\kappa(\tau) := \frac{\mathcal{W}(\tau)}{\left(\tau - t_2^*\right)^2}.$$

is strictly monotonic. The first derivative of $\kappa(\tau)$ can be obtained as follows:

$$\dot{\kappa}(\tau) = \frac{\dot{\mathcal{W}}(\tau)\left(\tau - t_2^*\right) - 2\mathcal{W}(\tau)}{\left(\tau - t_2^*\right)^3} =: \frac{F(\tau)}{\left(\tau - t_2^*\right)^3}. \tag{4.20}$$

Assume that there exists $\eta_1 \in (t_1^*, t_2^*)$ such that $F(\eta_1) = 0$. Since $F(\eta_1) = F(t_2^*) = 0$, the Rolle's Theorem implies that there is $\eta_2 \in (\eta_1, t_2^*)$ such that

$$\dot{F}(\eta_2) = \ddot{\mathcal{W}}(\eta_2)\left(\eta_2 - t_2^*\right) - \dot{\mathcal{W}}(\eta_2) = 0.$$

Furthermore, $\dot{F}(t_2^*)$ is also zero. Hence, from the Rolle's Theorem, there exists $\eta_3 \in (\eta_2, t_2^*)$ such that

$$\ddot{F}(\eta_3) = \dot{w}(\eta_3)\left(\eta_3 - t_2^*\right) = 0,$$

which follows that $\dot{w}(\eta_3) = 0$, and this contradicts hypothesis H1. Therefore, $\kappa(\tau)$ is strictly monotonic and, as a consequence, the equation $p_4^{max}(\tau; p_2^{max}) = 0$ has at most one root in the interval $[t_1^*, t_2^*)$.

Let $\bar{\tau} \in (t_1^*, t_2^*)$ be the root of the equation $p_4^{\max}(\tau; p_2^{\max}) = 0$. Substituting equation (4.19) into $p_4^{\max}(\tau; p_2^{\max})$ yields

$$\dot{p}_4^{\max}\left(\bar{\tau}; p_2^{\max}\right) = \frac{-\dot{W}(\bar{\tau})\left(\bar{\tau} - t_2^*\right) + 2W(\bar{\tau})}{\bar{\tau} - t_2^*}$$

$$= \frac{F(\bar{\tau})}{t_2^* - \bar{\tau}} \neq 0.$$

Therefore, the condition $\dot{p}_4^{\max}(\bar{\tau}; p_2^{\max}) < 0$ that can be expressed as

$$\frac{2}{t_2^* - \bar{\tau}} \int_{\bar{\tau}}^{t_2^*} \int_s^{t_2^*} w(\eta) \, d\eta \, ds > \int_{\bar{\tau}}^{t_2^*} w(s) \, ds \tag{4.21}$$

implies that $p_4^{\max}(\tau; p_2^{\max}) > 0$ for any $\tau \in [t_1^*, \bar{\tau})$ and $p_4^{\max}(\tau; p_2^{\max}) < 0$ for any $\tau \in (\bar{\tau}, t_2^*]$. If condition (4.21) is not satisfied, then $p_4^{\max}(\tau; p_2^{\max}) < 0$ for any $\tau \in [t_1^*, \bar{\tau})$ and $p_4^{\max}(\tau; p_2^{\max}) > 0$ for any $\tau \in (\bar{\tau}, t_2^*]$.

Since $m < x_3^0 < M$ and $x^{\max}(\tau)$ is the optimal trajectory, $S_1(x^0)$ and $S_2(x^0)$ are negative that result in $\lambda_1^{\max}(t_1^*) = \lambda_2^{\max}(t_1^*) = 0$. Without loss of generality, assume that $p_4^{\max}(t_1^*; p_2^{\max}) \neq 0$.[3] Then there are four possible cases.

Case 1: Assume that there exists $\bar{\tau} \in (t_1^*, t_2^*)$ such that $p_4^{\max}(\bar{\tau}; p_2^{\max}) = 0$ and inequality (4.21) holds. In this case, on the interval $[t_1^*, \bar{\tau})$, $v^{\max}(\tau) = L_1$ and consequently,

$$x_4^{\max}(\tau) = x_4^0 + L_1\left(\tau - t_1^*\right)$$

$$x_3^{\max}(\tau) = x_3^0 + x_4^0\left(\tau - t_1^*\right) + \frac{L_1}{2}\left(\tau - t_1^*\right)^2$$

$$x_2^{\max}(\tau) = x_3^0\left(\tau - t_1^*\right) + \frac{x_4^0}{2}\left(\tau - t_1^*\right)^2 + \frac{L_1}{6}\left(\tau - t_1^*\right)^3.$$

Because $p_4^{\max}(\tau; p_2^{\max}) > 0$ on the interval $[t_1^*, \bar{\tau})$, the trajectory $x^{\max}(\tau)$, $t_1^* \leq \tau < \bar{\tau}$ cannot enter onto the boundary $S_2 = 0$. From Ref. [99, p. 118], since the control of S_1 is obtained only by changing $\dddot{S}_1$, no finite control can keep the optimal trajectory of the system Σ_a on the constraint boundary $S_1 = 0$, unless the following tangency constraints[4] hold:

$$N_1(x^{\max}(\tau)) := \begin{bmatrix} S_1(x^{\max}(\tau)) \\ \dot{S}_1(x^{\max}(\tau)) \end{bmatrix} = \begin{bmatrix} m - x_3^{\max}(\tau) \\ -x_4^{\max}(\tau) \end{bmatrix} = \begin{bmatrix} 0 \\ 0 \end{bmatrix}. \tag{4.22}$$

---

[3] If $p_4^{\max}(t_1^*; p_2^{\max}) = 0$, then $\bar{\tau} = t_1^*$. Since the nonlinear equation $p_4^{\max}(\tau; p_2^{\max}) = 0$ has at most one root in the interval $[t_1^*, t_2^*)$, on the open set (t_1^*, t_2^*), $p_4^{\max}(\tau; p_2^{\max}) \neq 0$. In addition, $v^{\max}(t_1^*)$ is finite and as a consequence, without loss of generality, we can assume that $p_4^{\max}(t_1^*; p_2^{\max}) \neq 0$.

[4] The terminology of a *tangency constraint* is taken from Ref. [99, p. 118].

Because $L_1 < 0$, $x_3^{\max}(\tau)$, $t_1^* \leq \tau < \bar{\tau}$ is a quadratic and concave function with respect to τ. Thus, the condition $m < x_3^0 < M$ implies that the tangency constraints (4.22) cannot be satisfied on the interval $[t_1^*, \bar{\tau})$. Next, define

$$\bar{x}_4 := x_4^{\max}(\bar{\tau}) = x_4^0 + L_1(\bar{\tau} - t_1^*)$$

$$\bar{x}_3 := x_3^{\max}(\bar{\tau}) = x_3^0 + x_4^0(\bar{\tau} - t_1^*) + \frac{L_1}{2}(\bar{\tau} - t_1^*)^2$$

$$\bar{x}_2 := x_2^{\max}(\bar{\tau}) = x_3^0(\bar{\tau} - t_1^*) + \frac{x_4^0}{2}(\bar{\tau} - t_1^*)^2 + \frac{L_1}{6}(\bar{\tau} - t_1^*)^3.$$

Also, on the interval $(\bar{\tau}, t_2^*]$, $v^{\max}(\tau) = L_2$ and thus,

$$x_4^{\max}(\tau) = \bar{x}_4 + L_2(\tau - \bar{\tau})$$

$$x_3^{\max}(\tau) = \bar{x}_3 + \bar{x}_4(\tau - \bar{\tau}) + \frac{L_2}{2}(\tau - \bar{\tau})^2$$

$$x_2^{\max}(\tau) = \bar{x}_2 + \bar{x}_3(\tau - \bar{\tau}) + \frac{\bar{x}_4}{2}(\tau - \bar{\tau})^2 + \frac{L_2}{6}(\tau - \bar{\tau})^3.$$

Since $p_4^{\max}(\tau; p_2^{\max}) < 0$ on $(\bar{\tau}, t_2^*]$, the optimal trajectory $x^{\max}(\tau)$, $\bar{\tau} < \tau \leq t_2^*$ cannot enter onto the boundary $S_1 = 0$. Moreover, the tangency constraints to remain on the boundary $S_2 = 0$ can be expressed as

$$N_2(x^{\max}(\tau)) := \begin{bmatrix} S_2(x^{\max}(\tau)) \\ \dot{S}_2(x^{\max}(\tau)) \end{bmatrix} = \begin{bmatrix} x_3^{\max}(\tau) - M \\ x_4^{\max}(\tau) \end{bmatrix} = \begin{bmatrix} 0 \\ 0 \end{bmatrix}. \qquad (4.23)$$

The fact that $x_3^{\max}(\tau)$, $\bar{\tau} < \tau \leq t_2^*$ is a quadratic and convex function with respect to τ in combination with the condition[5] $m < \bar{x}_3 < M$ implies that the tangency constraints in equation (4.23) cannot be satisfied on the interval $(\bar{\tau}, t_2^*]$. Consequently, $S_1(x^{\max}(\tau))$, $S_2(x^{\max}(\tau)) < 0$ for any $\tau \in [t_1^*, t_2^*]$.

The final constraint $x_2^{\max}(t_2^*) = x_2^f$ can also be expressed as the following third-degree equation:

$$\frac{L_1 - L_2}{6}(\bar{\tau} - t_2^*)^3 + x_3^0 l_{\max} + \frac{x_4^0}{2} l_{\max}^2 + \frac{L_1}{6} l_{\max}^3 = x_2^f, \qquad (4.24)$$

---

[5] Since the optimal trajectory is feasible (i.e., $m \leq x_3^{\max}(\tau) \leq M$) and $L_1 < 0$, $\bar{x}_3 = m$ will result in $\bar{x}_4 < 0$, which, in turn, implies the existence of $\hat{\tau} \in (\bar{\tau}, t_2^*]$ such that $x_3^{\max}(\tau) < m$ for any $\tau \in (\bar{\tau}, \hat{\tau})$. This contradicts the feasibility of the optimal trajectory $x^{\max}(\tau)$. In a similar manner, it can be shown that $\bar{x}_3 \neq M$.

where $l_{\max} := t_2^* - t_1^*$. From equation (4.24), $\bar{\tau} \in \mathbb{R}$ is unique and can be calculated as follows:

$$\bar{\tau} = t_2^* + \sqrt[3]{\frac{6}{L_1 - L_2}\left(x_2^f - x_3^0 l_{\max} - \frac{x_4^0}{2}l_{\max}^2 - \frac{L_1}{6}l_{\max}^3\right)}.$$

If $t_1^* < \bar{\tau} < t_2^*$ and inequality (4.21) is satisfied, then $\bar{\tau}$ is a feasible solution and as a consequence, the validity of Case 1 is confirmed.

Case 2: There exists $\bar{\tau} \in (t_1^*, t_2^*)$ such that $p_4^{\max}(\bar{\tau}; p_2^{\max}) = 0$ and

$$\frac{2}{t_2^* - \bar{\tau}}\int_{\bar{\tau}}^{t_2^*}\int_s^{t_2^*} w(\eta)\,d\eta\,ds < \int_{\bar{\tau}}^{t_2^*} w(s)\,ds. \qquad (4.25)$$

An analysis similar to that presented for Case 1 can be performed. However, the third-degree equation in (4.24) is given by

$$\frac{L_2 - L_1}{6}\left(\bar{\tau} - t_2^*\right)^3 + x_3^0 l_{\max} + \frac{x_4^0}{2}l_{\max}^2 + \frac{L_2}{6}l_{\max}^3 = x_2^f. \qquad (4.26)$$

Equation (4.26) has the following real and unique root:

$$\bar{\tau} = t_2^* + \sqrt[3]{\frac{6}{L_2 - L_1}\left(x_2^f - x_3^0 l_{\max} - \frac{x_4^0}{2}l_{\max}^2 - \frac{L_2}{6}l_{\max}^3\right)}$$

that is feasible if $t_1^* < \bar{\tau} < t_2^*$ and inequality (4.25) is satisfied.

Case 3: The function $p_4^{\max}(\tau; p_2^{\max})$ is positive on the interval $[t_1^*, t_2^*]$. This implies that $p_2^{\max}$ is not unique. In addition, $v^{\max}(\tau) \equiv L_1$ and similar to the analysis performed for the interval $[t_1^*, \bar{\tau})$ in Case 1, it can be shown that $S_1(x^{\max}(\tau)), S_2(x^{\max}(\tau)) < 0$. Also, the final condition $x_2^{\max}(t_2^*) = x_2^f$ can be satisfied only for the following specific value of x_2^f:

$$x_2^f = x_3^0 l_{\max} + \frac{x_4^0}{2}l_{\max}^2 + \frac{L_1}{6}l_{\max}^3.$$

Case 4: If the function $p_4^{\max}(\tau; p_2^{\max})$ is negative on the interval $[t_1^*, t_2^*]$, $p_2^{\max}$ is not unique and, moreover, $p_4^{\max}(\tau; p_2^{\max}) < 0$ and $L_2 > 0$ imply that $S_1(x^{\max}(\tau)), S_2(x^{\max}(\tau)) < 0$, $t_1^* \leq \tau \leq t_2^*$. The final condition $x_2^{\max}(t_2^*) = x_2^f$ can be satisfied only for the following specific value of x_2^f:

$$x_2^f = x_3^0 l_{\max} + \frac{x_4^0}{2} l_{\max}^2 + \frac{L_2}{6} l_{\max}^3.$$

The proof of Lemma 4.1 follows from the results obtained in Cases 1–4.   ∎

Remark 4.2 (Solutions of the Minimization Problem)   *In the minimization problem,*

$$p_3^{\min}(\tau; p_2^{\min}) = \int_{\tau}^{t_2^*} w(s)\, ds - p_2^{\min}(\tau - t_2^*)$$

$$p_4^{\min}(\tau; p_2^{\min}) = \int_{\tau}^{t_2^*}\int_s^{t_2^*} w(\eta)\, d\eta\, ds + \frac{p_2^{\min}}{2}(\tau - t_2^*)^2.$$

Moreover, the nonlinear equation $p_4^{\min}(\tau; p_2^{\min}) = 0$ has at most one root in the interval $[t_1^*, t_2^*)$. Let $\underline{\tau} \in [t_1^*, t_2^*)$ be such that $p_4^{\min}(\underline{\tau}, p_2^{\min}) = 0$. The validity of Cases 1 and 2 in the minimization problem are confirmed by

$$\frac{2}{t_2^* - \underline{\tau}} \int_{\underline{\tau}}^{t_2^*}\int_s^{t_2^*} w(\eta)\, d\eta\, ds < \int_{\underline{\tau}}^{t_2^*} w(s)\, ds$$

and

$$\frac{2}{t_2^* - \underline{\tau}} \int_{\underline{\tau}}^{t_2^*}\int_s^{t_2^*} w(\eta)\, d\eta\, ds > \int_{\underline{\tau}}^{t_2^*} w(s)\, ds,$$

respectively. Cases 3 and 4 of the minimization problem are similar to the ones presented in the maximization problem.

Remark 4.3 (Infeasible Cases of the Optimization Problems)   *As discussed previously, by hypothesis H1, the function $F(\tau)$ defined in equation (4.20) is nonzero on the interval (t_1^*, t_2^*). Thus, without loss of generality, we will assume that $F(\tau) < 0$ on the interval (t_1^*, t_2^*). This assumption imposes that Case 2 of the maximization problem and Case 1 of the minimization problem are not feasible.*

Now let $m < x_3^0 < M$ and $x_4^0 \in \mathbb{R}$. Moreover, define $\Omega_{m,M,L_1,L_2}^{\max}(x_3^0, x_4^0)$ and $\Omega_{m,M,L_1,L_2}^{\min}(x_3^0, x_4^0)$ to be the sets of all $x_2^f \in \mathbb{R}$ for which the optimal solutions of the maximization and minimization problems starting from the initial point $(0, 0, x_3^0, x_4^0)'$ exist. Denote the solutions of the maximization and minimization problems corresponding to x_2^f by $x^{\max}(\tau; x_2^f)$ and $x^{\min}(\tau; x_2^f)$, respectively. The functions $\pi^{\max}: \Omega_{m,M,L_1,L_2}^{\max}(x_3^0, x_4^0) \to \mathbb{R}$ and $\pi^{\min}: \Omega_{m,M,L_1,L_2}^{\min}(x_3^0, x_4^0) \to \mathbb{R}$ are introduced by

$$\pi^{\max}(x_2^f) := \int_{t_1^*}^{t_2^*} w(s)\, x_3^{\max}\left(s; x_2^f\right) ds = x_1^{\max}\left(t_2^*; x_2^f\right)$$

$$\pi^{\min}(x_2^f) := \int_{t_1^*}^{t_2^*} w(s)\, x_3^{\min}\left(s; x_2^f\right) ds = x_1^{\min}\left(t_2^*; x_2^f\right)$$

(4.27)

(see Fig. 4.4). We claim that if the sets $\Omega^{\max}_{m,M,L_1,L_2}(x_3^0, x_4^0)$ and $\Omega^{\min}_{m,M,L_1,L_2}(x_3^0, x_4^0)$ are nonempty, they are connected sets. For this purpose, we present the following lemma for which it is shown that $\Omega^{\max}_{m,M,L_1,L_2}(x_3^0, x_4^0)$ is a connected set. A similar result can also be obtained for the set $\Omega^{\min}_{m,M,L_1,L_2}(x_3^0, x_4^0)$.

Lemma 4.2   *Let* $m < x_3^0 < M$ *and* $x_4^0 \in \mathbb{R}$. *Assume that* $\alpha < \beta$ *are two scalars such that* $\alpha, \beta \in \Omega^{\max}_{m,M,L_1,L_2}(x_3^0, x_4^0)$. *Then, for any* $\gamma \in (\alpha, \beta)$, $\gamma \in \Omega^{\max}_{m,M,L_1,L_2}(x_3^0, x_4^0)$.

The proof is given in Appendix B.1. Now we are in a position to present the main result of this section. This result is expressed as the following theorem that determines the C^1 open-loop control input μ transferring the state of the system Σ from the origin at t_1^* to the final point $(x_1^f, x_2^f)' \in A_{m,M,L_1,L_2}(x_3^0, x_4^0)$ at t_2^*.

Theorem 4.1 (Reachable Set from the Origin)   *Let* $m < x_3^0 < M$ *and* $x_4^0 \in \mathbb{R}$. *Assume that* $L_1 < 0$ *and* $L_2 > 0$ *are such that*

$$\min\left(x_3^0,\, x_3^0 + x_4^0 l_{\max} + \frac{L_1}{2} l^2_{\max}\right) > m$$

$$\max\left(x_3^0,\, x_3^0 + x_4^0 l_{\max} + \frac{L_2}{2} l^2_{\max}\right) < M.$$

(4.28)

Then, the set $A_{m,M,L_1,L_2}(x_3^0, x_4^0)$ *is given by*

$$A_{m,M,L_1,L_2}\left(x_3^0, x_4^0\right) = \left\{ \left(x_1^f, x_2^f\right)' \in \mathbb{R}^2 \,|\, \underline{x}_2^f \le x_2^f \le \bar{x}_2^f,\, \pi^{\min}\left(x_2^f\right) \right.$$

$$\left. \le x_1^f \le \pi^{\max}(x_2^f) \right\},$$

where

$$\underline{x}_2^f := x_3^0 l_{\max} + \frac{x_4^0}{2} l^2_{\max} + \frac{L_1}{6} l^3_{\max}$$

$$\bar{x}_2^f := x_3^0 l_{\max} + \frac{x_4^0}{2} l^2_{\max} + \frac{L_2}{6} l^3_{\max}.$$

Proof. From Remark 4.3, define

$$
v^{\max}\left(\tau; x_2^f\right) := \begin{cases} L_1 & t_1^* \le \tau < \bar{\tau}\left(x_2^f\right) \\ L_2 & \bar{\tau}\left(x_2^f\right) < \tau \le t_2^* \end{cases}
\tag{4.29}
$$

and

$$
v^{\min}\left(\tau; x_2^f\right) := \begin{cases} L_2 & t_1^* \le \tau < \underline{\tau}\left(x_2^f\right) \\ L_1 & \underline{\tau}\left(x_2^f\right) < \tau \le t_2^*, \end{cases}
\tag{4.30}
$$

where

$$
\bar{\tau}\left(x_2^f\right) := t_2^* + \sqrt[3]{\frac{6}{L_1 - L_2}\left(x_2^f - \underline{x}_2^f\right)}
$$

$$
\underline{\tau}\left(x_2^f\right) := t_2^* + \sqrt[3]{\frac{6}{L_2 - L_1}\left(x_2^f - \bar{x}_2^f\right)}.
$$

Since from part (c) of Lemma 4.1, the sets $\mathcal{T}_0^{\max}$ and $\mathcal{T}_0^{\min}$ are Lebesgue negligible, we shall leave the functions $v^{\max}$ and $v^{\min}$ undefined on them. For any $\tau \in [t_1^*, t_2^*]$, $v^{\max}(\tau; \underline{x}_2^f) = v^{\min}(\tau; \underline{x}_2^f) \equiv L_1$ (see point A in Fig. 4.4) and $v^{\max}(\tau; \bar{x}_2^f) = v^{\min}(\tau; \bar{x}_2^f) \equiv L_2$ (see point B in Fig. 4.4), which, in turn, imply that

$$
x_3^{\max}\left(\tau; \underline{x}_2^f\right) = x_3^{\min}\left(\tau; \underline{x}_2^f\right) = x_3^0 + x_4^0\left(\tau - t_1^*\right) + \frac{L_1}{2}\left(\tau - t_1^*\right)^2
$$

$$
x_3^{\max}\left(\tau; \bar{x}_2^f\right) = x_3^{\min}\left(\tau; \bar{x}_2^f\right) = x_3^0 + x_4^0\left(\tau - t_1^*\right) + \frac{L_2}{2}\left(\tau - t_1^*\right)^2,
$$

and, consequently, $x_3^{\max}(\tau; \underline{x}_2^f) \le x_3^{\max}(\tau; \bar{x}_2^f)$. This fact in combination with inequality (4.28), while considering $L_1 < 0 < L_2$, implies that for every $\tau \in [t_1^*, t_2^*]$,

$$
m < x_3^{\max}\left(\tau; \underline{x}_2^f\right) \le x_3^{\max}\left(\tau; \bar{x}_2^f\right) < M.
$$

Therefore, $\underline{x}_2^f, \bar{x}_2^f \in \Omega_{m,M,L_1,L_2}^{\max}(x_3^0, x_4^0), \Omega_{m,M,L_1,L_2}^{\min}(x_3^0, x_4^0)$. Moreover, from Lemma 4.2,

$$
\Omega_{m,M,L_1,L_2}^{\max}\left(x_3^0, x_4^0\right) = \Omega_{m,M,L_1,L_2}^{\min}\left(x_3^0, x_4^0\right) = \left[\underline{x}_2^f, \bar{x}_2^f\right].
$$

Also, $\pi^{\max}(\underline{x}_2^f) = \pi^{\min}(\underline{x}_2^f)$ and $\pi^{\max}(\bar{x}_2^f) = \pi^{\min}(\bar{x}_2^f)$.

Next we claim that the functions $\pi^{\max}$ an $\pi^{\min}$ are strictly increasing functions with respect to x_2^f. To show this, from the definition of $\pi^{\max}(x_2^f)$ in equation (4.27), $\pi^{\max}(x_2^f)$ can be rewritten as follows:

$$\pi^{\max}\left(x_2^f\right) = \int_{t_1^*}^{\bar{\tau}\left(x_2^f\right)} w(s)\, x_3^{\max}\left(s; x_2^f\right) ds + \int_{\bar{\tau}\left(x_2^f\right)}^{t_2^*} w(s)\, x_3^{\max}\left(s; x_2^f\right) ds. \quad (4.31)$$

Differentiating equation (4.31) with respect to x_2^f results in

$$\frac{\partial \pi^{\max}}{\partial x_2^f}\left(x_2^f\right) = \int_{\bar{\tau}\left(x_2^f\right)}^{t_2^*} w(s) \frac{\partial x_3^{\max}}{\partial \bar{\tau}}\left(s; x_2^f\right) ds \, \frac{\partial \bar{\tau}}{\partial x_2^f}\left(x_2^f\right)$$

$$= \frac{2}{\left(\bar{\tau}\left(x_2^f\right) - t_2^*\right)^2} \int_{\bar{\tau}\left(x_2^f\right)}^{t_2^*} w(s)\left(s - \bar{\tau}\left(x_2^f\right)\right) ds$$

$$= \frac{2}{\left(\bar{\tau}\left(x_2^f\right) - t_2^*\right)^2} \int_{\bar{\tau}\left(x_2^f\right)}^{t_2^*} \int_s^{t_2^*} w(\eta)\, d\eta\, ds$$

$$= \frac{2}{\left(\bar{\tau}\left(x_2^f\right) - t_2^*\right)^2} \mathcal{W}\left(\bar{\tau}\left(x_2^f\right)\right) > 0.$$

In the above derivation, we have made use of the integration by parts for the third equality. In a similar way, $\frac{\partial \pi^{\min}}{\partial x_2^f}(x_2^f) = \frac{2}{(\underline{\tau}(x_2^f) - t_2^*)^2}\mathcal{W}(\underline{\tau}(x_2^f)) > 0$. Also,[6]

$$\frac{\partial^2 \pi^{\max}}{\partial x_2^{f2}}\left(x_2^f\right) = \frac{\partial}{\partial \bar{\tau}} \frac{2}{\left(\bar{\tau} - t_2^*\right)^2} \mathcal{W}(\bar{\tau}) \frac{\partial \bar{\tau}}{\partial x_2^f}\left(x_2^f\right)$$

$$= \frac{4F\left(\bar{\tau}\left(x_2^f\right)\right)}{(L_1 - L_2)\left(\bar{\tau}\left(x_2^f\right) - t_2^*\right)^5} < 0$$

and

$$\frac{\partial^2 \pi^{\min}}{\partial x_2^{f2}}\left(x_2^f\right) = \frac{4F\left(\underline{\tau}\left(x_2^f\right)\right)}{(L_2 - L_1)\left(\underline{\tau}\left(x_2^f\right) - t_2^*\right)^5} > 0.$$

[6] From Remark 4.3, it is assumed that $F(\tau) < 0$ for any $\tau \in [t_1^*, t_2^*)$.

Next we show that $\pi^{\min}(x_2^f) < \pi^{\max}(x_2^f)$ for any $x_2^f \in (\underline{x}_2^f, \bar{x}_2^f)$. Introduce the error function

$$e\left(x_2^f\right) := \pi^{\max}\left(x_2^f\right) - \pi^{\min}\left(x_2^f\right)$$

and assume that there exists $\tilde{x}_2^f \in (\underline{x}_2^f, \bar{x}_2^f)$ such that $e(\tilde{x}_2^f) = 0$. Since $e(\underline{x}_2^f) = e(\tilde{x}_2^f) = e(\bar{x}_2^f) = 0$, by the Rolle's Theorem there exist $\xi_1 \in (\underline{x}_2^f, \tilde{x}_2^f)$ and $\xi_2 \in (\tilde{x}_2^f, \bar{x}_2^f)$ such that $\frac{d}{dx_2^f} e(\xi_1) = \frac{d}{dx_2^f} e(\xi_2) = 0$. However, $\frac{d^2}{dx_2^{f2}} e(x_2^f) = \frac{\partial^2 \pi^{\max}}{\partial x_2^{f2}}(x_2^f) - \frac{\partial^2 \pi^{\min}}{\partial x_2^{f2}}(x_2^f) < 0$, and hence $\frac{d}{dx_2^f} e(x_2^f)$ is strictly monotonic that contradicts the assumed existence of $\tilde{x}_2^f$. This result in combination with

$$\frac{\partial^2 \pi^{\max}}{\partial x_2^{f2}}\left(x_2^f\right) < 0 \,, \qquad \frac{\partial^2 \pi^{\min}}{\partial x_2^{f2}}\left(x_2^f\right) > 0$$

implies that $\pi^{\min}(x_2^f) < \pi^{\max}(x_2^f)$ for any $x_2^f \in (\underline{x}_2^f, \bar{x}_2^f)$.

We show that for any $(x_1^f, x_2^f)' \in \mathcal{A}_{m,M,L_1,L_2}(x_3^0, x_4^0)$, there exists an admissible open-loop control input $v \in \mathcal{V}_{L_1,L_2}$ that transfers the state of the system Σ_a from the initial point $(0, 0, x_3^0, x_4^0)'$ (at t_1^*) to the point $(x_1^f, x_2^f, x_3^f, x_4^f)'$ (at t_2^*). For this purpose, choose $\vartheta \in [0, 1]$ such that

$$x_1^f = \vartheta \, \pi^{\min}\left(x_2^f\right) + (1 - \vartheta) \pi^{\max}\left(x_2^f\right),$$

(see point E in Fig. 4.4) and define the following open-loop control input:

$$v\left(\tau; x_1^f, x_2^f\right) := \vartheta \, v^{\min}\left(\tau; x_2^f\right) + (1 - \vartheta) v^{\max}\left(\tau; x_2^f\right).$$

Since[7] $\vartheta \in [0, 1]$, $v(\tau; x_1^f, x_2^f) \in [L_1, L_2]$. Moreover, due to the fact that[8]

$$x_3(\tau) = \vartheta \, x_3^{\min}\left(\tau; x_2^f\right) + (1 - \vartheta) x_3^{\max}\left(\tau; x_2^f\right) \in (m, M),$$

the open-loop control $v(\tau; x_1^f, x_2^f)$ transfers the state of the system Σ_a from the initial point $(0, 0, x_3^0, x_4^0)'$ (at t_1^*) to the point $(x_1^f, x_2^f, x_3^f, x_4^f)'$ (at t_2^*) such that $m < x_3(\tau) < M$, $t_1^* \le \tau \le t_2^*$.

Finally, it can be shown that the Hamilton–Jacobi–Bellman Equation is satisfied along optimal trajectories of the system Σ_a. The proof of this statement will not be presented here (a detailed proof of a similar rechability problem is to be presented in

---

[7] We remark that from the assumptions of Theorem 4.1, the set $\mathcal{A}_{m,M,L_1,L_2}(x_3^0, x_4^0)$ is convex.
[8] Note that the system Σ_a is linear. In addition, from part (a) of Lemma 4.1, $x_3^{\min}(\tau; x_2^f), x_3^{\max}(\tau; x_2^f) \in (m, M)$ for every $\tau \in [t_1^*, t_2^*]$.

Chapter 5, Lemma 5.2). Thus, Theorem 5.12 of Ref. [101, p. 357] implies that the sufficient conditions for optimality are satisfied, and hence, the developed open-loop control laws (i.e., $v^{\max}$ and $v^{\min}$) are indeed optimal. ∎

Using the constructive proof of Theorem 4.1, we develop an online algorithm for reconfiguration. For this purpose, suppose that the assumptions of Theorem 4.1 are satisfied.

Algorithm 4.1 *Online reconfiguration algorithm*

Step 1: Define $x_1^f := \frac{\sigma_{cm}^*}{\sigma_{cm}} \int_{t_1^*}^{t_2^*} w(s)ds$ and $x_2^f := t_2 - t_1$, and suppose that $(x_1^f, x_2^f)' \in \mathcal{A}_{m,M,L_1,L_2}(x_3^0, x_4^0)$. By assuming that the values of the functions $\pi^{\max}(x_2^f)$ and $\pi^{\min}(x_2^f)$ on the interval $[\underline{x}_2^f, \bar{x}_2^f]$ are precomputed and stored in a lookup table, choose ϑ such that $x_1^f = \vartheta \pi^{\min}(x_2^f) + (1 - \vartheta)\pi^{\max}(x_2^f)$ and for any $\tau \in [t_1^*, t_2^*]$, let

$$v\left(\tau; x_1^f, x_2^f\right) := \vartheta\, v^{\min}\left(\tau; x_2^f\right) + (1 - \vartheta)\, v^{\max}\left(\tau; x_2^f\right)$$
$$x_3\left(\tau; x_1^f, x_2^f\right) := \vartheta\, x_3^{\min}\left(\tau; x_2^f\right) + (1 - \vartheta)\, x_3^{\max}\left(\tau; x_2^f\right) \qquad (4.32)$$
$$x_4\left(\tau; x_1^f, x_2^f\right) := \vartheta\, x_4^{\min}\left(\tau; x_2^f\right) + (1 - \vartheta)\, x_4^{\max}\left(\tau; x_2^f\right).$$

Note that the functions $v(\tau; x_1^f, x_2^f)$, $x_3(\tau; x_1^f, x_2^f)$, and $x_4(\tau; x_1^f, x_2^f)$ can be computed in an online manner.

Step 2: Introduce the state vector

$$z(t) := \begin{bmatrix} z_1(t) \\ z_2(t) \\ z_3(t) \end{bmatrix} := \begin{bmatrix} \tau(t) \\ \dot{\tau}(t) \\ \ddot{\tau}(t) \end{bmatrix}$$

for $t_1 \leq t \leq t_2$ to augment the state of the mechanical system. The augmented system over the time interval $[t_1, t_2]$ is given by

$$\begin{aligned} \dot{x}_f &= f_f(x_f) + g_f(x_f)u \\ \dot{z}_1 &= z_2 \\ \dot{z}_2 &= z_3 \\ \dot{z}_3 &= \frac{-v\left(z_1; x_1^f, x_2^f\right) x_3\left(z_1; x_1^f, x_2^f\right) + 3x_4^2\left(z_1; x_1^f, x_2^f\right)}{x_3^5\left(z_1; x_1^f, x_2^f\right)}. \end{aligned} \qquad (4.33)$$

The initial condition for $z(t)$ is also defined as $z(t_1) = (t_1^*, \frac{1}{x_3^0}, -\frac{x_4^0}{x_3^{03}})'$.

Step 3: Define

$$\varphi(t) := \varphi^*(z_1(t)), \quad t_1 \le t \le t_2$$

as the reference trajectory for the joint angles.

Remark 4.4 *From the definition of μ, $\dot{\tau}(t) = \frac{1}{\mu(\tau(t))}$, and thus,*

$$\frac{d}{dt}\dot{\tau}(t) = -\frac{\frac{\partial \mu}{\partial \tau}(\tau(t))}{\mu^2(\tau(t))}\dot{\tau}(t) = -\frac{\frac{\partial \mu}{\partial \tau}(\tau(t))}{\mu^3(\tau(t))}$$

$$\frac{d}{dt}\ddot{\tau}(t) = \frac{-\frac{\partial^2 \mu}{\partial \tau^2}(\tau(t))\mu^3(\tau(t)) + 3\mu^2(\tau(t))\left(\frac{\partial \mu}{\partial \tau}(\tau(t))\right)^2}{\mu^6(\tau(t))}\dot{\tau}(t)$$

$$= \frac{-\frac{\partial^2 \mu}{\partial \tau^2}(\tau(t))\mu(\tau(t)) + 3\left(\frac{\partial \mu}{\partial \tau}(\tau(t))\right)^2}{\mu^5(\tau(t))},$$

which follows the system introduced in equation (4.33).

Remark 4.5 (The First and Second Time Derivatives of φ) *Since we will make use of $\varphi(t)$ as the reference trajectory for the body angles, the states z_2 and z_3 are introduced so that the first and second time derivatives of φ can be calculated as follows:*

$$\dot{\varphi}(t) = \dot{\varphi}^*(z_1)z_2$$
$$\ddot{\varphi}(t) = \ddot{\varphi}^*(z_1)z_2^2 + \dot{\varphi}^*(z_1)z_3.$$

Remark 4.6 (Planar Multilink Systems Composed of $N \ge 3$ Links) *The proposed online reconfiguration algorithm can be used for motion planning problem of planar multilink systems composed of $N \ge 3$ rigid links which conserve angular momentum.*

Figure 4.5 illustrates a block diagram for the online reconfiguration algorithm over the time interval $[t_1, t_2]$ during flight phases of monopedal running. The following theorem is an extended version of the latter algorithm by which the problem of monopedal running can be treated in Section 4.4.

Theorem 4.2 (Online Reconfiguration Algorithm for Landing in a Fixed Configuration During Flight Phases of Monopedal Running) *Let t_f^* be a positive real number. Assume that $0 < t_1^* < t_2^* < t_f^*$ is a partition of the interval $[0, t_f^*]$. Let $\varphi^* : [0, t_f^*] \to \mathcal{Q}_b$ be a nominal C^2 trajectory in the configuration space satisfying hypothesis H1 such that*

$$\begin{aligned}\dot{\varphi}^*\left(t_1^*\right) = \ddot{\varphi}^*\left(t_1^*\right) = 0 \\ \dot{\varphi}^*\left(t_2^*\right) = \ddot{\varphi}^*\left(t_2^*\right) = 0.\end{aligned} \tag{4.34}$$

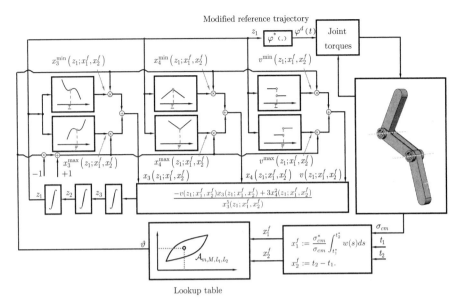

Lookup table

Figure 4.5 Block diagram of the online reconfiguration algorithm over the time interval $[t_1, t_2]$ during flight phases of monopedal running. In this diagram, for a given $t_1 < t_2$ and σ_{cm}, the quantities $x_1^f := \frac{\sigma_{cm}^*}{\sigma_{cm}} \int_{t_1^*}^{t_2^*} w(s)ds$ and $x_2^f := t_2 - t_1$ are computed. Then, by assuming that (i) $(x_1^f, x_2^f)' \in \mathcal{A}_{m,M,L_1,L_2}(x_3^0, x_4^0)$ and (ii) the values of the functions $\pi^{max}(x_2^f)$ and $\pi^{min}(x_2^f)$ on the interval $[\underline{x}_2^f, \bar{x}_2^f]$ are precomputed and stored in a lookup table, we choose ϑ such that $x_1^f = \vartheta\pi^{min}(x_2^f) + (1 - \vartheta)\pi^{max}(x_2^f)$. Using equation (4.32), ϑ is used to construct the functions $v(z_1; x_1^f, x_2^f)$, $x_3(z_1; x_1^f, x_2^f)$, and $x_4(z_1; x_1^f, x_2^f)$. Finally, on the basis of equation (4.33), the augmented states z_1, z_2, and z_3 are introduced to construct the desired trajectory $\varphi^d(t) = \varphi^*(z_1(t))$, $t_1 \le t \le t_2$ and its derivatives up to the second order. (See the color version of this figure in color plates section.)

Furthermore, suppose that when the angular momentum of the mechanical system about its COM is identical to σ_{cm}^*, the nominal absolute orientation, $\theta^*(t)$, satisfies the boundary conditions $\theta^*(0) = \theta_0$, $\theta^*(t_1^*) = \theta_1$, $\theta^*(t_2^*) = \theta_2$ and $\theta^*(t_f^*) = \theta_f$. For a given $t_f > 0$ and σ_{cm} with the property $\sigma_{cm}\sigma_{cm}^* > 0$, let $x_1^f := \frac{\sigma_{cm}^*}{\sigma_{cm}} \int_{t_1^*}^{t_2^*} w(s)ds$ and $x_2^f := t_2 - t_1$, where

$$t_1 := \frac{\sigma_{cm}^*}{\sigma_{cm}} t_1^*$$

$$t_2 := t_f - \frac{\sigma_{cm}^*}{\sigma_{cm}}(t_f^* - t_2^*).$$

Choose $x_3^0 \in (m, M)$, $x_4^0 \in \mathbb{R}$, $L_1 < 0$, and $L_2 > 0$ such that the assumptions of Theorem 4.1 hold. If $(x_1^f, x_2^f)' \in \mathcal{A}_{m,M,L_1,L_2}(x_3^0, x_4^0)$, then the following statements hold:

(a) The trajectory $\varphi^d : [0, t_f] \to \mathcal{Q}_b$ by

$$
\varphi^d(t) := \begin{cases} \varphi^*\left(\frac{\sigma_{\mathrm{cm}}}{\sigma_{\mathrm{cm}}^*}t\right) & 0 \le t \le t_1 \\ \varphi^*(z_1(t)) & t_1 \le t \le t_2 \\ \varphi^*\left(\frac{\sigma_{\mathrm{cm}}}{\sigma_{\mathrm{cm}}^*}(t - t_f) + t_f^*\right) & t_2 \le t \le t_f \end{cases}
$$

is C^2, where $z_1(t)$ was introduced in equation (4.33), and fulfills the following boundary conditions

$$
\begin{aligned}
\varphi^d(0) &= \varphi^*(0) \\
\varphi^d(t_f) &= \varphi^*(t_f^*) \\
\dot{\varphi}^d(0) &= \dot{\varphi}^*(0)\frac{\sigma_{\mathrm{cm}}}{\sigma_{\mathrm{cm}}^*} \\
\dot{\varphi}^d(t_f) &= \dot{\varphi}^*(t_f^*)\frac{\sigma_{\mathrm{cm}}}{\sigma_{\mathrm{cm}}^*}.
\end{aligned}
\tag{4.35}
$$

(b) The trajectory $\varphi^d(t)$ satisfies the boundary conditions $\theta(t_1) = \theta_1$, $\theta(t_2) = \theta_2$ and $\theta(t_f) = \theta_f$ when the initial condition of the absolute orientation and the angular momentum about the COM are $\theta(0) = \theta_0$ and σ_{cm}, respectively.

(c) If $x_3^0 = 1$, $x_4^0 = 0$, $0 < m < 1 < M$, and $L_2 = -L_1$ with the following property

$$
-\frac{2}{l_{\mathrm{max}}^2}(1 - m) < L_1 < 0 < L_2 < \frac{2}{l_{\mathrm{max}}^2}(M - 1),
$$

then for $\sigma_{\mathrm{cm}} = \sigma_{\mathrm{cm}}^*$ and $t_f = t_f^*$, $\varphi^d(t) = \varphi^*(t)$ for every $t \in [0, t_f]$.

The proof is given in Appendix B.2.

4.4   CONTROL LAWS FOR STANCE AND FLIGHT PHASES

This section presents a design method for determining the control laws during the stance and flight phases to realize a desired periodic trajectory as an asymptotically stable orbit. Let $\mathcal{O} := \mathcal{O}_s \cup \mathcal{O}_f$ denote a desired period-one orbit of the open-loop hybrid model of running in equation (4.5), in which $\mathcal{O}_s$ and $\mathcal{O}_f$ are the stance and flight phases of the periodic orbit, respectively. Reference [73] proposes a method based on a finite-dimensional optimization problem for generating the time trajectory of $\mathcal{O}$.[9]

4.4.1 Stance Phase Control Law

Following the ideas of Refs. [46, 52, 55], the stance phase controller is assumed to be a continuous time-invariant feedback law based on zeroing a *parameterized* holonomic output function with the uniform vector relative degree 2. This control law creates a parameterized finite-time attractive two-dimensional zero dynamics manifold in the corresponding state manifold (i.e., $\mathcal{X}_s$). To make this notion precise, introduce the following holonomic output function for the dynamical system of equation (4.4):

$$y_s(x_s; \alpha) := h_s(q_s; \alpha) := h_{\mathcal{O}_s}(q_s) + \mathfrak{B}(s(q_s); \alpha), \qquad (4.36)$$

where $h_{\mathcal{O}_s} : \mathcal{Q}_s \to \mathbb{R}^2$ is at least a C^2 function vanishing on the orbit $\mathcal{O}_s$. For determining $h_{\mathcal{O}_s}$, the sample-based virtual constraints method introduced in Ref. [94] can be used. In addition, the function $\mathfrak{B} : [0, 1] \times \mathcal{A} \to \mathbb{R}^2$ is an augmentation function that is expressed as a Bézier polynomial of degree N

$$\mathfrak{B}(s(q_s); \alpha) := \sum_{k=0}^{N} \frac{N!}{k!(N-k)!} \alpha_k s^k (1-s)^{N-k},$$

where $\alpha := [\alpha_0 \, \alpha_1 \, ... \, \alpha_{N-1} \, \alpha_N] \in \mathcal{A}$, and $\mathcal{A} \subset \mathbb{R}^{2 \times (N+1)}$ is an open set. Also, $s(q_s)$ is defined as the normalized value of the angle of the virtual leg, that is, $s(q_s) := \frac{\gamma_s(q_s) - \gamma_s^+}{\gamma_s^- - \gamma_s^+}$, in which γ_s^+ and γ_s^- are the initial and final values of the angle of the virtual leg on $\mathcal{O}_s$, respectively (see Fig. 4.1). It is assumed that the set

$$Z_{s,\alpha} := \{x_s \in T\mathcal{Q}_s | h_s(x_s; \alpha) = 0_{2 \times 1}, L_{f_s} h_s(x_s; \alpha) = 0_{2 \times 1}\}$$

is an embedded two-dimensional submanifold of $T\mathcal{Q}_s$. Moreover, suppose that $\mathcal{S}_s^f \cap Z_{s,\alpha}$ is an embedded one-dimensional submanifold of $T\mathcal{Q}_s$. By properties of Bézier polynomials, see Remark 3.16, for $\alpha_{N-1} = \alpha_N = 0_{2 \times 1}$, the manifold $\mathcal{S}_s^f \cap Z_{s,\alpha}$ is independent of α. Moreover, following the results of Ref. [18, p. 125], it can be expressed as

$$\mathcal{S}_s^f \cap Z_{s,\alpha} = \left\{ (q', \dot{q}')' | q = q_s^{-*}, \dot{q} = \dot{q}_s^{-*} \frac{\sigma_s^-}{\sigma_s^{-*}}, \sigma_s^- \in \mathbb{R} \right\}, \qquad (4.37)$$

---

[9] In this chapter, the design method introduced in Ref. [73] for generating $\mathcal{O}$ is modified such that hypothesis H1, condition (4.34), and HO3–HO4 of [18, p. 162] are satisfied. Specifically, for this purpose, some constraints are added to the proposed nonlinear optimization problem in Ref. [73].

where q_s^{-*} and $\dot{q}_s^{-*}$ are the final configuration and velocity of the robot on $\mathcal{O}_s$, respectively. σ_s^- represents the final value of the angular momentum of the mechanical system about the leg end. In addition, σ_s^{-*} denotes the value of σ_s^- on the orbit $\mathcal{O}_s$.[10] In the coordinates (γ_s, σ_s) for $Z_{s,\alpha}$, the stance phase zero dynamics can be given by

$$
\begin{aligned}
\dot{\gamma}_s &= \kappa_1(\gamma_s; \alpha)\,\sigma_s \\
\dot{\sigma}_s &= \kappa_2(\gamma_s; \alpha),
\end{aligned}
\tag{4.38}
$$

where σ_s is the angular momentum of the monoped robot about the leg end that can be obtained as $\sigma_s = D_3(q_s)\,\dot{q}_s$ [18, Proposition B.11, p. 430]. Furthermore, the stance phase feedback law is chosen as the *parameterized version* of the finite-time controller proposed in Refs. [18, p. 134, 46].

Remark 4.7 *In contrast to the approach of Ref. [55], the stance phase controller of our strategy is parameterized. The main reasons for this difference can be expressed as follows.*

(1) *In Ref. [55], a parameterized flight phase controller is used to achieve hybrid invariance and configuration determinism at landing. Specifically, due to hybrid invariance, an additional constraint is imposed on the vector of generalized velocities at the end of flight phases. Thus, by the Implicit Function Theorem and a numerical constrained optimization problem, the parameters of the flight phase controller are updated (at the beginning of the flight phase) to satisfy hybrid invariance and configuration determinism at landing.*

(2) *In our approach, this latter constraint on the final velocity is relaxed and an analytical reconfiguration algorithm is proposed. Instead, for creation of hybrid invariance, the parameter α_1 of the stance phase controller should be updated at the beginning of the stance phase (see equation (4.41)).*

4.4.2 Flight Phase Control Law

In this chapter, the flight phase control law is designed as a continuous feedback law to track the modified reference trajectories generated by Theorem 4.2 (i.e., $\varphi^d(t)$). Define σ_{cm}^*, $\varphi^*(t)$, and $\theta^*(t)$ as the angular momentum of the mechanical system about its COM, the time evolution of the joint angles, and the time evolution of the absolute orientation on the orbit $\mathcal{O}_f$, respectively. Now assume that $x_s^- \in \mathcal{S}_s^f \cap Z_{s,\alpha}$ is the state of the closed-loop hybrid system immediately before the takeoff. From equation (4.37), $x_s^- = (q_s^{\prime-*}, \dot{q}_s^{\prime-})'$, where $\dot{q}_s^- = \dot{q}_s^{-*} \frac{\sigma_s^-}{\sigma_s^{-*}}$. This latter result in combination with the fact that the position and velocity remain continuous during the takeoff implies that the joint angles and velocities at the beginning of the flight phase

---

[10] Equations (6.72) and (6.73) and Proposition 6.1 of Ref. [18, p. 158] in combination with hypotheses HO3 and HO4 of Ref. [18, p. 162] imply that $\sigma_s^{-*} \neq 0$.

can be given by $\varphi(0) = \varphi^*(0)$ and $\dot{\varphi}(0) = \dot{\varphi}^*(0)\frac{\sigma_s^-}{\sigma_s^{-*}}$. We remark that

$$\varphi^*(0) = [I_{2\times2} \ 0_{2\times1}]q_s^{-*}$$
$$\dot{\varphi}^*(0) = [I_{2\times2} \ 0_{2\times1}]\dot{q}_s^{-*}.$$

Moreover, following the notation of Ref. [55], on $\mathcal{S}_s^f \cap Z_{s,\alpha}$, the position and velocity of the COM are given by $p_{cm,s}^- = f_1(q_s^{-*})$

$$\dot{p}_{cm,s}^- = \begin{bmatrix} \lambda_x(q_s^{-*}) \\ \lambda_y(q_s^{-*}) \end{bmatrix} \sigma_s^-,$$

where $[\lambda_x(q_s^{-*}) \ \lambda_y(q_s^{-*})]' := \frac{\partial f_1}{\partial q}(q_s^{-*})\frac{\dot{q}_s^{-*}}{\sigma_s^{-*}}$. Continuity of the position and velocity during the takeoff and conservation of angular momentum about the COM during the flight phase in combination with equation (C.57) of Ref. [18, p. 454] imply that the angular momentum about the COM in the flight phase can be expressed as $\sigma_{cm} = \chi\sigma_s^-$, where $\chi := 1 + m_{tot}y_{cm,s}^-\lambda_x(q_s^{-*}) - m_{tot}x_{cm,s}^-\lambda_y(q_s^{-*})$. If $\chi \neq 0$,[11] $\frac{\sigma_{cm}}{\sigma_{cm}^*} = \frac{\sigma_s^-}{\sigma_s^{-*}}$, and, consequently, from part (a) of Theorem 4.2, $\varphi(0) = \varphi^d(0)$ and $\dot{\varphi}(0) = \dot{\varphi}^d(0)$.

Next, let $\Omega(\varphi, \dot{q}) := \bar{C}(\varphi, \dot{q})\dot{q}$. Then, the static feedback law

$$u = \bar{A}_{\varphi\varphi}(\varphi)\bar{u} + \bar{\Omega}_{\varphi}(\varphi, \dot{q}), \tag{4.39}$$

where $\bar{A}_{\varphi\varphi}(\varphi) := A_{\varphi\varphi} - A_{\varphi3}A_{33}^{-1}A_{3\varphi}$ and $\bar{\Omega}_{\varphi}(\varphi, \dot{q}) := \Omega_{\varphi} - A_{\varphi3}A_{33}^{-1}\Omega_3$, yields the following partially feedback linearized result that is known as the Spong normal form [95]:

$$\ddot{\varphi} = \bar{u}$$
$$\dot{\theta} = \frac{\sigma_{cm}}{A_{33}(\varphi)} - J(\varphi)\dot{\varphi}$$
$$\ddot{x}_{cm} = 0$$
$$\ddot{y}_{cm} = -g_0.$$

Furthermore, since $\varphi(0) = \varphi^d(0)$ and $\dot{\varphi}(0) = \dot{\varphi}^d(0)$, the feedback law

$$\bar{u} := \ddot{\varphi}^d(t) - K_1(\dot{\varphi} - \dot{\varphi}^d(t)) - K_0(\varphi - \varphi^d(t)),$$

---

[11] It is assumed that on the desired periodic trajectory $\mathcal{O}$, $\chi \neq 0$. The condition $\chi \neq 0$ can be imposed through an inequality constraint in the nonlinear optimization problem of Ref. [73].

where $K_1, K_0 \in \mathbb{R}^{2 \times 2}$ are diagonal and positive definite matrices, imposes that $\varphi(t) = \varphi^d(t), 0 \le t \le t_f$, which, in turn, from part (b) of Theorem 4.2, implies that $\theta(t_f) = \theta^*(t_f^*) = \theta_f$. Hence, at the end of the flight phase, the values of the joint angles and absolute orientation are specified and equal to the desired values, that is, $[\varphi(t_f) \ \theta(t_f)]' = q_s^{+*}$, where q_s^{+*} is the initial configuration on $\mathcal{O}_s$ (configuration determinism at landing). Since the impact map preserves positions during the transition from flight to stance, q_s^{+*} will be the initial configuration of the mechanical system at the beginning of the stance phase. Thus, to achieve hybrid invariance, it is necessary to choose $\alpha_0 = 0_{2 \times 1}$ as in this case,

$$h_s(q_s^{+*}; \alpha) = h_{\mathcal{O}_s}(q_s^{+*}) + \mathfrak{B}(0; \alpha) = 0_{2 \times 1}.$$

As in Ref. [55], the configuration determinism at landing implies that the height of the COM at the beginning of the stance phase , $y_{cm,s}^+$, is predetermined. Therefore, the flight time t_f satisfies the following quadratic equation:

$$y_{cm,s}^+ = y_{cm,s}^- + \lambda_y(q_s^{-*}) \sigma_s^- t_f - \frac{1}{2} g_0 t_f^2 \tag{4.40}$$

from which, t_f can be computed as a function of σ_s^-.

Remark 4.8 (Configuration Determinism at Landing)   *Since (i) the modified reference trajectories generated by Theorem 4.2 are C^2 and (ii) for every $x_s^- \in \mathcal{S}_s^f \cap Z_{s,\alpha}$, the projection of the trajectory onto TQ_b at the beginning of the flight phase is identical to that of the orbit $\mathcal{O}_f$ (i.e., $\varphi(0) = \varphi^d(0)$ and $\dot{\varphi}(0) = \dot{\varphi}^d(0)$), the feedback law of equation (4.39) together with parts (b) and (c) of Theorem 4.2 result in configuration determinism at landing. Moreover, since the sets $Z_{s,\alpha}$ are locally continuously finite-time attractive (see Definition 2.8, Section 2.3), it can be concluded that there exist open sets $V_{s,\alpha}$ containing $Z_{s,\alpha}$ such that for every initial condition $x_s^0 \in V_{s,\alpha}$, the solution of the closed-loop hybrid model of running through x_s^0 at time $t = 0$ satisfies configuration determinism at landing.*

4.4.3 Event-Based Update Law

The event-based update law updates the coefficients of the augmentation function $\mathfrak{B}(s; \alpha)$ at each impact event (i.e., transition from flight to stance) to achieve hybrid invariance and asymptotic stabilization of the desired periodic orbit $\mathcal{O}$ as described in Section 2.4. We remark that these coefficients are held constant during the stance phase. As mentioned previously, $\alpha_{N-1} = \alpha_N = 0_{2 \times 1}$ implies that $\mathcal{S}_s^f \cap Z_{s,\alpha}$ is independent of α. The parameter α_0 was also chosen as zero in the previous section. Here, we obtain an update law for α_1 in terms of σ_s^- (i.e., the value of the angular momentum about the leg end at the end of previous stance phase) to render the family of manifolds $\mathbf{Z}_s := \{Z_{s,\alpha} | \alpha \in \mathcal{A}\}$ invariant under the transition map $\Delta : \mathcal{S}_s^f \to \mathcal{X}_s$ defined by $\Delta(x_s) := \Delta_f^s \circ \mathcal{F}_f \circ \Delta_s^f(x_s)$, where $\mathcal{F}_f$ represents the flow map of the flight phase. In particular, the parameter α_1 is updated in a stride-to-stride manner such that $\Delta(\mathcal{S}_s^f \cap Z_{s,\alpha}) \subset \mathbf{Z}_s$. The update laws of $\alpha_2, \cdots, \alpha_{N-2}$ that stabilize the desired periodic trajectory will be addressed in Section 4.5.

At the end of the flight phase, according to the definition of the flight time t_f as a function of σ_s^- in equation (4.40), the generalized velocity of the mechanical system at the end of the flight phase, $\dot{q}_f^-$, can be obtained as a function of σ_s^- [12]. Moreover, the impact map in equation (4.5) yields immediately the initial velocity in the stance phase, $\dot{q}_s^+$, in terms of σ_s^-, that is, $\dot{q}_s^+(\sigma_s^-)$. As discussed previously, $\alpha_0 = 0_{2\times1}$ implies that $h_s(q_s^+; \alpha) = 0_{2\times1}$. Let us define $x_s^+(\sigma_s^-) := (q_s^{'+*}, \dot{q}_s^{'+}(\sigma_s^-))'$. To create hybrid invariance, at the beginning of the stance phase, the event-based law should update α such that $x_s^+(\sigma_s^-) \in Z_{s,\alpha}$. To achieve this goal, α is updated so that

$$L_{f_s} h_s(x_s^+(\sigma_s^-); \alpha) = \frac{\partial h_s}{\partial q_s}(q_s^{+*}; \alpha)\, \dot{q}_s^+(\sigma_s^-) = 0_{2\times1}.$$

In particular,

$$L_{f_s} h_s(x_s^+(\sigma_s^-); \alpha) = \frac{\partial h_{\mathcal{O}_s}}{\partial q_s}(q_s^{+*})\, \dot{q}_s^+(\sigma_s^-) + \frac{N(\alpha_1 - \alpha_0)}{\gamma_s^- - \gamma_s^+} \frac{\partial \gamma_s}{\partial q_s}(q_s^{+*})\, \dot{q}_s^+(\sigma_s^-)$$

$$= 0_{2\times1},$$

and as a consequence, α_1 should be updated by the following law[13]:

$$\alpha_1(\sigma_s^-) = -\frac{\gamma_s^- - \gamma_s^+}{N} \frac{\partial h_{\mathcal{O}_s}}{\partial q_s}(q_s^{+*})\, \dot{q}_s^+(\sigma_s^-) \left(\frac{\partial \gamma_s}{\partial q_s}(q_s^{+*})\, \dot{q}_s^+(\sigma_s^-)\right)^{-1}. \qquad (4.41)$$

4.5 HYBRID ZERO DYNAMICS AND STABILIZATION

To obtain HZD for the closed-loop hybrid model of monopedal running, let $N \geq 3$ be an integer number and $\alpha \in \mathcal{A}$. Assume that $\alpha_0 = \alpha_{N-1} = \alpha_N = 0_{2\times1}$ and α_1 is updated as in equation (4.41) at the beginning of the stance phase. Then, the angular momentum of the mechanical system about the leg end at the beginning of the stance phase is given by $\sigma_s^+ = \omega(\sigma_s^-)$, where

$$\omega(\sigma_s^-) := D_3(q_s^{+*})\, \dot{q}_s^+(\sigma_s^-).$$

---

[12] Note that since the flight phase controller is based on the reconfiguration algorithm of Theorem 4.2, $\dot{q}_f^-(\sigma_s^-)$ is different from that presented in equation (9.51) of Ref. [18, p. 271].

[13] Hypothesis HO3 of Ref. [18, p. 162] implies that on the stance phase of the periodic orbit, $\dot{\gamma}_s \neq 0$. Hence, there exists an open neighborhood $\mathcal{N}(\sigma_s^{-*})$ such that for every $\sigma_s^- \in \mathcal{N}(\sigma_s^{-*})$, $\frac{\partial \gamma_s}{\partial q_s}(q_s^{+*})\dot{q}_s^+(\sigma_s^-) \neq 0$.

Thus, hybrid invariance reduces the analysis of the full-order model to the analysis of the following reduced-order system:

$$
\Sigma_{\text{zero}} : \begin{cases} \begin{bmatrix} \dot{\gamma}_s \\ \dot{\sigma}_s \end{bmatrix} = \begin{bmatrix} \kappa_1(\gamma_s; \alpha) \, \sigma_s \\ \kappa_2(\gamma_s; \alpha) \end{bmatrix} & \gamma_s \neq \gamma_s^- \\[3mm] \begin{bmatrix} \gamma_s^+ \\ \sigma_s^+ \end{bmatrix} = \begin{bmatrix} \gamma_s^+ \\ \omega(\sigma_s^-) \end{bmatrix} & \gamma_s = \gamma_s^-, \end{cases} \tag{4.42}
$$

which is referred to as HZD.

By extending the results of Ref. [52] to HZD of equation (4.42) and also assuming that σ_s is negative during the stance phase, we define the *restricted Poincaré return map* as $\rho_{\bar{\alpha}} : S_s^f \cap Z_{s,\alpha} \to S_s^f \cap Z_{s,\alpha}$ by

$$
\rho_{\bar{\alpha}}(\sigma_s^-) := -\left(\omega^2(\sigma_s^-) + 2 \int_{\gamma_s^+}^{\gamma_s^-} \frac{\kappa_2(\gamma_s; \bar{\alpha}, \alpha_1(\sigma_s^-))}{\kappa_1(\gamma_s; \bar{\alpha}, \alpha_1(\sigma_s^-))} \, d\gamma_s \right)^{\frac{1}{2}},
$$

where $\bar{\alpha} := [\alpha_2' \, \alpha_3' \, \cdots \, \alpha_{N-2}']' \in \mathbb{R}^{p \times 1}$ and $p := 2(N-3)$.

Remark 4.9 *Since in this chapter, (i) the constraint of Ref. [55] on the final velocities of the robot at the end of flight phases is relaxed, and (ii) instead, hybrid invariance is created by the parameter update law given in equation (4.41), the resultant restricted Poincaré return map is different from that of Ref. [55]. Specifically, the first derivative of $\rho_{\bar{\alpha}}$ with respect to σ_s^- should be calculated numerically.*

Let $\rho(\sigma_s^-; \bar{\alpha}) := \rho_{\bar{\alpha}}(\sigma_s^-)$. Then, the following discrete-time system

$$
\sigma_s^-[k+1] = \rho(\sigma_s^-[k]; \bar{\alpha}[k]) \tag{4.43}
$$

with the one-dimensional state space $S_s^f \cap Z_{s,\alpha}$ and input $\bar{\alpha}$ can be considered to analyze stabilization. From the constructive procedure of $h_{\mathcal{O}_s}(q_s)$ and determination of the parameters of the stance and flight phase controllers on the basis of the periodic orbit $\mathcal{O}$, σ_s^{-*} is an equilibrium point of the discrete-time system in equation (4.43) when the input $\bar{\alpha}$ is zero (i.e., $\bar{\alpha} = \bar{\alpha}^* = 0_{p \times 1}$).

Theorem 4.3 (Asymptotic Stability of the Periodic Orbit) *Suppose that the assumptions of part (c) of Theorem 4.2 hold. Define $a := \frac{\partial \rho}{\partial \sigma_s^-}(\sigma_s^{-*}; \bar{\alpha}^*)$ and $b := \frac{\partial \rho}{\partial \bar{\alpha}}(\sigma_s^{-*}; \bar{\alpha}^*)$. If $b \neq 0_{1 \times p}$, then there exists a gain matrix $K \in \mathbb{R}^{p \times 1}$ such that using the within-stride controllers and the following static update law*

$$
\bar{\alpha}(\sigma_s^-) = -K(\sigma_s^- - \sigma_s^{-*}), \tag{4.44}
$$

O is an asymptotically stable periodic orbit for the closed-loop hybrid model of running.

Proof. If $b \neq 0_{1 \times p}$, the pair (a, b) is controllable, which, in turn, implies the existence of $K \in \mathbb{R}^{p \times 1}$ such that $|a_{\mathrm{cl}}| < 1$, where $a_{\mathrm{cl}} := a - bK$. Hence, σ_s^{-*} is a locally exponentially stable equilibrium point for the closed-loop discrete-time system[14] $\sigma_s^{-}[k + 1] = \rho_{\mathrm{cl}}(\sigma_s^{-}[k])$, where $\rho_{\mathrm{cl}}(\sigma_s^{-}) := \rho(\sigma_s^{-}; \bar{\alpha}(\sigma_s^{-}))$. Next, denote the right-hand side of the closed-loop augmented system of equation (4.33) by $f_a(x_f, z)$, which is discontinuous at the following hypersurfaces:

$$\mathcal{Z}_1^{\max} := \left\{ x_a := \left(x_f', z' \right)' \in \mathcal{X}_f \times \mathbb{R}^3 \big| z_1 = \bar{\tau} \right\}$$
$$\mathcal{Z}_1^{\min} := \left\{ x_a := \left(x_f', z' \right)' \in \mathcal{X}_f \times \mathbb{R}^3 \big| z_1 = \underline{\tau} \right\}.$$

Since the vector field f_a is transversal to $\mathcal{Z}_1^{\max}$ and $\mathcal{Z}_1^{\min}$ at every point in

$$\mathcal{X}_a := \left\{ x_a := \left(x_f', z' \right)' \in \mathcal{X}_f \times \mathbb{R}^3 \big| z_2 \neq 0 \right\},$$

it follows from [102, Lemma 2, p. 107] that there exists an open set $\tilde{\mathcal{X}}_a \subset \mathcal{X}_a$ such that the closed-loop ordinary differential equation of equation (4.33) for every initial condition in $\tilde{\mathcal{X}}_a$ has a unique solution in forward time. Moreover, from [102, Corollary, p. 93], this solution depends continuously on the parameters of $f_a(x_f, z)$, that is, σ_{cm} and t_f. From $\sigma_{\mathrm{cm}} = \chi \sigma_s^{-}$, equation (4.40) and $|a_{\mathrm{cl}}| < 1$, this latter fact in combination with Theorem 2.5 of Chapter 2 and part (c) of Theorem 4.2 guarantees that the proposed control scheme realizes $\mathcal{O}$ as an asymptotically stable periodic orbit for the closed-loop hybrid system. ∎

4.6 NUMERICAL RESULTS

This section presents a numerical example for the proposed online reconfiguration algorithm and control scheme. It is assumed that all masses of the three-link monoped robot are lumped. The torques u_1 and u_2 are applied between the femur and tibia, and the torso and femur, respectively. The physical parameters of the monoped robot are given in Table 4.1.[15] Similar to the motion planning algorithm presented in Ref. [73], a modified algorithm can be developed for designing a feasible period-one trajectory $\mathcal{O}$ satisfying hypothesis H1, condition (4.34), and hypotheses HO3–HO4 of

---

[14] Since $a_{\mathrm{cl}} = \frac{\partial \rho_{\mathrm{cl}}}{\partial \sigma_s^{-}}(\sigma_s^{-})\big|_{\sigma_s^{-} = \sigma_s^{-*}}$, $|a_{\mathrm{cl}}| < 1$ implies that σ_s^{-*} is locally exponentially stable equilibrium point for the closed-loop discrete-time system.

[15] The fourth row of Table 4.1 represents the distance between the COM of the links and joints. Note that for the torso and tibia links, the position of the COM is measured with respect to the hip and knee joints, respectively. Furthermore, the position of the COM of the femur is measured with respect to the hip joint.

TABLE 4.1 Physical Parameters of the Monoped Robot

	Femur	Tibia	Torso
Length in m	0.5	0.5	0.5
Mass in kg	2	2	10
Mass center in m	0.1667	0.2500	0.3333

Ref. [18, p. 162] and also to minimize the cost function

$$J := \frac{1}{L_s} \int_0^T \|u(t)\|_2^2 \, dt, \tag{4.45}$$

in which T and L_s denote the period of $\mathcal{O}$ and the step length, respectively. The cost function (4.45) roughly represents electrical motor energy in the body joints per distance traveled. Using the fmincon function of the MATLAB's Optimization Toolbox on the optimal trajectory, the joint paths during the stance and flight phases are defined by polynomial evolutions (i.e., φ_s^* and φ_f^*) with respect to t, specifically,

$$\varphi_s^*(t) := \sum_{i=0}^{6} a_s^i t^i, \qquad 0 \le t < t_s^*$$

$$\varphi_f^*(t) := \sum_{i=0}^{10} a_f^i t^i, \qquad 0 \le t < t_f^*,$$

where t_s^* and t_f^* are the stance and flight times, respectively. The coefficients $a_s^i, i = 0, \cdots, 6$ and $a_f^i, i = 0, \cdots, 10$ are given in Tables 4.2 and 4.3, respectively. Furthermore, the initial position and velocity for the absolute orientation at the beginning of the stance phase are $\theta_s^*(0) = 0.8376(\text{rad})$ and $\dot{\theta}_s^*(0) = 0.5555(\text{rad/s})$, which, in turn, in combination with $\varphi_s^*(t), \varphi_f^*(t)$ and the open-loop hybrid system in equation (4.5) completely determine the reference trajectory $\mathcal{O}$. This trajectory has a period of $T = t_f^* + t_s^* = 0.2073 + 0.2356 = 0.4429(\text{s})$, a step length of $L_s = 0.4429(\text{m})$, and

TABLE 4.2 Coefficients $a_s^i, i = 0, \ldots, 6$ for the Joint Paths During the Stance Phase

$a_s^{0'}$	0.6315	1.7704
$a_s^{1'}$ (10^1)	1.0000	−0.3770
$a_s^{2'}$ (10^2)	−2.6992	1.7708
$a_s^{3'}$ (10^3)	2.7612	−2.5305
$a_s^{4'}$ (10^4)	−1.1920	1.9725
$a_s^{5'}$ (10^4)	1.9346	−7.8882
$a_s^{6'}$ (10^5)	−0.0379	1.2286

TABLE 4.3   Coefficients a^i_f, $i = 0, \ldots, 10$ for the Joint Paths During the Flight Phase

$a^{0'}_f$	0.7840	2.1446
$a^{1'}_f$	0.5988	9.9929
$a^{2'}_f$ (10^3)	1.3096	−0.5331
$a^{3'}_f$ (10^4)	−8.1640	1.7127
$a^{4'}_f$ (10^6)	2.1984	−0.3909
$a^{5'}_f$ (10^7)	−3.1680	0.6625
$a^{6'}_f$ (10^8)	2.6560	−0.7794
$a^{7'}_f$ (10^9)	−1.3335	0.5823
$a^{8'}_f$ (10^9)	3.9338	−2.5843
$a^{9'}_f$ (10^9)	−6.2405	6.1852
$a^{10'}_f$ (10^9)	4.0564	−6.1401

an average running speed of 1(m/s). On the trajectory, the robot will not slip for a coefficient of friction greater than 0.55. Table 4.4 presents the reference trajectory statistics that will be used in the control law. Desired state trajectories corresponding to two steps of the mechanical system are depicted in Fig. 4.6, where the discontinuities

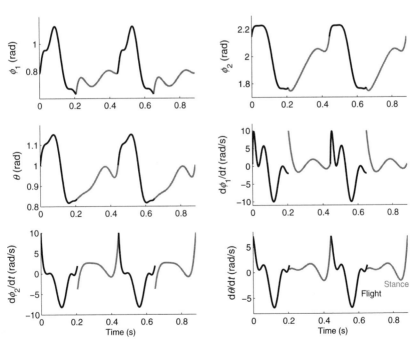

Figure 4.6   Plot of the state trajectories corresponding to two consecutive steps of the desired periodic orbit. The discontinuities in velocity are due to the impact. (See the color version of this figure in color plates section.)

TABLE 4.4   Periodic Trajectory Statistics

γ_s^+	γ_s^-	σ_s^{-*}	σ_{cm}^*	$x_{cm,s}^-$
1.2486	1.5342	-13.7227	0.9021	0.1215
$y_{cm,s}^-$	$y_{cm,s}^+$	$\lambda_x(q_s^{-*})$	$\lambda_y(q_s^{-*})$	χ
1.0041	0.9663	-0.0832	-0.0608	-0.0657

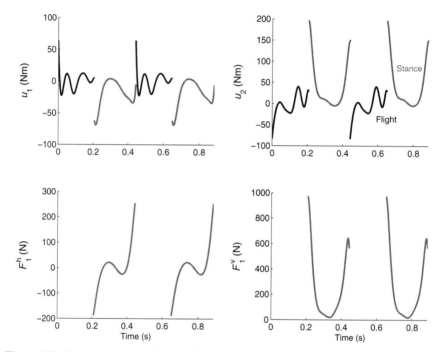

Figure 4.7   Plot of commanded control inputs and ground reaction force during two consecutive steps of the desired periodic orbit. The discontinuities are due to the transitions between
the stance and flight phases. (See the color version of this figure in color plates section.)

in velocity are due to impact. The control signals and components of the ground reaction force at the leg end during two steps of the desired periodic orbit are also shown
in Fig. 4.7. The discontinuities of the open-loop control signals are due to transitions
between the stance and flight phases.

   The results of the stability analysis performed for the desired trajectory with a
fourth-degree Bézier polynomial as an augmentation function are given in Table 4.5.

TABLE 4.5   Stability Analysis of the Desired Periodic Trajectory

a	b	K'	a_{cl}
0.9576	[0.1194 0.1838]	[1.1372 1.7556]	0.4981

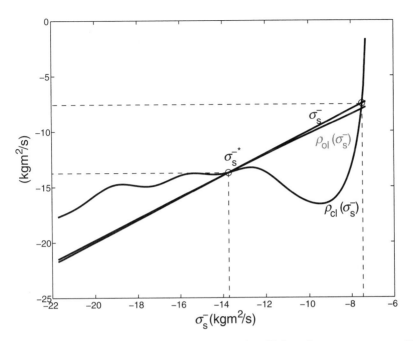

Figure 4.8 Plot of the open-loop and closed-loop restricted Poincaré return maps ρ_{ol}, ρ_{cl}. The plot is truncated at $-7.3227(\text{kgm}^2/\text{s})$ because this point is an upper bound for the domain of definition of ρ_{cl}. For $|\sigma_s^-|$ sufficiently large, the ground reaction force at the leg end will not be in the static friction cone. The mapping ρ_{cl} has two fixed points. One fixed point ($\sigma_s^- = \sigma_s^{-*} = -13.7227(\text{kgm}^2/\text{s})$) is asymptotically stable and corresponds to the desired periodic trajectory, while the other fixed point is unstable and occurs at approximately $\sigma_s^- = -7.4964(\text{kgm}^2/\text{s})$. (See the color version of this figure in color plates section.)

Since $a = 0.9576 \in (-1, 1)$, this trajectory is asymptotically stable orbit for the closed-loop system. However, to improve the convergence rate, we will make use of the static update law in equation (4.44). Moreover, the gain of this update law can be calculated via DLQR.[16] In this design method, the gain K is obtained such that by the static update law in equation (4.44), the cost function

$$\mathcal{J} := \frac{1}{2}\sum_{k=0}^{\infty}\left\{q\left(\delta\sigma_s^-[k]\right)^2 + r\delta\bar{\alpha}'[k]\delta\bar{\alpha}[k]\right\}$$

subject to the linearization of the system in equation (4.43) about $(\sigma_s^{-*}, \bar{\alpha}^*)$ is minimized, where $q \geq 0$ and $r > 0$. Calculation for $q = 10$ and $r = 1$ by the

---

[16] In Ref. [61], the DLQR design method has been used in control of walking of an underactuated 3D biped robot.

TABLE 4.6   Parameters of the Online Reconfiguration Algorithm

m	M	L_1	L_2	t_1^*	t_2^*	$l_{\max}$
0.001	1000	-104	104	0.0345	0.1727	0.1382

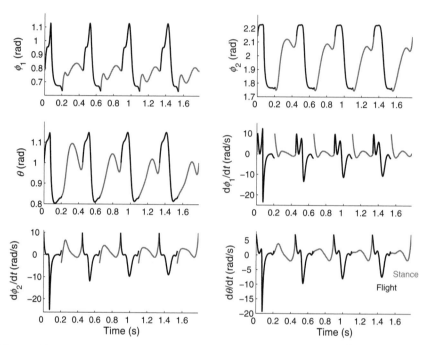

Figure 4.9   Plot of the state trajectories corresponding to four consecutive steps of the monoped robot. The discontinuities in velocity are due to the impact. (See the color version of this figure in color plates section.)

$\texttt{dlqr}$ function of MATLAB yields $K = \begin{bmatrix} 1.1372 & 1.7556 \end{bmatrix}'$ and, as a consequence, $a_{\mathrm{cl}} = a - bK = 0.4981$. The open-loop and closed-loop restricted Poincaré return maps (i.e., ρ_{ol} and ρ_{cl}) are shown in Fig. 4.8.[17] From part (c) of Theorem 4.2, we choose $x_3^0 = 1$, $x_4^0 = 0$, $m = 0.001$, $M = 1000$, and $L_2 = -L_1 = 104$. Moreover, $l_{\max} = t_2^* - t_1^* = 0.1382(\mathrm{s})$ (see Table 4.6).

To illustrate the convergence to the desired periodic orbit, the simulation of the closed-loop hybrid model of running is started at the end of the stance phase with an initial velocity 4% higher than the value on $\mathcal{O}$. State trajectories corresponding to four steps of the mechanical system are depicted in Fig. 4.9. Discontinuities in velocity are

---

[17] $\rho_{\mathrm{ol}}(\sigma_s^-)$ is identical to $\rho_{\bar{\alpha}}(\sigma_s^-)$ when $\bar{\alpha} = \bar{\alpha}^* = 0_{p \times 1}$.

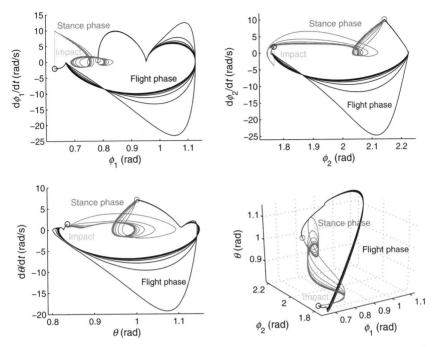

Figure 4.10 Phase–plane plots and projection of the state trajectories during 10 consecutive steps onto $(\varphi_1, \varphi_2, \theta)$. The convergence to the desired periodic trajectory can be seen. (See the color version of this figure in color plates section.)

due to the impact. Figure 4.10 represents the phase portraits and projection of the state trajectories onto $(\varphi_1, \varphi_2, \theta)$. The effect of the impact with the ground is illustrated by jumps of the velocity in the phase portraits. Commanded control inputs during four consecutive steps of running are also shown in Fig. 4.11. The discontinuities in the control inputs are due to the transitions between the stance and flight phases. Finally, Fig. 4.12 depicts the desired trajectories for the joint angles (i.e., $\varphi^d(t)$) generated by the algorithm of Theorem 4.2 and absolute orientation during the flight phases of 10 consecutive steps. As mentioned in Chapter 1, most of the past work in the literature of legged locomotion is based on quasistatic stability criteria and flat-footed walking and running motions such as ZMP criterion and its extensions [2–16]. The monopedal model studied in this chapter has a point foot, and hence, the ZMP criterion cannot be applied. By using the approach of this chapter, the periodic orbit is asymptotically stable in the sense of Lyapunov. In addition, the configuration of the mechanical system during discrete transitions (i.e., impact and takeoff) are predetermined, and hence, the step length of the robot during consecutive steps of running is equal to that on the periodic orbit $\mathcal{O}$. In the literature of monopedal and bipedal gait control, the

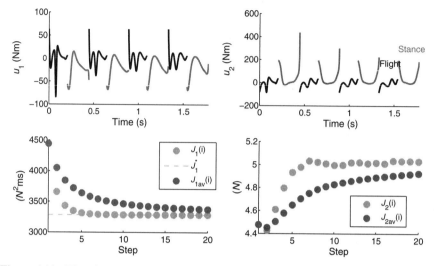

Figure 4.11   Plot of commanded control inputs during four consecutive steps of running (top graphs). The discontinuities in the control inputs are due to the transitions between the stance and flight phases. The bottom graphs present the plot of the cost function $J_1(i)$, $J_2(i)$, $J_{1,av}(i)$ and $J_{2,av}(i)$ for $i = 1, 2, \ldots, 20$. The periodic orbit $\mathcal{O}$ is designed to minimize the cost function (4.45). On this trajectory, $J = J_1^* = 3.2836 \times 10^3 (\text{N}^2\text{ms})$. From the figure, the value of J_1 after a short transient period (four steps) is approximately equal to J_1^*, which, in turn, illustrates the efficiency of the algorithm in the sense of electric motor energy per distance traveled. (See the color version of this figure in color plates section.)

two most popular cost functions are [18, 52, 73]

$$J_1(i) := \frac{1}{L_s} \int_{\text{step}\,(i)} \|u(t)\|_2^2 \, dt, \quad i = 1, 2, \ldots$$

$$J_2(i) := \frac{1}{L_s} \int_{\text{step}\,(i)} \langle \dot{q}, B u \rangle \, dt, \quad i = 1, 2, \ldots,$$

(4.46)

in which L_s is the common step length, B is the input matrix, $\langle x, y \rangle := x'y$, and step (i) represents the ith step, $i = 1, 2, \cdots$. In equation (4.46), $J_1(i)$ and $J_2(i)$ denote the electric motor energy and the integral of instantaneous mechanical power, per distance traveled during the ith step, respectively. Figure 4.11 (bottom graphs) illustrates the value of $J_1(i)$, $J_2(i)$, $J_{1,av}(i)$, and $J_{2,av}(i)$ during 20 consecutive steps of running, where $J_{1,av}$ and $J_{2,av}$ are the average values of J_1 and J_2, respectively, that is, $J_{1,av}(i) := \frac{1}{i} \sum_{j=1}^{i} J_1(j)$ and $J_{2,av}(i) := \frac{1}{i} \sum_{j=1}^{i} J_2(j)$. As mentioned previously, the periodic orbit $\mathcal{O}$ is designed to minimize the cost function (4.45). From Fig. 4.11, it can be concluded that by applying the proposed feedback scheme, the value of J_1 after a short transient period (four steps) is approximately equal to that on $\mathcal{O}$, which, in turn, illustrates the efficiency of the algorithm in the sense of control effort (electric motor

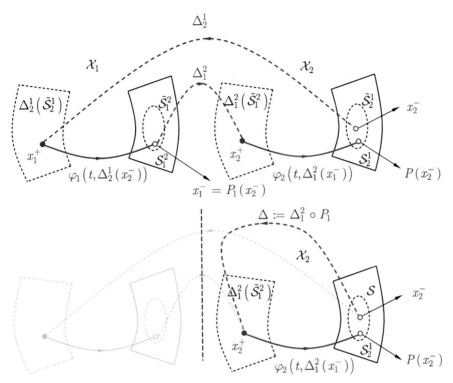

Figure 2.1 Geometric description of Theorem 2.1. The Poincaré return map of the autonomous hybrid system $\Sigma(\mathcal{X}_1, \mathcal{X}_2, \mathcal{S}_1^2, \mathcal{S}_2^1, \Delta_1^2, \Delta_2^1, f_1, f_2)$, $P : \tilde{\mathcal{S}}_2^1 \to \mathcal{S}_2^1$, is also the Poincaré return map for the autonomous system with impulse effects $\Sigma_{ie}(\mathcal{X}_2, \mathcal{S}, \Delta, f_2)$, where $\mathcal{S} := \tilde{\mathcal{S}}_2^1$ and $\Delta(x_2) := \Delta_1^2 \circ P_1(x_2)$.

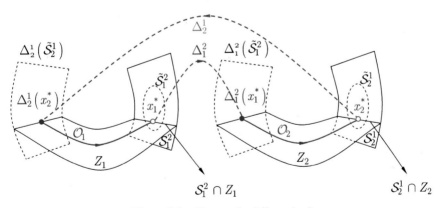

Figure 2.2 *(See text for full caption.)*

Hybrid Control and Motion Planning of Dynamical Legged Locomotion, Nasser Sadati, Guy A. Dumont, Kaveh Akbari Hamed, and William A. Gruver.
© 2012 by the Institute of Electrical and Electronics Engineers, Inc. Published 2012 by John Wiley & Sons, Inc.

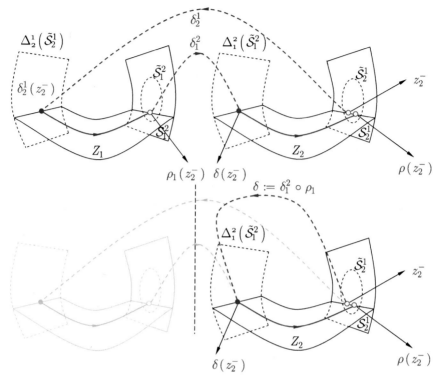

Figure 2.3 Geometric description of the restricted Poincaré return map $\rho : \tilde{\mathcal{S}}_2^1 \cap Z_2 \to \mathcal{S}_2^1 \cap Z_2$. By hypotheses H2–H5 and the construction of $\Sigma|_Z$, $\rho(z_2) = P|_Z(z_2)$, where $P|_Z$ is the restriction of the Poincaré return map of the full-dimensional hybrid system Σ to Z. By applying Theorem 2.1, it follows that ρ is also the Poincaré return map for the reduced-order system with impulse effects $\Sigma_{ie}|_{Z_2}(Z_2, \mathcal{S} \cap Z_2, \delta, f_2|_{Z_2})$, where $\delta(z_2) := \delta_1^2 \circ \rho_1(z_2)$.

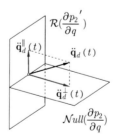

Figure 3.3 Geometrical description of the motion planning algorithm during double support. In equation (3.66), it is assumed that $\text{rank}(\frac{\partial p_2}{\partial q}(\mathbf{q}_d(t))) = 2$ and $\ddot{\mathbf{q}}_d(t)$ can be expressed as $\ddot{\mathbf{q}}_d(t) = \ddot{\mathbf{q}}_d^{\parallel}(t) + \ddot{\mathbf{q}}_d^{\perp}(t)$.

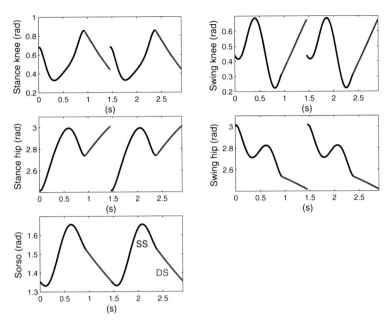

Figure 3.6 Angular positions of the knee, hip, and torso joints during two consecutive steps of the optimal motion. The discontinuities are due to the coordinate relabling for swapping the role of the legs.

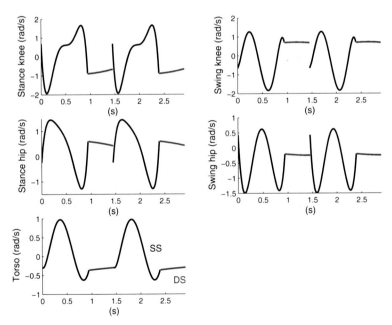

Figure 3.7 Angular velocities of the knee, hip and torso joints during two consecutive steps of the optimal motion. The discontinuities are due to the coordinate relabling for swapping the role of the legs.

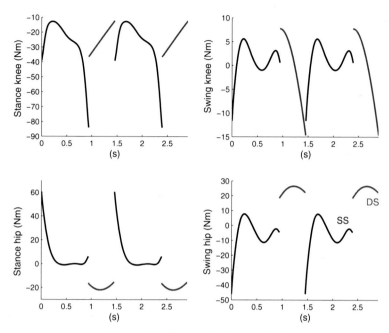

Figure 3.8 (*See text for full caption.*)

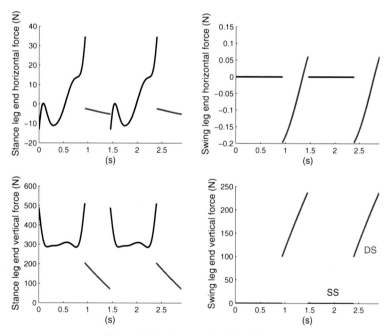

Figure 3.9 (*See text for full caption.*)

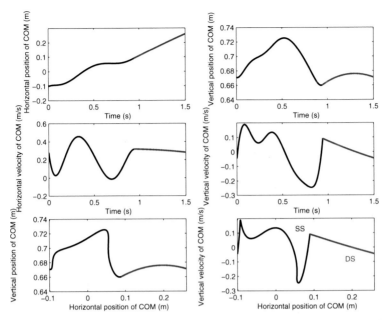

Figure 3.11 Plot of the vertical height and velocity of the COM versus time and the horizontal position of the COM on $\mathcal{O}$. At the impact, the velocity of the COM is not pointed downward.

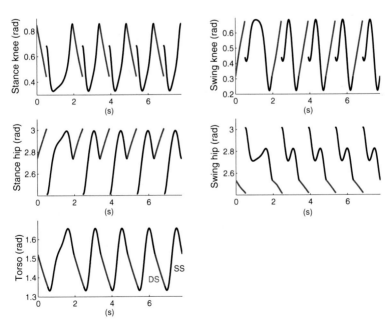

Figure 3.12 Configuration variables during five steps of the closed-loop simulation. Discontinuities in the graphs are due to coordinate relabling.

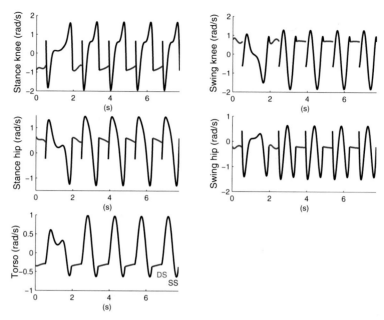

Figure 3.13 Velocity variables during five steps of the closed-loop simulation. Discontinuities in the graphs are due to coordinate relabling.

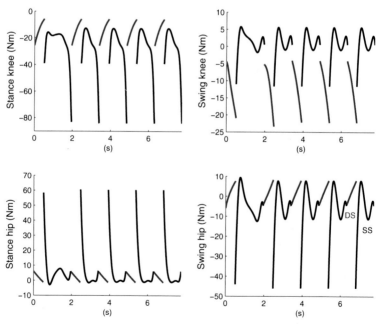

Figure 3.14 Control inputs (i.e., joint torques) during five steps of the closed-loop simulation. Discontinuities in the graphs are due to transition between the continuous phases.

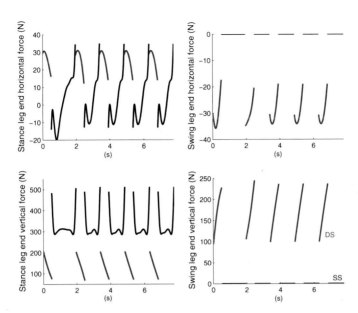

Figure 3.15 (*See text for full caption.*)

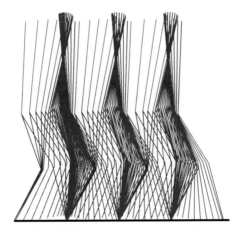

Figure 3.16 (*See text for full caption.*)

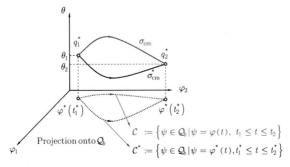

Figure 4.2 (*See text for full caption.*)

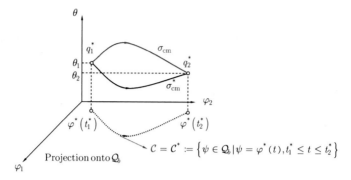

Figure 4.3 *(See text for full caption.)*

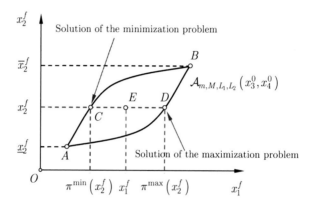

Figure 4.4 *(See text for full caption.)*

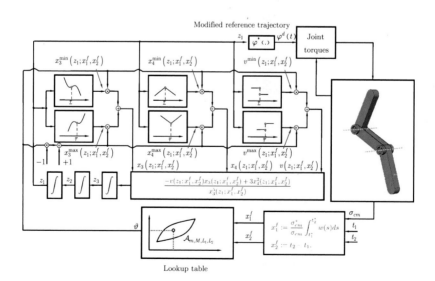

Figure 4.5 *(See text for full caption.)*

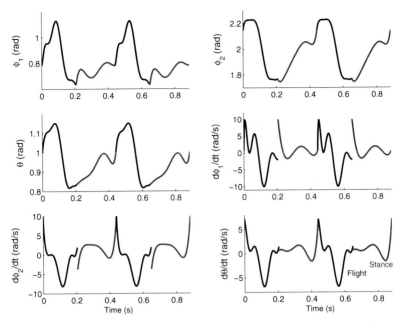

Figure 4.6 Plot of the state trajectories corresponding to two consecutive steps of the desired periodic orbit. The discontinuities in velocity are due to the impact.

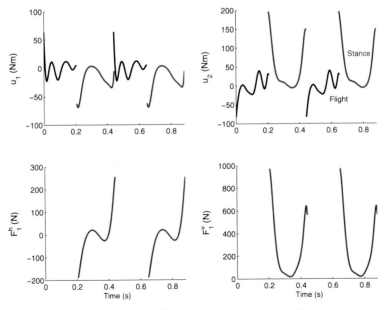

Figure 4.7 Plot of commanded control inputs and ground reaction force during two consecutive steps of the desired periodic orbit. The discontinuities are due to the transitions between the stance and flight phases.

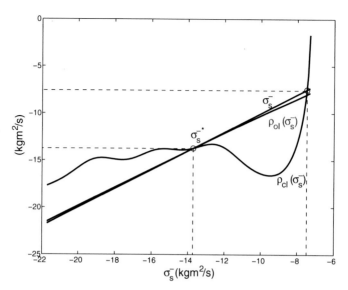

Figure 4.8 Plot of the open-loop and closed-loop restricted Poincaré return maps ρ_{ol}, ρ_{cl}. The plot is truncated at $-7.3227(\text{kgm}^2/\text{s})$ because this point is an upper bound for the domain of definition of ρ_{cl}. For $|\sigma_s^-|$ sufficiently large, the ground reaction force at the leg end will not be in the static friction cone. The mapping ρ_{cl} has two fixed points. One fixed point ($\sigma_s^- = \sigma_s^{-*} = -13.7227(\text{kgm}^2/\text{s})$) is asymptotically stable and corresponds to the desired periodic trajectory, while the other fixed point is unstable and occurs at approximately $\sigma_s^- = -7.4964(\text{kgm}^2/\text{s})$.

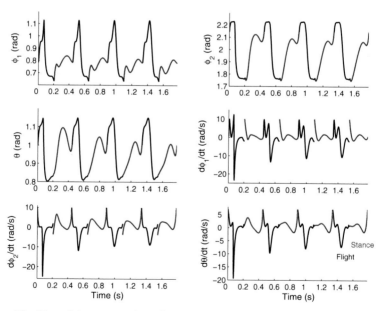

Figure 4.9 Plot of the state trajectories corresponding to four consecutive steps of the monoped robot. The discontinuities in velocity are due to the impact.

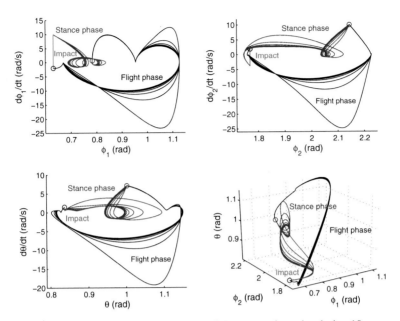

Figure 4.10 Phase–plane plots and projection of the state trajectories during 10 consecutive steps onto $(\varphi_1, \varphi_2, \theta)$. The convergence to the desired periodic trajectory can be seen.

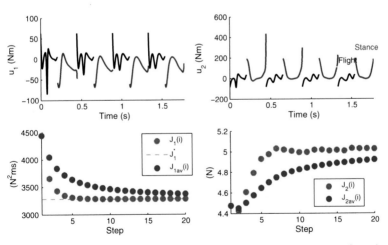

Figure 4.11 Plot of commanded control inputs during four consecutive steps of running (top graphs). The discontinuities in the control inputs are due to the transitions between the stance and flight phases. The bottom graphs present the plot of the cost function $J_1(i)$, $J_2(i)$, $J_{1,av}(i)$ and $J_{2,av}(i)$ for $i = 1, 2, \ldots, 20$. The periodic orbit $\mathcal{O}$ is designed to minimize the cost function (4.45). On this trajectory, $J = J_1^* = 3.2836 \times 10^3 (\text{N}^2\text{ms})$. From the figure, the value of J_1 after a short transient period (four steps) is approximately equal to J_1^*, which, in turn, illustrates the efficiency of the algorithm in the sense of electric motor energy per distance traveled.

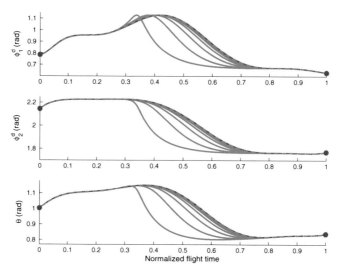

Figure 4.12 *(See text for full caption.)*

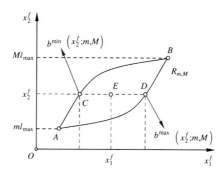

Figure 5.2 *(See text for full caption.)*

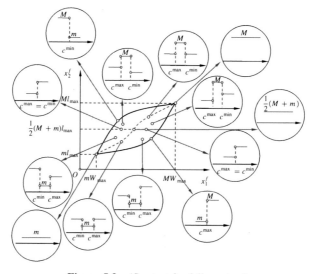

Figure 5.3 *(See text for full caption.)*

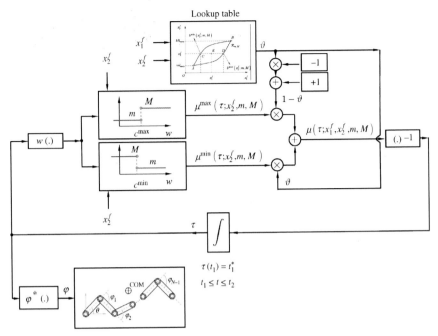

Figure 5.4 Block diagram of the online motion planning algorithm for generation of continuous joint motion $\varphi(t) = \varphi^*(\tau(t))$, $t_1 \leq t \leq t_2$ to solve configuration determinism.

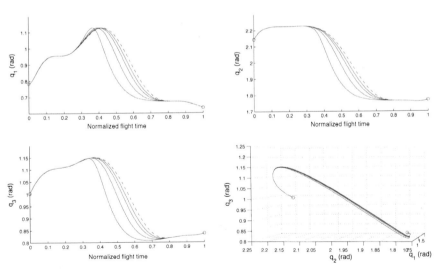

Figure 5.5 Plot of the desired trajectories for the joint angles (i.e., q_1 and q_2) generated by the online motion planning algorithm of Theorem 5.2, the absolute orientation (q_3) versus normalized time during the flight phases of four consecutive steps (solid curves) and the projection of the state variables onto the configuration space. The nominal trajectory is depicted by dashed curves. The circles at the both ends represent the initial and final predetermined configurations.

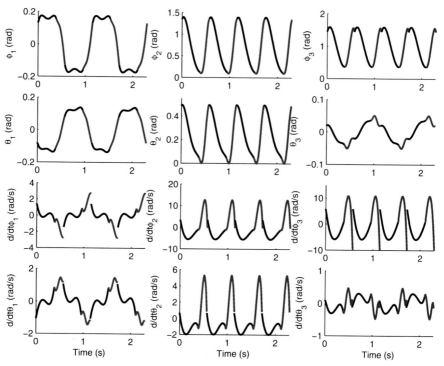

Figure 6.2 (*See text for full caption.*)

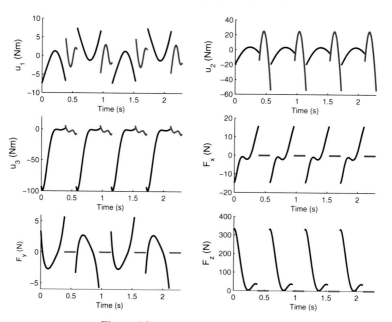

Figure 6.3 (*See text for full caption.*)

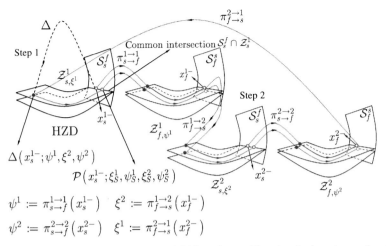

$$\psi^1 := \pi_{s \to f}^{1 \to 1}\left(x_s^{1-}\right) \quad \xi^2 := \pi_{f \to s}^{1 \to 2}\left(x_f^{1-}\right)$$

$$\psi^2 := \pi_{s \to f}^{2 \to 2}\left(x_s^{2-}\right) \quad \xi^1 := \pi_{f \to s}^{2 \to 1}\left(x_f^{2-}\right)$$

Figure 6.4 Geometric description of hybrid invariance. The plot depicts that under the 4-tuple event-based update law ($\pi_{s \to f}^{1 \to 1}, \pi_{f \to s}^{1 \to 2}, \pi_{s \to f}^{2 \to 2}, \pi_{f \to s}^{2 \to 1}$), the family of the zero dynamics manifolds for the first stance phase $\mathbf{Z}_s^1$ is hybrid invariant, that is, $\Delta(x_s^{1-}; \psi^1, \xi^2, \psi^2) \in \mathcal{Z}_{s, \xi^1}^1$, where $\Delta(x_s^{1-}; \psi^1, \xi^2, \psi^2) := \Delta_f^s(x_f^{2-})$ is the two-step reset map. In addition, $a_{N_s-1}^j = a_{N_s}^j = 0_{3 \times 1}$ results in the common intersection $\mathcal{S}_s^f \cap \mathcal{Z}_s^j$. Plot also illustrates the five-dimensional restricted Poincaré return map $\mathcal{P}(x_s^{1-}; \xi_S^1, \psi_S^1, \xi_S^2, \psi_S^2)$ and the HZD.

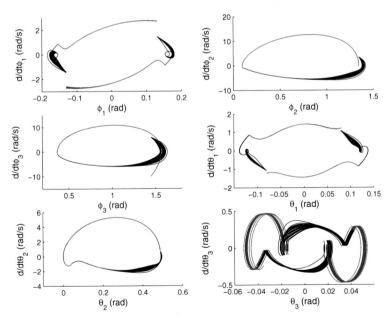

Figure 6.5 Phase portraits of the state trajectories during 40 consecutive steps of running. The stance and flight phases are shown by bold and light curves, respectively. In the figure, the effect of the impact with the ground is illustrated by jumps in the velocity. The convergence to the desired limit cycle $\mathcal{O}$ can be seen.

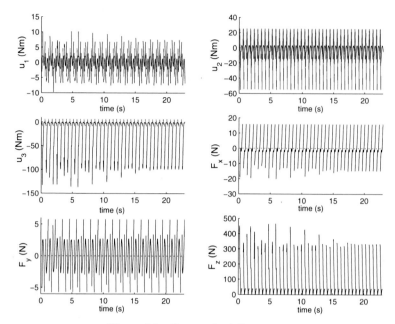

Figure 6.6 (*See text for full caption.*)

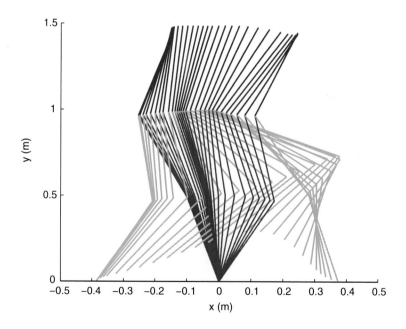

Figure 7.2 Stick animation of the bipedal robot during one step of the optimal motion.

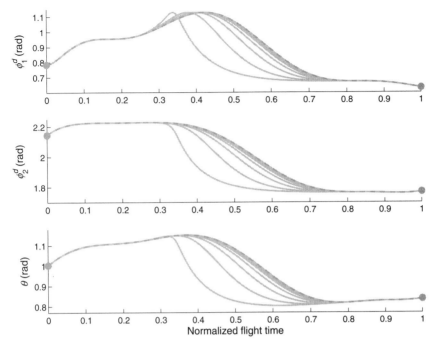

Figure 4.12   Plot of the desired trajectories for the joint angles (i.e., $\varphi^d(t)$) generated by the algorithm of Theorem 4.2 (two top graphs) and absolute orientation (bottom graph) versus normalized time during the flight phases of 10 consecutive steps. In three graphs, the nominal trajectory is depicted by dashed curves. The circles represent the initial and final configurations. (See the color version of this figure in color plates section.)

energy) per distance traveled. In addition, by defining dimensionless cost function $J_3(i) := \frac{J_2(i)}{m_{tot}g_0} = \frac{1}{m_{tot}g_0 L_s} \int_{\text{step}(i)} \langle \dot{q}, B\,u \rangle\, dt$, as the mechanical energy per unit weight per unit distance traveled during the ith step, where m_{tot} is the total mass of the robot, it can be observed that $J_3(i) \in [0.0322, 0.0367]$, $i = 1, 2, \ldots$, and $\lim_{i \to \infty} J_3(i) = 0.0366$. Comparing these dimensionless results with those reported in Ref. [73], which describe desired reference trajectories during running by another robot, demonstrates that the resultant running locomotion on the desired trajectory $\mathcal{O}$ and the closed-loop trajectory is energy efficient.

Online Generation of Joint Motions During Flight Phases of Planar Running

5.1 INTRODUCTION

The motion of a planar bipedal robot during running can be described by a hybrid system with two continuous phases, a stance phase (one leg on the ground) and a flight phase (no leg on the ground), and discrete transitions between the continuous phases for take-off and landing. An offline motion planning algorithm, based on a finite-dimensional nonlinear optimization problem with equality and inequality constraints, has been proposed in Ref. [73] to generate time trajectories of a desired periodic orbit for the hybrid model of bipedal running. Following the results of Chapter 4, to asymptotically stabilize the desired periodic orbit for the hybrid model of running using a one-dimensional restricted Poincaré return map and HZD approach, the configuration of the mechanical system should be transferred from a predetermined initial pose (immediately after take-off) to a predetermined final pose (immediately before landing) during the flight phases of running. As mentioned in Chapter 4, this problem is referred to as *configuration determinism at transitions*. The objective of this chapter is to present *modified* online motion planning algorithms for generation of continuous (C^0) and continuously differentiable (C^1) open-loop trajectories in the body configuration space of the mechanical system such that the reconfiguration problem is solved [82, 83]. The algorithms presented here are extensions of that presented in Chapter 4. In particular, the generated trajectories in Chapter 4 were twice continuously differentiable (C^2) while the reachable sets associated with the algorithms of the present chapter are larger than that of Chapter 4. We address the motion planning problem for general planar open kinematic chains composed of $N \geq 3$ rigid links interconnected with frictionless and rotational joints. The configuration of this *multilink* system is specified by the absolute position and orientation of the mechanical system with respect to an inertial world frame, and by the joint angles determining the shape of the robot. It is assumed that the angular momentum of the mechanical system about its COM is conserved.

Hybrid Control and Motion Planning of Dynamical Legged Locomotion, Nasser Sadati, Guy A. Dumont, Kaveh Akbari Hamed, and William A. Gruver.
© 2012 by the Institute of Electrical and Electronics Engineers, Inc. Published 2012 by John Wiley & Sons, Inc.

The main contribution of this chapter is to present online motion planning algorithms based on *virtual time* for generation of joint motions to satisfy configuration determinism at transitions. Since the flight time and angular momentum about the COM may differ during consecutive steps, the reconfiguration problem must be solved online. Control problems for reconfiguring multilink mechanisms with zero angular momentum have been treated in the literature, for example, Refs. [74–78]. Reference [80] proposed a method based on the averaging theorem [79, Theorem 2.1] such that for any value of the angular momentum, joint motions can reorient a multilink mechanism over an arbitrary time interval. Since this latter approach is based on determining roots of nonlinear equations, it cannot be employed online. We assume that the time trajectory of a desired joint motion, precomputed offline, solves the reconfiguration problem. By replacing the time argument of the desired motion by a strictly increasing function of time called the *virtual time*, we show how to determine continuous and continuously differentiable joint motions online during consecutive steps of running so that they also solve the reconfiguration problem. In this chapter, it is shown that the reconfiguration problem can be viewed in terms of reachability and an optimal control model for a linear time-varying system with input constraints.

5.2 MECHANICAL MODEL OF A PLANAR OPEN KINEMATIC CHAIN

Throughout this chapter, we treat a planar multilink system comprised of $N \geq 3$ rigid links that are connected by frictionless rotational joints and constituting an open kinematic chain conserving angular momentum about its COM (see Fig. 5.1). The joints have internal actuators such as dc motors. Assume that a coordinate frame called the world frame is attached to the ground. To represent the configuration of the multilink system, a convenient choice of coordinates consists of the body angles, the absolute orientation, and the absolute position of the mechanical system with respect to the world frame. The body angles consist of the relative angles $\varphi := (\varphi_1, \ldots, \varphi_{N-1})' \in \mathcal{Q}_b$ describing the shape of the multilink system, where

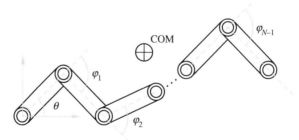

Figure 5.1 A planar multilink system composed of $N \geq 3$ rigid links interconnected with frictionless rotational joints in the form of an open kinematic chain. The configuration of the mechanical system is completely determined by the body angles $\varphi := (\varphi_1, \ldots, \varphi_{N-1})' \in \mathcal{Q}_b$, the absolute orientation $\theta \in \mathbb{S}^1$, and the absolute position $p_{cm} := (x_{cm}, y_{cm})' \in \mathbb{R}^2$.

prime denotes matrix transpose. Furthermore, $\mathcal{Q}_b$ is a simply connected and open subset of

$$\mathbb{S}^{N-1} := \underbrace{\mathbb{S}^1 \times \cdots \times \mathbb{S}^1}_{N-1}$$

referred to as the *body configuration space*, in which $\mathbb{S}^1 := [0, 2\pi)$ denotes the unit circle. The absolute orientation of the system is represented by θ, whereas the absolute position is represented by the Cartesian coordinates of its COM, $p_{cm} := (x_{cm}, y_{cm})' \in \mathbb{R}^2$. Hence, the generalized coordinates for the mechanical system can be expressed as $q_e := (\varphi', \theta, p'_{cm})' = (q', p'_{cm})'$, where $q := (\varphi', \theta)'$. Moreover, the configuration space, $\mathcal{Q}_e$, is chosen as $\mathcal{Q}_e := \mathcal{Q}_b \times \mathbb{S}^1 \times \mathbb{R}^2$.

By introducing the Lagrangian of the mechanical system as the real-valued function $\mathcal{L}_e : T\mathcal{Q}_e \to \mathbb{R}$ by

$$\mathcal{L}_e(q_e, \dot{q}_e) := \mathcal{K}_e(\varphi, \dot{q}_e) - \mathcal{V}_e(q_e),$$

where $\mathcal{K}_e$ and $\mathcal{V}_e$ represent the total kinetic and potential energy, respectively, a dynamical model for describing the motions of the multilink system can be obtained. To this end, the potential energy can be expressed as $\mathcal{V}_e(q_e) := m_{tot} g_0 y_{cm}$, in which m_{tot} is the total mass of the multilink system and g_0 is the gravitational constant. Moreover, the total kinetic energy of the system can be represented by the positive definite quadratic function $\mathcal{K}_e(\varphi, \dot{q}_e) := \frac{1}{2}\dot{q}'_e D_e(\varphi)\dot{q}_e$, where

$$D_e(\varphi) := \begin{bmatrix} A(\varphi) & 0_{N \times 2} \\ 0_{2 \times N} & m_{tot} I_{2 \times 2} \end{bmatrix} \in \mathbb{R}^{(N+2) \times (N+2)} \tag{5.1}$$

is a block-diagonal mass-inertia matrix. By applying Lagrange's equation, the dynamical model of the multilink mechanism can be expressed as a nonlinear, ordinary differential equation

$$D_e(\varphi)\ddot{q}_e + C_e(\varphi, \dot{q}_e)\dot{q}_e + G_e(q_e) = B_e u, \tag{5.2}$$

where $C_e \in \mathbb{R}^{(N+2) \times (N+2)}$ is a matrix containing Coriolis and centrifugal terms, $G_e \in \mathbb{R}^{N+2}$ is a gravity vector, $u := (u_1, \ldots, u_{N-1})' \in \mathbb{R}^{N-1}$ is a vector of actuator torques, and

$$B_e := \begin{bmatrix} I_{(N-1) \times (N-1)} \\ 0_{3 \times (N-1)} \end{bmatrix}$$

is the input matrix. Using the block diagonal form of the mass-inertia matrix in equation (5.1) and equations (7.60) and (7.62) of Ref. [90, p. 256], C_e can be expressed as

$$C_e(\varphi, \dot{q}_e) = \begin{bmatrix} \bar{C}(\varphi, \dot{q}) & 0_{N \times 2} \\ 0_{2 \times N} & 0_{2 \times 2} \end{bmatrix}, \tag{5.3}$$

and consequently, the equation of motions in equation (5.2) can be decomposed as follows:

$$A(\varphi)\ddot{q} + \bar{C}(\varphi, \dot{q})\dot{q} = Bu \tag{5.4}$$

$$\ddot{x}_{cm} = 0 \tag{5.5}$$

$$\ddot{y}_{cm} + g_0 = 0, \tag{5.6}$$

where $B := [I_{(N-1)\times(N-1)} \ 0_{(N-1)\times 1}]'$. By introducing $x_e := (q'_e, \dot{q}'_e)'$ as the state vector of the mechanical system, equation (5.2) can be expressed in following state space form:

$$\dot{x}_e = f_e(x_e) + g_e(x_e)u.$$

Moreover, the state manifold is taken as the tangent bundle of Q_e, that is,

$$\mathcal{X}_e := TQ_e := \left\{ x_e := (q'_e, \dot{q}'_e)' | q_e \in Q_e, \dot{q}_e \in \mathbb{R}^{N+2} \right\}.$$

5.3 MOTION PLANNING ALGORITHM TO GENERATE CONTINUOUS JOINT MOTIONS

Following the results of Section 4.3, the configuration of the mechanical system should be transferred from a specified initial pose to a specified final pose during the flight phases of running. In other words, we desire that the take-off and landing occur in fixed configurations. Regulating the robot's shape and absolute orientation during flight phases is referred to as the *reconfiguration problem*. As mentioned in Chapter 4, during running of the robot, the angular momentum about the COM and flight time may differ during consecutive steps. Consequently, the reconfiguration problem must be solved online. The *conservation of angular momentum* about the COM of the mechanical system studied here is expressed in the Nth row of matrix equation (5.4) that can be rewritten as[1]

$$\dot{\theta} = \frac{\sigma_{cm}}{A_{N,N}(\varphi)} - \sum_{i=1}^{N-1} \frac{A_{N,i}(\varphi)}{A_{N,N}(\varphi)} \dot{\varphi}_i \tag{5.7}$$

$$= \frac{\sigma_{cm}}{A_{N,N}(\varphi)} - J(\varphi)\dot{\varphi},$$

where σ_{cm} is a constant representing the angular momentum of the mechanical system about its COM and

$$J(\varphi) := \frac{1}{A_{N,N}(\varphi)} [A_{N,1}(\varphi), \ldots, A_{N,N-1}(\varphi)] \in \mathbb{R}^{1\times N-1}.$$

---

[1] Since the matrix $A(\varphi)$ is positive definite, $A_{N,N}(\varphi) > 0$ for any $\varphi \in Q_b$.

Remark 5.1 *Since θ is a cyclic variable [1], for the Lagrangian $\mathcal{L}_e$, the mass-inertia and Coriolis matrices in equations (5.1) and (5.3) are independent of θ. Hence, the right-hand side of equation (5.7) is expressed as a function of φ and $\dot{\varphi}$.*

Assume that the twice continuously differentiable nominal trajectory φ^* : $[t_1^*, t_2^*] \to \mathcal{Q}_b$ can transfer the state of the system given in equation (5.7) from the initial condition θ_1 to the final condition θ_2 when the angular momentum about the COM is identical to σ_{cm}^*, that is,

$$\theta_2 = \theta_1 + \int_{t_1^*}^{t_2^*} \left(\frac{\sigma_{cm}^*}{A_{N,N}(\varphi^*(s))} - J(\varphi^*(s))\, \dot{\varphi}^*(s) \right) ds.$$

Next we let the angular momentum about the COM be σ_{cm}, where $\sigma_{cm} \neq \sigma_{cm}^*$. The objective of this section is to develop an online algorithm for generating the trajectory $\varphi : [t_1, t_2] \to \mathcal{Q}_b$ such that (i) $\varphi(t_1) = \varphi^*(t_1^*)$, (ii) $\varphi(t_2) = \varphi^*(t_2^*)$, and (iii)

$$\theta_2 = \theta_1 + \int_{t_1}^{t_2} \left(\frac{\sigma_{cm}}{A_{N,N}(\varphi(s))} - J(\varphi(s))\, \dot{\varphi}(s) \right) ds,$$

where $t_1 \neq t_1^*$ and $t_2 \neq t_2^*$. Let

$$C^* := \{\psi \in \mathcal{Q}_b | \psi = \varphi^*(t),\ t_1^* \leq t \leq t_2^*\}$$
$$C := \{\psi \in \mathcal{Q}_b | \psi = \varphi(t),\ \ t_1 \leq t \leq t_2\}$$

be the projections of the nominal and generated trajectories onto the body configuration space $\mathcal{Q}_b$. Attention is now turned to online generation of the trajectory $\varphi(t), t_1 \leq t \leq t_2$. For this purpose, integration of equation (5.7) over the time interval $[t_1, t_2]$ implies that

$$\theta(t_2) = \theta_1 + \int_{t_1}^{t_2} \frac{\sigma_{cm}}{A_{N,N}(\varphi(t))}\, dt - \int_C J(\varphi)\, d\varphi. \tag{5.8}$$

Analogous to the approach of Chapter 4, by assuming $\varphi(t) := \varphi^*(\tau(t))$, where τ : $[t_1, t_2] \to [t_1^*, t_2^*]$ fulfills the following constraints:

(i) $\tau(t_1) = t_1^*$

(ii) $\tau(t_2) = t_2^*$ $\hspace{2cm}$ (5.9)

(iii) $\inf_{t_1 \leq t \leq t_2} \dot{\tau}(t) > 0,$

$\mathcal{C} = \mathcal{C}^*$, and equation (5.8) can be rewritten as follows:

$$\theta(t_2) = \theta_1 + \int_{t_1}^{t_2} \frac{\sigma_{cm}}{A_{N,N}(\varphi^*(\tau(t)))} \, dt - \int_{\mathcal{C}^*} J(\varphi^*) \, d\varphi^*$$

$$= \theta_1 + \int_{t_1^*}^{t_2^*} \frac{\sigma_{cm}}{A_{N,N}(\varphi^*(s))} \frac{ds}{\dot{\tau} \circ \tau^{-1}(s)} - \int_{\mathcal{C}^*} J(\varphi^*) \, d\varphi^*.$$

Hence,

$$\theta(t_2) - \theta_2 = \int_{t_1^*}^{t_2^*} \frac{1}{A_{N,N}(\varphi^*(s))} \left(\frac{\sigma_{cm}}{\dot{\tau} \circ \tau^{-1}(s)} - \sigma_{cm}^* \right) ds.$$

Since τ can be viewed as the argument of $\varphi^*(.)$, it is called the *virtual time*. By defining $\mu(s) := \frac{1}{\dot{\tau} \circ \tau^{-1}(s)} > 0$ and $w(s) := \frac{1}{A_{N,N}(\varphi^*(s))} > 0$ for $s \in [t_1^*, t_2^*]$, and assuming $\sigma_{cm} \neq 0$, the condition $\theta(t_2) = \theta_2$ can be expressed as the following equality constraint:

$$\int_{t_1^*}^{t_2^*} w(s) \, \mu(s) \, ds = \frac{\sigma_{cm}^*}{\sigma_{cm}} \int_{t_1^*}^{t_2^*} w(s) \, ds. \tag{5.10}$$

Moreover, from the definition of $\mu(s)$, $\dot{\tau}(t) = \frac{1}{\mu(\tau(t))}$, $t_1 \leq t \leq t_2$, and hence,

$$\int_{t_1^*}^{t_2^*} \mu(s) \, ds = t_2 - t_1. \tag{5.11}$$

Determination of the piecewise continuous function $\mu(\tau) > 0$, $t_1^* \leq \tau \leq t_2^*$ such that equality constraints in equations (5.10) and (5.11) are satisfied is equivalent to solving for the open-loop control $\mu : [t_1^*, t_2^*] \to \mathbb{R}^{>0}$ transferring the state of the following system in the virtual time domain:

$$\Sigma : \begin{aligned} \dot{x}_1 &= w(\tau) \, \mu \\ \dot{x}_2 &= \mu \end{aligned} \tag{5.12}$$

from the initial condition $(x_1(t_1^*), x_2(t_1^*))' = (0, 0)'$ to the final condition $(x_1(t_2^*), x_2(t_2^*))' = (x_1^f, x_2^f)'$, where $(\dot{.}) := \frac{d}{d\tau}(.)$ and

$$x_1^f := \frac{\sigma_{cm}^*}{\sigma_{cm}} \int_{t_1^*}^{t_2^*} w(s) \, ds \tag{5.13}$$

$$x_2^f := t_2 - t_1.$$

Due to the fact that $w(\tau)$ is continuous on the compact set $[t_1^*, t_2^*]$, $m_w := \min_{t_1^* \leq \tau \leq t_2^*} w(\tau)$ and $M_w := \max_{t_1^* \leq \tau \leq t_2^*} w(\tau)$ exist and are positive scalars.

Moreover, since $\mu(\tau) > 0$, the following inequality holds:

$$m_w \int_{t_1^*}^{t_2^*} \mu(s)\,ds \leq \int_{t_1^*}^{t_2^*} w(s)\,\mu(s)\,ds \leq M_w \int_{t_1^*}^{t_2^*} \mu(s)\,ds,$$

which, in turn, follows $0 < m_w x_2^f \leq x_1^f \leq M_w x_2^f$. Thus, the state of the system Σ cannot be transferred to any arbitrary final point $(x_1^f, x_2^f)' \in \mathbb{R}^2$ by positive open-loop control μ. In the following, we assume that $\frac{\sigma_{cm}^*}{\sigma_{cm}} > 0$.

5.3.1 Determining of the Reachable Set from the Origin

The purpose of this subsection is to determine the reachable set from the origin at t_1^* at time t_2^* for the system Σ. For this purpose, we present the following definitions.

Definition 5.1 (The Admissible Open-Loop Control Inputs) *The set of admissible open-loop control inputs for system Σ is denoted by $\mathcal{U}_{m,M}$ and defined to be the set of all piecewise continuous functions $\tau \mapsto \mu(\tau) \in [m, M]$ defined on the interval $[t_1^*, t_2^*]$, where $0 < m < M$.*

Definition 5.2 (The Reachable Set from the Origin) *The reachable set from the origin (at t_1^*) with respect to $\mathcal{U}_{m,M}$ at time t_2^* is denoted by $\mathcal{R}_{m,M}$ and defined to be the set of all points $(x_1^f, x_2^f)' \in \mathbb{R}^2$ for which there exists an admissible open-loop control μ (i.e., $\mu \in \mathcal{U}_{m,M}$) such that there is a trajectory of the system Σ with the property $(x_1(t_1^*), x_2(t_1^*))' = (0, 0)'$ and $(x_1(t_2^*), x_2(t_2^*))' = (x_1^f, x_2^f)'$.*

To determine the set $\mathcal{R}_{m,M}$, we first formulate an optimal control problem, in which the optimal admissible open-loop control input, $\mu^{\max}(\tau) \in \mathcal{U}_{m,M}$, $t_1^* \leq \tau \leq t_2^*$ transfers the system Σ from the origin to the final point $(x_1(t_2^*), x_2(t_2^*))'$, while the performance measure

$$\mathcal{I}(\mu) := x_1(t_2^*) \tag{5.14}$$

is maximized.[2] In this problem, we remark that $x_2(t_2^*) = x_2^f$ is specified (see point D in Fig. 5.2). By introducing the Hamiltonian

$$\mathcal{H}(x_1, x_2, p_1, p_2, \mu, \tau) := p_1\, w(\tau)\, \mu + p_2\, \mu,$$

where p_1 and p_2 denote the costate variables, the costate equations are

$$\dot{p}_1^{\max}(\tau) = -\frac{\partial \mathcal{H}}{\partial x_1} \left(x_1^{\max}, x_2^{\max}, p_1^{\max}, p_2^{\max}, \mu^{\max}, \tau \right) = 0$$

$$\dot{p}_2^{\max}(\tau) = -\frac{\partial \mathcal{H}}{\partial x_2} \left(x_1^{\max}, x_2^{\max}, p_1^{\max}, p_2^{\max}, \mu^{\max}, \tau \right) = 0.$$

---

[2] This problem is equivalent to the minimization of the performance $-\mathcal{I}(\mu) = -x_1(t_2^*)$.

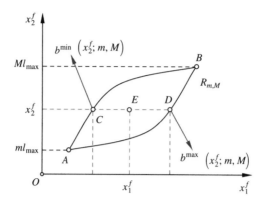

Figure 5.2   The reachable set $\mathcal{R}_{m,M}$. Solutions of the minimization and maximization problems for a given x_2^f are represented by points C and D, respectively. The values of the cost function $\mathcal{I}$ for the minimization and maximization problems are also denoted by $b^{\min}(x_2^f; m, M)$ and $b^{\max}(x_2^f; m, M)$. (See color version of this figure in color plates section.)

Thus, the costate variables $p_1^{\max}(\tau)$ and $p_2^{\max}(\tau)$ are constant valued functions of the virtual time. Note that the superscript "*max*" denotes the solutions of the maximization problem. Since the final value of x_1 (i.e., $x_1(t_2^*)$) is free, from Table 5.1 of Ref. [100, p. 200], $p_1^{\max}(\tau) = p_1^{\max}(t_2^*) = -1$. Moreover, by Pontryagin's Minimum Principle [101] and defining $c := p_2^{\max}$, the optimal open-loop control input is given by

$$\mu^{\max}(\tau) := \begin{cases} m & w(\tau) < c \\ M & w(\tau) > c \\ \text{undetermined} & w(\tau) = c. \end{cases} \tag{5.15}$$

Let

$$\mathcal{T}_-(c) := \left\{ \tau \in \left[t_1^*, t_2^* \right] \mid w(\tau) \leq c \right\}$$
$$\mathcal{T}_+(c) := \left\{ \tau \in \left[t_1^*, t_2^* \right] \mid w(\tau) \geq c \right\},$$

and assume that $w(\tau)$ satisfies the following hypothesis:

(H1)   For any $c \in [m_w, M_w]$, the set $\mathcal{T}_0(c) := \mathcal{T}_+(c) \cap \mathcal{T}_-(c)$ is Lebesgue negligible.

If the function $w(\tau)$ passes through c, a switching of the optimal control input $\mu^{\max}(\tau)$ is indicated. However, hypothesis H1 implies that for any $c \in [m_w, M_w]$, $w(\tau)$ is not equal to c on any finite time interval, and thus, the singular condition does not occur. For the later purposes, define $W_-(c)$ and $W_+(c)$ as follows:

$$W_-(c) := \int_{t_1^*}^{t_2^*} w(s)\, \mathbf{1}(c - w(s))\, ds$$

$$W_+(c) := \int_{t_1^*}^{t_2^*} w(s)\, \mathbf{1}(w(s) - c)\, ds,$$

where $\mathbf{1}(.)$ is the unit Heaviside step function. From hypothesis H1, $W_-(c) + W_+(c) = W_{\max}$, where $W_{\max} := \int_{t_1^*}^{t_2^*} w(s)ds$ because Lebesgue integration over an interval with zero measure results in zero. Furthermore, define the functions $l_-, l_+ : \mathbb{R} \to \mathbb{R}^{\geq 0}$ by

$$l_-(c) := \int_{t_1^*}^{t_2^*} \mathbf{1}(c - w(s))\, ds$$

$$l_+(c) := \int_{t_1^*}^{t_2^*} \mathbf{1}(w(s) - c)\, ds,$$

which can be viewed as the Lebesgue measures of the sets $\mathcal{T}_-(c)$ and $\mathcal{T}_+(c)$, respectively. Hypothesis H1 also implies that $l_-(c) + l_+(c) = l_{\max}$, where $l_{\max} := t_2^* - t_1^*$. Therefore, the final constraint $x_2^{\max}(t_2^*) = x_2^f$ can be expressed as

$$x_2^{\max}(t_2^*) = \int_{t_1^*}^{t_2^*} \mu(s)\, ds$$

$$= ml_-(c) + M(l_{\max} - l_-(c))$$

$$= x_2^f,$$

and consequently, c is the solution of the equation

$$l_-(c) = \frac{Ml_{\max} - x_2^f}{M - m}.$$

Since $0 \leq l_-(c) \leq l_{\max}$, x_2^f satisfies the constraint $ml_{\max} \leq x_2^f \leq Ml_{\max}$ (see Fig. 5.2).

Lemma 5.1 (Existence and Uniqueness of Optimal Solutions) *Let $\varphi^* : [t_1^*, t_2^*] \to \mathcal{Q}_b$ be a C^2 nominal trajectory such that hypothesis H1 is met. Then, for every $0 < \bar{l} < l_{\max}$, the equation $l_-(c) = \bar{l}$ has a unique solution.*

Proof. $l_-(c)$ is a well-defined function for any $c \in \mathbb{R}$. It can be shown that[3]

$$\frac{dl_-(c)}{dc} = \sum_{\tau \in \mathcal{T}_0(c)} \frac{1}{|\dot{w}(\tau)|}.$$

Thus, for every $m_w \leq c \leq M_w$ for which $\dot{w}(\tau) \neq 0$ on the set $\mathcal{T}_0(c)$, $l_-(c)$ is differentiable at c and $\frac{dl_-(c)}{dc} > 0$. From hypothesis H1, for any $c \leq m_w, l_-(c) = 0$. Moreover, for any $c \geq M_w, l_-(c) = l_{\max}$. Hence, $l_-(c)$ is a strictly increasing function of c for every $m_w < c < M_w$ that completes the proof. ∎

---

[3] We remark that since the functions $F(s, c) := \mathbf{1}(c - w(s))$ and $\frac{\partial F}{\partial c}(s, c)$ are discontinuous in c, the Leibniz integral rule cannot be applied to obtain $\frac{dl_-(c)}{dc}$.

From Lemma 5.1, for every $ml_{max} < x_2^f < Ml_{max}$, there exists a unique solution for the equation $l_-(c) = \frac{Ml_{max} - x_2^f}{M - m}$, and hence, the final value of x_1^{max} can be expressed as follows:

$$x_1^{max}(t_2^*) = \int_{t_1^*}^{t_2^*} w(s)\, \mu^{max}(s)\, ds$$

$$= MW_{max} - (M - m)\, W_- \circ l_-^{-1}\left(\frac{Ml_{max} - x_2^f}{M - m}\right)$$

$$=: b^{max}\left(x_2^f; m, M\right).$$

Moreover, the optimal open-loop control input can be described in terms of x_2^f, m and M as

$$\mu^{max}\left(\tau; x_2^f, m, M\right) := \begin{cases} m & w(\tau) < c^{max}\left(x_2^f\right) \\ M & w(\tau) > c^{max}\left(x_2^f\right), \end{cases} \tag{5.16}$$

where

$$c^{max}\left(x_2^f\right) := l_-^{-1}\left(\frac{Ml_{max} - x_2^f}{M - m}\right).$$

We remark that the optimal open-loop control input μ^{max} is not defined on the discontinuity points $\tau \in T_0(c^{max}(x_2^f))$. In other words, since the set $T_0(c^{max}(x_2^f))$ is Lebesgue negligible, it therefore makes sense for the function $\mu^{max}(\tau; x_2^f, m, M)$ to be undefined on the points of discontinuity.

If the optimal control problem is defined as the minimization of the performance measure in equation (5.14) (see point C in Fig. 5.2), an analogous analysis can be performed and it can be shown that

$$x_1^{min}(t_2^*) = mW_{max} + (M - m)\, W_- \circ l_-^{-1}\left(\frac{x_2^f - ml_{max}}{M - m}\right)$$

$$=: b^{min}\left(x_2^f; m, M\right).$$

In addition, the optimal open-loop control input is given by

$$\mu^{min}\left(\tau; x_2^f, m, M\right) := \begin{cases} M & w(\tau) < c^{min}\left(x_2^f\right) \\ m & w(\tau) > c^{min}\left(x_2^f\right), \end{cases} \tag{5.17}$$

where

$$c^{min}\left(x_2^f\right) := l_-^{-1}\left(\frac{x_2^f - ml_{max}}{M - m}\right).$$

Next, we show that sufficient conditions for optimality are satisfied along the optimal trajectories of the system Σ. For this purpose, the following result is presented.

Lemma 5.2 (Sufficient Conditions for Optimality) *Let hypothesis H1 hold. Then, for every $x_2^f \in [ml_{max}, Ml_{max}]$, the functions $\mu^{max}(\tau; x_2^f, m, M)$ and $\mu^{min}(\tau; x_2^f, m, M)$, $t_1^* \leq \tau \leq t_2^*$ given in equations (5.16) and (5.17) are optimal open-loop control inputs for the maximization and minimization problems, respectively.*

Proof. To verify sufficiency, we remark that the minimization and maximization problems for system Σ are equivalent to the minimization and maximization of the performance index

$$\mathcal{J}(x_0, t_0) := \int_{t_0}^{t_2^*} w(s)\, \mu(s)\, ds$$

subject to the system $\Sigma_e : \dot{x} = \mu, \mu \in \mathcal{U}_{m,M}$ transfers the state of the system Σ_e from the initial pair $(x_0, t_0)' = (0, t_1^*)'$ to the final pair $(x_2^f, t_2^*)'$. We shall verify that the Hamilton–Jacobi–Bellman Equation is satisfied along the optimal trajectories of the system Σ_e. In the proof of Lemma 5.2, sufficient conditions for the minimization problem are verified. A similar reasoning can also be presented for the maximization problem. Introduce the Hamiltonian

$$\mathcal{H}(x, p, \mu, \tau) := (w(\tau) + p)\, \mu,$$

where p denotes the costate variable. From Definition 5.12 of Ref. [101, p. 357], $\mathcal{H}$ is normal relative to $\mathcal{X} := \tilde{\mathcal{X}} \times [t_1^*, t_2^*]$, where $\tilde{\mathcal{X}} \subset \mathbb{R}$ is a connected set containing the points 0 and x_2^f. Hypothesis H1 implies that the control

$$\mu^{min}(\tau) := \begin{cases} m & w(\tau) > c \\ M & w(\tau) < c \end{cases}$$

is the $\mathcal{H}$-minimal control relative to $\mathcal{X}$, where $c = -p$. By defining

$$l_-(c, t_0) := \int_{t_0}^{t_2^*} \mathbf{1}(c - w(s))\, ds$$

$$W_-(c, t_0) := \int_{t_0}^{t_2^*} w(s)\mathbf{1}(c - w(s))\, ds$$

and also considering hypothesis H1, the final constraint $x^{min}(t_2^*) = x_2^f$ can be expressed as

$$(M - m)l_-(c, t_0) + m(t_2^* - t_0) = x_2^f - x_0. \tag{5.18}$$

Moreover, the performance index along the optimal trajectory of the minimization problem is given by

$$\mathcal{J}^{min}(x_0, t_0) = m \int_{t_0}^{t_2^*} w(s)\, ds + (M - m)W_-(c, t_0).$$

From equation (5.18),

$$(M - m)\frac{\partial l_-}{\partial c}(c, t_0)\frac{\partial c}{\partial x_0}(x_0, t_0) = -1$$

$$(M - m)\frac{\partial l_-}{\partial c}(c, t_0)\frac{\partial c}{\partial t_0}(x_0, t_0) + (M - m)\frac{\partial l_-}{\partial t_0}(c, t_0) - m = 0,$$

which, in turn, in combination with $\frac{\partial W_-}{\partial c}(c, t_0) = c\frac{\partial l_-}{\partial c}(c, t_0)$, yield

$$\frac{\partial \mathcal{J}^{\min}}{\partial x_0}(x_0, t_0) = -c$$

$$\frac{\partial \mathcal{J}^{\min}}{\partial t_0}(x_0, t_0) = -mw(t_0) + cm + (M - m)c\,\mathbf{1}(c - w(t_0))$$

$$- (M - m)w(t_0)\,\mathbf{1}(c - w(t_0)).$$

From the definition of the $\mathcal{H}$-minimal control $\mu^{\min}(\tau)$, we deduce that the Hamilton–Jacobi–Bellman Equation is satisfied along the optimal trajectory of the system Σ_e, that is,

$$\frac{\partial \mathcal{J}^{\min}}{\partial t_0}(x_0, t_0) + \mathcal{H}\left(x^{\min}(t_0), \frac{\partial \mathcal{J}^{\min}}{\partial x_0}(x_0, t_0), \mu^{\min}(t_0), t_0\right) = 0,$$

for all $(x_0, t_0)' \in \mathcal{X}$. This fact together with Theorem 5.12 of Ref. [101, p. 357] implies that sufficient conditions for optimality are satisfied, and hence, the control law $\mu^{\min}(\tau; x_2^f, m, M)$ is optimal. ∎

Now we can present the main result of this section as follows.

Theorem 5.1 (Reachable Set $\mathcal{R}_{m,M}$) *Let hypothesis H1 hold. Then, for any arbitrary m, M with the property $0 < m < M$, the reachable set at t_2^* (from the origin at t_1^*) can be represented by*

$$\mathcal{R}_{m,M} = \left\{\left(x_1^f, x_2^f\right)' \in \mathbb{R}^2 | ml_{\max} \leq x_2^f \leq Ml_{\max},\right.$$

$$\left. b^{\min}\left(x_2^f; m, M\right) \leq x_1^f \leq b^{\max}\left(x_2^f; m, M\right)\right\}.$$

Proof. If $\mu(\tau) \equiv m$, the trajectory of the system Σ is transferred from the origin at t_1^* to the final point $(mW_{\max}, ml_{\max})'$ at t_2^* (see point A in Fig. 5.2). Moreover, $\mu(\tau) \equiv M$ transfers the trajectory of the system from the origin to the point $(MW_{\max}, Ml_{\max})'$ (see point B in Fig. 5.2). Thus,

$$b^{\max}(ml_{\max}; m, M) = b^{\min}(ml_{\max}; m, M) = mW_{\max}$$

$$b^{\max}(Ml_{\max}; m, M) = b^{\min}(Ml_{\max}; m, M) = MW_{\max}.$$

The fact that $\frac{dW_-(c)}{dc} = c\frac{dl_-(c)}{dc}$ for any $m_w \leq c \leq M_w$ implies that for every $x_2^f \in (ml_{\max}, Ml_{\max})$,

$$\frac{\partial b^{\max}}{\partial x_2^f}\left(x_2^f; m, M\right) = -(M - m)\frac{dW_-}{dc}(c^{\max})\frac{\partial c^{\max}}{\partial x_2^f}\left(x_2^f\right)$$

$$= c^{\max}\left(x_2^f\right) > 0,$$

and in a similar manner, $\frac{\partial b^{\min}}{\partial x_2^f}(x_2^f; m, M) = c^{\min}(x_2^f) > 0$. Moreover, from the proof of Lemma 5.1, the derivative of the function $l_-^{-1}(.)$ with respect to its argument is positive, and hence,

$$\frac{\partial^2 b^{\max}}{\partial x_2^{f2}}\left(x_2^f; m, M\right) = \frac{\partial c^{\max}}{\partial x_2^f}\left(x_2^f\right) < 0$$

$$\frac{\partial^2 b^{\min}}{\partial x_2^{f2}}\left(x_2^f; m, M\right) = \frac{\partial c^{\min}}{\partial x_2^f}\left(x_2^f\right) > 0. \tag{5.19}$$

We show that $b^{\min}(x_2^f; m, M) < b^{\max}(x_2^f; m, M)$ for every $x_2^f \in (ml_{\max}, Ml_{\max})$. For this purpose, introduce the error function $e : [ml_{\max}, Ml_{\max}] \to \mathbb{R}$ by

$$e\left(x_2^f\right) := b^{\max}\left(x_2^f; m, M\right) - b^{\min}\left(x_2^f; m, M\right).$$

Assume that there exists $\tilde{x}_2^f \in (ml_{\max}, Ml_{\max})$ such that $e(\tilde{x}_2^f) = 0$. Since $e(ml_{\max}) = e(\tilde{x}_2^f) = e(Ml_{\max}) = 0$, Rolle's Theorem implies that there exist $\xi_1 \in (ml_{\max}, \tilde{x}_2^f)$ and $\xi_2 \in (\tilde{x}_2^f, Ml_{\max})$ such that

$$\frac{d}{dx_2^f}e(\xi_1) = \frac{d}{dx_2^f}e(\xi_2) = 0.$$

However,

$$\frac{d^2}{dx_2^{f2}}e\left(x_2^f\right) = \frac{\partial^2 b^{\max}}{\partial x_2^{f2}}\left(x_2^f; m, M\right) - \frac{\partial^2 b^{\min}}{\partial x_2^{f2}}\left(x_2^f; m, M\right) < 0$$

for any $x_2^f \in (ml_{\max}, Ml_{\max})$. Therefore, $\frac{d}{dx_2^f}e(x_2^f)$ is strictly monotonic, which, in turn, contradicts the assumed existence of $\tilde{x}_2^f$. This result in combination with equation (5.19) follows that

$$b^{\min}\left(x_2^f; m, M\right) < b^{\max}\left(x_2^f; m, M\right)$$

for every $x_2^f \in (ml_{\max}, Ml_{\max})$.

Next, we claim that for any $(x_1^f, x_2^f)' \in \mathcal{R}_{m,M}$ there exists an admissible open-loop control input that transfers the state of the system Σ from the origin at t_1^* to the final point $(x_1^f, x_2^f)'$ at t_2^*. To show this, choose[4] $\vartheta \in [0, 1]$ such that

$$x_1^f = \vartheta \, b^{\min}\left(x_2^f; m, M\right) + (1 - \vartheta)\, b^{\max}\left(x_2^f; m, M\right)$$

(see point E in Fig. 5.2) and define the following open-loop control input:

$$\mu\left(\tau; x_1^f, x_2^f, m, M\right) := \vartheta \, \mu^{\min}\left(\tau; x_2^f, m, M\right) + \left(1 - \vartheta\right) \mu^{\max}(\tau; x_2^f, m, M). \quad (5.20)$$

Since $\mu^{\min}, \mu^{\max} \in \{m, M\}$, μ is also admissible. Furthermore,

$$
\begin{aligned}
x_1(t_2^*) &= \vartheta \int_{t_1^*}^{t_2^*} \mu^{\min}\left(s; x_2^f, m, M\right) w(s)\, ds + (1 - \vartheta) \int_{t_1^*}^{t_2^*} \mu^{\max}\left(s; x_2^f, m, M\right) w(s)\, ds \\
&= \vartheta \, b^{\min}\left(x_2^f; m, M\right) + (1 - \vartheta)\, b^{\max}\left(x_2^f; m, M\right) \\
&= x_1^f
\end{aligned}
$$

and

$$
\begin{aligned}
x_2(t_2^*) &= \vartheta \int_{t_1^*}^{t_2^*} \mu^{\min}\left(s; x_2^f, m, M\right) ds + (1 - \vartheta) \int_{t_1^*}^{t_2^*} \mu^{\max}\left(s; x_2^f, m, M\right) ds \\
&= \vartheta \, x_2^f + (1 - \vartheta)\, x_2^f \\
&= x_2^f.
\end{aligned}
$$

Finally, use of the sufficient conditions for optimality, as verified in Lemma 5.2, completes the proof. ∎

5.3.2 Motion Planning Algorithm

From the constructive proof of Theorem 5.1, an online reconfiguration algorithm to satisfy configuration determinism at transitions can be expressed as follows.

Assume that $0 < m < M$ and hypothesis H1 holds.

Step 1: For the given σ_{cm}, t_1, and t_2, calculate x_1^f and x_2^f as follows:

$$x_1^f := \frac{\sigma_{\text{cm}}^*}{\sigma_{\text{cm}}} \int_{t_1^*}^{t_2^*} w(s)\, ds$$

$$x_2^f := t_2 - t_1.$$

---

[4] We remark that the set $\mathcal{R}_{m,M}$ is convex.

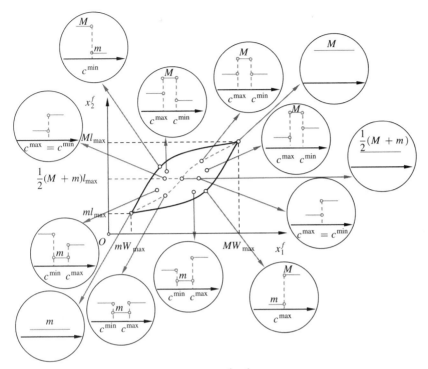

Figure 5.3 Plot of the open-loop control $\mu(\tau; x_1^f, x_2^f, m, M)$ versus $w(\tau)$ to transfer the state of the system Σ from the initial condition $(x_1(t_1^*), x_2(t_1^*))' = (0, 0)'$ to the final condition $(x_1(t_2^*), x_2(t_2^*))' = (x_1^f, x_2^f)' \in \mathcal{R}_{m,M}$ for 13 different typical cases. It is clear that the control μ can switch at most twice. (See color version of this figure in color plates section.)

Step 2: Suppose that $(x_1^f, x_2^f)' \in \mathcal{R}_{m,M}$. By assuming that the values of functions $b^{\min}(x_2^f; m, M)$ and $b^{\max}(x_2^f; m, M)$ on the interval $[ml_{\max}, Ml_{\max}]$ are precomputed and stored in a lookup table, choose $\vartheta \in [0, 1]$ such that

$$x_1^f = \vartheta\, b^{\min}\left(x_2^f; m, M\right) + (1 - \vartheta)\, b^{\max}\left(x_2^f; m, M\right).$$

For every $\tau \in [t_1^*, t_2^*]$, let (see Fig. 5.3)

$$\mu\left(\tau; x_1^f, x_2^f, m, M\right) := \vartheta\, \mu^{\min}\left(\tau; x_2^f, m, M\right) + (1 - \vartheta)\, \mu^{\max}\left(\tau; x_2^f, m, M\right).$$

Step 3: Over the time interval $[t_1, t_2]$, integrate the following ordinary differential equation

$$\dot{\tau}(t) = \frac{1}{\mu\left(\tau(t); x_1^f, x_2^f, m, M\right)} \tag{5.21}$$

with the initial condition $\tau(t_1) = t_1^*$.

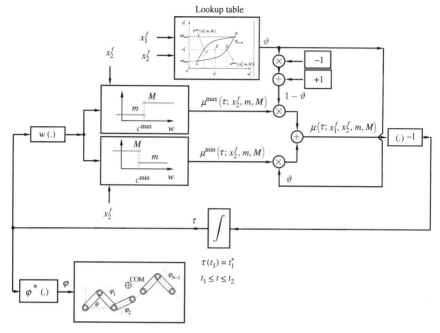

Figure 5.4 Block diagram of the online motion planning algorithm for generation of continuous joint motion $\varphi(t) = \varphi^*(\tau(t))$, $t_1 \leq t \leq t_2$ to solve configuration determinism. (See color version of this figure in color plates section.)

Step 4: Define

$$\varphi(t) := \varphi^*(\tau(t)), \quad t_1 \leq t \leq t_2$$

as the continuous open-loop joint motion that transfers the configuration of the mechanical system from q_1^* to q_2^*.

Figure 5.4 represents a block diagram for the proposed motion planning algorithm to generate continuous joint motions.

5.4 MOTION PLANNING ALGORITHM TO GENERATE CONTINUOUSLY DIFFERENTIABLE JOINT MOTIONS

Based on the results of Section 5.3, this section presents an alternative online modification of the nominal trajectory $\varphi^* : [t_1^*, t_2^*] \to \mathcal{Q}_b$ to construct a continuously differentiable (C^1) trajectory $\varphi : [t_1, t_2] \to \mathcal{Q}_b$ transferring θ from $\theta(t_1) = \theta_1$ to $\theta(t_2) = \theta_2$. Without loss of generality, we shall assume that $t_1^* = 0$, $t_2^* = T^*$, $t_1 = 0$, and $t_2 = T$. To present the main results, as given in preceding sections of Chapters 4 and 5, let $\varphi(t) = \varphi^*(\tau(t))$, where $\tau : [0, T] \to [0, T^*]$ satisfies the constraints (i) $\tau(0) = 0$,

(ii) $\tau(T) = T^*$, and (iii) $\frac{d}{dt}\tau(t) > 0$ for all $t \in [0, T]$. By defining

$$w(\xi) := \frac{1}{A_{N,N}(\varphi^*(\xi))} > 0$$

$$\mu(\xi) := \frac{1}{\dot{\tau} \circ \tau^{-1}(\xi)} > 0$$

for $\xi \in [0, T^*]$, and also considering the fact that

$$\int_0^T J(\varphi)\dot{\varphi}\,dt = \int_0^{T^*} J(\varphi^*)\dot{\varphi}^*\,dt,$$

the condition $\theta(T) = \theta_2$ can be expressed as the following equality constraint:

$$\int_0^{T^*} w(\xi)\,\mu(\xi)\,d\xi = \frac{\sigma_{cm}^*}{\sigma_{cm}} \int_0^{T^*} w(\xi)\,d\xi. \tag{5.22}$$

Moreover, condition (ii) can be rewritten as

$$\int_0^{T^*} \mu(\xi)\,d\xi = T. \tag{5.23}$$

Since we desire that the generated trajectory φ is continuously differentiable with respect to t, $\dot{\tau}(t)$ and equivalently $\mu(\xi)$ should be continuous. Consequently, we let

$$\mu(\xi) = 1 + \int_0^\xi v(\zeta)\,d\zeta, \quad \xi \in [0, T^*], \tag{5.24}$$

in which v is a piecewise continuous function. The number 1 is added to the right-hand side of equation (5.24) to guarantee that for $v(\xi) \equiv 0$, $\mu(\xi) \equiv 1$, which, in turn, results in $\tau(t) \equiv t$ (for $\sigma_{cm} = \sigma_{cm}^*$ and $T = T^*$). For the later purposes, assume that $\mathcal{V}_L$ is the set of all piecewise continuous functions $v : [0, T^*] \to [-L, L]$. Determination of $v \in \mathcal{V}_L$ such that the constraints (5.22)–(5.24) are satisfied is equivalent to determining the admissible control input $v \in \mathcal{V}_L$ such that the states of the following linear system

$$\frac{dz_1}{d\xi}(\xi) = w(\xi)\,z_3(\xi), \quad 0 \le \xi \le T^*$$

$$\frac{dz_2}{d\xi}(\xi) = z_3(\xi) \tag{5.25}$$

$$\frac{dz_3}{d\xi}(\xi) = v(\xi)$$

are transferred from $(z_1(0), z_2(0), z_3(0))' = (0, 0, 1)'$ to

$$(z_1(T^*), z_2(T^*), z_3(T^*))' = (z_{1,f}, z_{2,f}, z_{3,f})',$$

in which

$$z_{1,f} := \frac{\sigma^*_{cm}}{\sigma_{cm}} \int_0^{T^*} w(\xi)\,d\xi$$

$$z_{2,f} := T$$

and $z_{3,f}$ is free. For a given $L > 0$, the reachable set $\mathcal{R}_L$ is defined to be the set of all points $(\hat{z}_{1,f}, \hat{z}_{2,f})' \in \mathbb{R}^2$ for which there exists an admissible control $v \in \mathcal{V}_L$ transferring the state of system (5.25) from $z_0 := (0, 0, 1)'$ (at $\xi = 0$) to $\hat{z}_f := (\hat{z}_{1,f}, \hat{z}_{2,f}, \hat{z}_{3,f})'$ (at $\xi = T^*$), where $\hat{z}_3^f$ is assumed to be free. To determine the set $\mathcal{R}_L$, we study the following optimal control problems for a given $\hat{z}_{2,f}$:

$$\begin{cases} \max_{v \in \mathcal{V}_L} z_1(T^*) \\[2mm] \text{s.t.} \quad \text{system (5.25)} \\[1mm] \qquad z(0) = z_0 \\[1mm] \qquad z_2(T^*) = \hat{z}_{2,f} \end{cases} \qquad \begin{cases} \min_{v \in \mathcal{V}_L} z_1(T^*) \\[2mm] \text{s.t.} \quad \text{system (5.25)} \\[1mm] \qquad z(0) = z_0 \\[1mm] \qquad z_2(T^*) = \hat{z}_{2,f}. \end{cases} \qquad (5.26)$$

By introducing the Hamiltonian function

$$\mathcal{H}(\xi, z, p) := (p_1 w(\xi) + p_2) z_3 + p_3 v,$$

where $p := (p_1, p_2, p_3)'$ is the costate vector, and also using Table 5.1 of Ref. [100, p. 200], it can be concluded that the costate variables for the maximization problem are

$$p_1(\xi) \equiv -1$$

$$p_2(\xi) = \frac{1}{T^*} \int_0^{T^*} w(\zeta)\,d\zeta$$

$$p_3(\xi) = \int_0^{\xi} w(\zeta)\,d\zeta - \left(\frac{1}{T^*} \int_0^{T^*} w(\zeta)\,d\zeta \right) \xi.$$

Lemma 5.3 *Assume that the equation $\frac{d}{d\xi} w(\xi) = 0$ has at most one root in the open interval $(0, T^*)$. Then, the nonlinear equation $p_3(\xi) = 0$ has at most one root in $(0, T^*)$.*

Proof. Assume that there exist $\xi_1, \xi_2 \in (0, T^*)$ such that $\xi_1 \neq \xi_2$ ($\xi_1 < \xi_2$) and $p_3(\xi_1) = p_3(\xi_2) = 0$. Then, the Rolle's Theorem implies that there is $\xi_3 \in (\xi_1, \xi_2)$ such that $\frac{dp_3}{d\xi}(\xi_3) = 0$, or equivalently

$$w(\xi_3) = \frac{1}{T^*} \int_0^{T^*} w(\zeta)\,d\zeta.$$

In addition, $p_3(0)$ and $p_3(T^*)$ are also zero. Thus, there exist $\xi_4 \in (0, \xi_1)$ and $\xi_5 \in (\xi_2, T^*)$ to satisfy $\frac{dp_3}{d\xi}(\xi_4) = \frac{dp_3}{d\xi}(\xi_5) = 0$, which, in turn, results in

$$w(\xi_4) = w(\xi_5) = \frac{1}{T^*} \int_0^{T^*} w(\zeta) \, d\zeta.$$

Since $w(\xi_4) = w(\xi_3) = w(\xi_5)$, applying the Rolle's Theorem follows that there exist $\xi_6 \in (\xi_4, \xi_3)$ and $\xi_7 \in (\xi_3, \xi_5)$ such that $\frac{dw}{d\xi}(\xi_6) = \frac{dw}{d\xi}(\xi_7) = 0$, which contradicts the assumption of Lemma 5.3. ∎

The Pontryagin's Minimum Principle [101] results in

$$v = -L \operatorname{sign}(p_3).$$

Following Lemma 5.3 without loss of generality, we shall assume that $\frac{dw}{d\xi}(\xi = 0^+) < 0$. Next, let $\xi^{\max} \in (0, T^*)$ be the root of equation $p_3^{\max}(\xi) = 0$, in which the super-script "max" denotes the solution of the maximization problem. Then

$$v^{\max}(\xi) = L \operatorname{sign}(\xi - \xi^{\max})$$
$$z_3^{\max}(\xi) = 1 + L \, \sigma_1(\xi; \xi^{\max})$$
$$z_2^{\max}(\xi) = \xi + L \, \sigma_2(\xi; \xi^{\max}),$$

where

$$\sigma_1(\xi; \xi^{\max}) := \int_0^\xi \operatorname{sign}(\zeta - \xi^{\max}) \, d\zeta$$

$$\sigma_2(\xi; \xi^{\max}) := \int_0^\xi \sigma_1(\zeta; \xi^{\max}) \, d\zeta.$$

The boundary condition $z_2^{\max}(T^*) = \hat{z}_{2,f}$ also implies that

$$\xi^{\max}(\hat{z}_{2,f}) = T^* - \sqrt{\frac{1}{L}\left(\hat{z}_{2,f} - T^* + \frac{L}{2}T^{*2}\right)}. \tag{5.27}$$

An analogous analysis for the minimization problem results in

$$p_1^{\min}(\xi) \equiv 1$$

$$p_2^{\min}(\xi) = -\frac{1}{T^*} \int_0^{T^*} w(\zeta) \, d\zeta$$

$$p_3^{\min}(\xi) = -\int_0^\xi w(\zeta) \, d\zeta + \left(\frac{1}{T^*} \int_0^{T^*} w(\zeta) \, d\zeta\right) \xi$$

(see Table 5.1 of Ref. [100, p. 200]) that together with the assumption $\frac{dw}{d\xi}(\xi = 0^+) < 0$ yields

$$v^{\min}(\xi) = -L\,\mathrm{sign}(\xi - \xi^{\min})$$
$$z_3^{\min}(\xi) = 1 - L\,\sigma_1(\xi; \xi^{\min})$$
$$z_2^{\min}(\xi) = \xi - L\,\sigma_2(\xi; \xi^{\min}),$$

where

$$\xi^{\min}(\hat{z}_{2,f}) = T^* - \sqrt{\frac{1}{L}\left(\frac{L}{2}T^{*2} + T^* - \hat{z}_{2,f}\right)}. \tag{5.28}$$

From equations (5.27) and (5.28), it can be concluded that

$$\hat{z}_{2,f} \in \left[T^* - \frac{L}{2}T^{*2},\, T^* + \frac{L}{2}T^{*2}\right].$$

Furthermore, we define

$$Z_1^{\max}(\hat{z}_{2,f}) := \int_0^{T^*} w(\zeta)\,z_3^{\max}(\zeta)\,d\zeta$$

and

$$Z_1^{\min}(\hat{z}_{2,f}) := \int_0^{T^*} w(\zeta)\,z_3^{\min}(\zeta)\,d\zeta$$

as the values of the cost functions in the maximization and minimization problems, respectively. Since the system (5.25) is linear, $\mathcal{R}_L$ is convex. In addition, Theorem 5.12 of Ref. [101, p. 357] implies that the sufficient conditions for optimality are satisfied along the optimal solutions (see Section 5.3, Lemma 5.2 for a similar optimization problem). Thus, the reachable set can be expressed as

$$\mathcal{R}_L = \{(\hat{z}_{1,f}, \hat{z}_{2,f})' \in \mathbb{R}^2 \mid Z_1^{\min}(\hat{z}_{2,f}) \le \hat{z}_{1,f} \le Z_1^{\max}(\hat{z}_{2,f}),\, \underline{z}_{2,f} \le \hat{z}_{2,f} \le \overline{z}_{2,f}\},$$

in which $\underline{z}_{2,f} := T^* - \frac{L}{2}T^{*2}$ and $\overline{z}_{2,f} := T^* + \frac{L}{2}T^{*2}$. The following theorem presents the main result of this section.

Theorem 5.2 (Motion Planning Algorithm) *Assume that the functions $Z_1^{\max}$ and $Z_1^{\min}$ on the interval $[\underline{z}_{2,f}, \overline{z}_{2,f}]$ are precomputed and stored in a lookup table. For a given σ_{cm} and T, calculate $z_{1,f} := \frac{\sigma_{cm}^*}{\sigma_{cm}}\int_0^{T^*} w(\xi)\,d\xi$ and $z_{2,f} := T$. If $(z_{1,f}, z_{2,f})' \in \mathcal{R}_L$, let $\vartheta \in [0, 1]$ be such that*

$$z_{1,f} = \vartheta\,Z_1^{\min}(z_{2,f}) + (1 - \vartheta)\,Z_1^{\max}(z_{2,f}).$$

Then, the trajectory

$$\varphi(t) = \varphi^*(s_1(t)), \quad 0 \le t \le T$$

can transfer the state of system (5.7) form $\theta(0) = \theta_1$ to $\theta(T) = \theta_2$, where s_1 is the state of the following system with the initial condition $(s_1(0), s_2(0))' = (0, 1)'$

$$\dot{s}_1 = s_2$$
$$\dot{s}_2 = -\frac{\vartheta \, v^{\min}(s_1) + (1 - \vartheta) \, v^{\max}(s_1)}{\left(\vartheta z_3^{\min}(s_1) + (1 - \vartheta)z_3^{\max}(s_1)\right)^3}. \tag{5.29}$$

Proof. Let us define $s_1(t) := \tau(t) = \xi$ and $s_2(t) := \dot{\tau}(t)$. Since system (5.25) is linear,

$$v(s_1) = \vartheta \, v^{\min}(s_1) + (1 - \vartheta) \, v^{\max}(s_1) \tag{5.30}$$

implies that

$$\mu(s_1) = z_3(s_1) = \vartheta \, z_3^{\min}(s_1) + (1 - \vartheta) \, z_3^{\max}(s_1). \tag{5.31}$$

We also remark that

$$z_2(T^*) = \vartheta \, z_{2,f} + (1 - \vartheta) \, z_{2,f} = z_{2,f}$$
$$z_1(T^*) = \vartheta \, Z_1^{\min}(z_{2,f}) + (1 - \vartheta) \, Z_1^{\max}(z_{2,f}) = z_{1,f}.$$

By considering the fact that $\dot{\tau}(t) = \frac{1}{\mu(\tau(t))}$, it can be concluded that

$$\dot{s}_2(t) = -\frac{\partial\mu/\partial\tau(\tau(t))}{\mu^2(\tau(t))} \dot{\tau}(t) = -\frac{\partial\mu/\partial\tau(\tau(t))}{\mu^3(\tau(t))}, \tag{5.32}$$

which, in turn, results in equation (5.29). In addition, the right-hand side of system (5.29) is discontinuous with respect to s_1. However, from Lemma 2 of Ref. [102, p. 107], there exists a unique solution for system (5.29) through $(s_1(0), s_2(0))' = (0, 1)'$ (at $t = 0$). ∎

Figure 5.5 depicts the trajectories for the joint angles (q_1 and q_2) generated by the algorithm of Theorem 5.2 and the absolute orientation q_3 during the flight phases of four consecutive steps for the monopedal robot investigated in Section 4.6. It is clearly observed that configuration determinism at landing is satisfied.

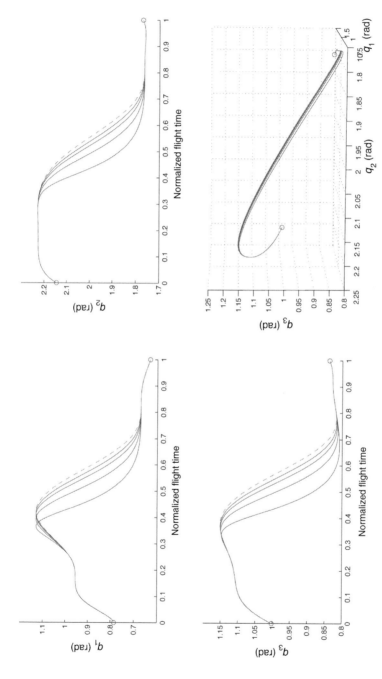

Figure 5.5 Plot of the desired trajectories for the joint angles (i.e., q_1 and q_2) generated by the online motion planning algorithm of Theorem 5.2, the absolute orientation (q_3) versus normalized time during the flight phases of four consecutive steps (solid curves) and the projection of the state variables onto the configuration space. The nominal trajectory is depicted by dashed curves. The circles at the both ends represent the initial and final predetermined configurations. (See color version of this figure in color plates section.)

CHAPTER 6

Stabilization of Periodic Orbits for 3D Monopedal Running

6.1 INTRODUCTION

This chapter presents a motion planning algorithm to generate periodic time trajectories for running by a 3D monopedal robot. In order to obtain a symmetric gait along a straight line, the overall open-loop model of running can be expressed as a hybrid system with four continuous phases consisting of two stance phases (the leg is on the ground) and two flight phases (the leg is off the ground), and discrete transitions between them (takeoff and impact). The robot studied here is a 3D, three-link, three-actuator, monopedal mechanism with a point foot. During the stance phases, the robot has three degrees of underactuation, whereas it has six degrees of underactuation in the flight phases. The motion planning algorithm is developed on the basis of a finite-dimensional nonlinear optimization problem with equality and inequality constraints. The main objective of this chapter is to develop time-invariant feedback scheme to exponentially stabilize a desired periodic orbit generated by the motion planning algorithm for the hybrid model of running. When the amount of underactuation during locomotion of legged robots is increased, it becomes difficult to create hybrid invariant manifolds. Reference [60] proposed a method to generate an open-loop augmented system with impulse effects, a parameterized holonomic output function for the resultant system, and an event-based update law for the parameters of the output such that the zero dynamics manifold associated with this output is hybrid invariant under the closed-loop augmented system. Recently, this approach has been used in the design of time-invariant controllers for walking of a 3D biped robot in Refs. [61, 62]. In this chapter, we show how to create hybrid invariant manifolds during 3D running [85]. By assuming that the control inputs of the mechanical system have discontinuities during discrete transitions between continuous phases, the takeoff switching hypersurface can be expressed as a zero-level set of a scalar holonomic function. In other words, takeoff occurs when a scalar quantity, a strictly increasing function of time on the desired gait, passes through a threshold value. The virtual constraints during stance

Hybrid Control and Motion Planning of Dynamical Legged Locomotion, Nasser Sadati, Guy A. Dumont,
Kaveh Akbari Hamed, and William A. Gruver.
© 2012 by the Institute of Electrical and Electronics Engineers, Inc. Published 2012 by John Wiley & Sons, Inc.

phases are defined as the summation of two terms including a nominal holonomic output function vanishing on the periodic orbit and an additive parameterized Bézier polynomial, both in terms of the latter strictly increasing scalar. By properties of Bézier polynomials, an update law for the parameters of the stance phase virtual constraints is developed, which, in turn, results in a common intersection of the parameterized stance phase zero dynamics manifolds and the takeoff switching hypersurface. By this approach, creation of hybrid invariance can be easily achieved by updating the other parameters of the Bézier polynomial. Consequently, a parameterized restricted Poincaré return map can be defined on the common intersection for studying the stabilization problem. Thus, the overall feedback scheme can be considered at two levels. At the first level, within-stride controllers including stance and flight phase controllers, which are continuous time-invariant and parameterized feedback laws, are employed to create a family of attractive zero dynamics manifolds in each of the continuous phases. At the second level, the parameters of the within-stride controllers are updated by event-based update laws during discrete transitions between continuous phases to achieve hybrid invariance and stabilization. By this means, the stability analysis of the periodic orbit for the full-order hybrid system can be treated in terms of a reduced-order hybrid system with a five-dimensional Poincaré return map.

6.2   OPEN-LOOP HYBRID MODEL OF A 3D RUNNING

In this section, we present an open-loop hybrid model for describing running by a 3D monopedal robot as illustrated in Fig. 6.1. The robot consists of three rigid links: a torso link and a leg with tibia and shin links. The links are connected by a set of actuated body joints including a two degree of freedom revolute hip joint and a one degree of freedom revolute knee joint. It is assumed that the robot has a point foot and cannot apply torques at the end of its leg. The body angles that determine the shape of the monoped are denoted by $\varphi := (\varphi_1, \varphi_2, \varphi_3)'$ as shown in Fig. 6.1, where prime represents matrix transpose. Assume that $o_0 x_0 y_0 z_0$ is an inertial *world frame* attached to the ground. Now attach the *torso frame* $o_t x_t y_t z_t$ rigidly to the torso link with the origin on its COM. The orientation of the torso frame relative to the world frame can be represented by the rotation matrix

$$R(\theta) = R_z(\theta_3)\, R_x(\theta_1)\, R_y(\theta_2),$$

in which $\theta := (\theta_1, \theta_2, \theta_3)'$ and R_x, R_y, and R_z are basic rotations about the x, y, and z axes, respectively. We assume that the *COM frame* $o_c x_c y_c z_c$ is a frame with the same orientation matrix $R(\theta)$ relative to the world frame such that its origin is on the COM of the robot. Let $p_{cm} := (x_{cm}, y_{cm}, z_{cm})'$ denote the position of the COM relative to the world frame. For describing the configuration of the mechanical system during the flight phase, the vector of generalized coordinates can be defined as $q_f := (\varphi', \theta', p'_{cm})' = (q', p'_{cm})'$, where $q := (\varphi', \theta')'$ and the subscript "f" will designate the flight phase. The configuration space for the flight phase is also denoted by $\mathcal{Q}_f$.

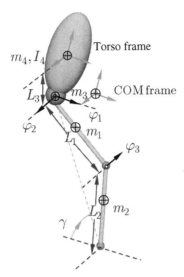

Figure 6.1   A three-link, three-actuator 3D monopedal robot with point foot. The body joints are denoted by $\varphi := (\varphi_1, \varphi_2, \varphi_3)'$. The orientation of the torso and COM frames relative to the world frame can be expressed by the rotation matrix $R(\theta) = R_z(\theta_3)R_x(\theta_1)R_y(\theta_2)$, where $\theta := (\theta_1, \theta_2, \theta_3)'$. R can be considered as the following sequence of basic rotations in the order specified: (i) a rotation of θ_3 about the fixed z-axis, (ii) a rotation of θ_1 about the current x-axis, and (iii) a rotation of θ_2 about the current y-axis. The virtual leg is depicted as a dashed line connecting the end of the leg and the hip joint. The angle of the virtual leg in the sagittal plane can be expressed as $\gamma = -\varphi_2 + \frac{\varphi_3}{2} + \theta_3 + \frac{\pi}{2}$.

Let $p_i \in \mathbb{R}^3$ for $i = 1, 2, 4$ represent the position of the COM of the tibia, shin, and torso links relative to the world frame, respectively. The masses of these links are also given by m_i for $i = 1, 2, 4$. Furthermore, $p_3 \in \mathbb{R}^3$ denotes the position of the lumped mass m_3 located at the hip joint. Let $\Gamma_i(q), i = 1, 2, 3, 4$ represent the Cartesian coordinates of the COM with respect to the mass m_i. Then, p_i can be expressed as

$$p_i = p_{\text{cm}} - \Gamma_i(q) = p_{\text{cm}} - R(\theta)\,\Upsilon_i(\varphi), \qquad i = 1, 2, 3, 4, \tag{6.1}$$

where $\Upsilon_i \in \mathbb{R}^3, i = 1, 2, 3, 4$ are smooth functions with respect to φ with the property $\sum_{i=1}^{4} m_i \Upsilon_i(\varphi) = 0$. Since $\sum_{i=1}^{4} m_i p_i / \sum_{i=1}^{4} m_i = p_{\text{cm}}$, it can be concluded that $\sum_{i=1}^{4} m_i \Gamma_i(q) = 0$ and $\sum_{i=1}^{4} m_i \Upsilon_i(\varphi) = 0$. In an analogous manner, the position of the leg end with respect to the world frame can be expressed as $p_l = p_{\text{cm}} - \Gamma_l(q) = p_{\text{cm}} - R(\theta)\Upsilon_l(\varphi)$, where the subscript "$l$" denotes the leg end. Now suppose that during the stance phase, the leg end is on the origin of the world frame, then p_{cm} can be expressed as $p_{\text{cm}} = \Gamma_l(q)$. Consequently, the generalized coordinates for the stance phase is given by $q_s := q = (\varphi', \theta')'$, where the subscript "$s$" denotes the stance phase.

6.2.1 Dynamics of the Flight Phase

Throughout this chapter, all of the link masses except the mass of the torso link are assumed to be lumped. Let I_4 denote the inertia tensor expressed in the torso frame. During the flight phase, the angular velocity vector of the torso link relative to the world frame is given by $\omega = (\omega_x, \omega_y, \omega_z)' = \Omega(\theta)\dot{\theta}$, where

$$\Omega(\theta) := \begin{bmatrix} \cos(\theta_3) & -\sin(\theta_3)\cos(\theta_1) & 0 \\ \sin(\theta_3) & \cos(\theta_3)\cos(\theta_1) & 0 \\ 0 & \sin(\theta_1) & 1 \end{bmatrix}. \tag{6.2}$$

Considering the fact that $\sum_{i=1}^{4} m_i \Gamma_i(q) = 0$, the positive definite quadratic kinetic energy of the mechanical system during the flight phase, $\mathcal{K}_f : T\mathcal{Q}_f \to \mathbb{R}$, can be expressed as

$$\mathcal{K}_f(q_f, \dot{q}_f) = \frac{1}{2}\sum_{i=1}^{4} m_i \, \dot{p}'_i \, \dot{p}_i + \frac{1}{2}\omega' \, R \, I_4 \, R' \, \omega = \frac{1}{2}\dot{q}'_f \, D_f(q) \, \dot{q}_f,$$

where $D_f(q) := \text{block diag}\{A(q), m_{\text{tot}} I_{3\times3}\} \in \mathbb{R}^{9\times9}$, $m_{\text{tot}} := \sum_{i=1}^{4} m_i$, and

$$A(q) := \sum_{i=1}^{4} m_i \frac{\partial \Gamma_i}{\partial q}' \frac{\partial \Gamma_i}{\partial q} + \begin{bmatrix} 0_{3\times3} & 0_{3\times3} \\ 0_{3\times3} & \Omega' \, R \, I_4 \, R' \, \Omega \end{bmatrix} \in \mathbb{R}^{6\times6}. \tag{6.3}$$

By introducing the Lagrangian of the flight phase as $\mathcal{L}_f(q_f, \dot{q}_f) := \mathcal{K}_f(q_f, \dot{q}_f) - \mathcal{V}_f(q_f)$, where $\mathcal{V}_f : \mathcal{Q}_f \to \mathbb{R}$ by

$$\mathcal{V}_f(q_f) := m_{\text{tot}} \, g_0 \, z_{\text{cm}}$$

is the total potential energy of the robot and g_0 is the gravitational constant, the second-order dynamical equation of motions during the flight phase can be expressed as

$$D_f(q)\ddot{q}_f + C_f(q, \dot{q}_f)\dot{q}_f + G_f(q_f) = B_f u, \tag{6.4}$$

in which, $C_f(q, \dot{q}_f)$ is a (9×9) Coriolis and centrifugal matrix, $G_f(q_f)$ is a (9×1) gravity vector, and $u := (u_1, u_2, u_3)' \in \mathbb{R}^{3\times1}$ is the vector of applied torques at the body joints. Moreover,

$$B_f := \begin{bmatrix} B \\ 0_{3\times3} \end{bmatrix} := \begin{bmatrix} \begin{bmatrix} I_{3\times3} \\ 0_{3\times3} \end{bmatrix} \\ 0_{3\times3} \end{bmatrix} \in \mathbb{R}^{9\times3}.$$

Using the Christoffel symbols [90, p. 256] and the block-diagonal form of the mass-inertia matrix, it can be shown that the Coriolis matrix C_f also has the block diagonal form as

$$C_f(q, \dot{q}_f) = \text{block diag}\{C(q, \dot{q}), 0_{3\times 3}\}.$$

Due to the block diagonal form of the mass-inertia matrix (i.e., D_f) and Coriolis and centrifugal (i.e., C_f) matrix, $q_f = (q', p'_{cm})'$ and

$$G_f(q_f) = \frac{\partial V_f'}{\partial q_f} = m_{tot}\, g_0 \begin{bmatrix} 0_{8\times 1} \\ 1 \end{bmatrix},$$

the equation of motion in (6.4) can be decomposed as follows:

$$A(q)\ddot{q} + C(q, \dot{q})\dot{q} = Bu$$

$$m_{tot}\,\ddot{p}_{cm} = m_{tot} \begin{bmatrix} 0 \\ 0 \\ -g_0 \end{bmatrix}. \tag{6.5}$$

Remark 6.1 (Cyclic Variables of the Flight Phase) *During the flight phase, θ_3 (the orientation about the z-axis), x_{cm}, y_{cm}, and z_{cm} are cyclic variables [1] in the sense that $\frac{\partial \mathcal{K}_f}{\partial \theta_3} = \frac{\partial \mathcal{K}_f}{\partial x_{cm}} = \frac{\partial \mathcal{K}_f}{\partial y_{cm}} = \frac{\partial \mathcal{K}_f}{\partial z_{cm}} = 0$. Thus, D_f and C_f in equation (6.4) are independent of them. By introducing the state vector $x_f := (q'_f, \dot{q}'_f)'$ for the flight phase, equation (6.4) can be expressed in a state space form $\dot{x}_f = f_f(x_f) + g_f(x_f)u$. Moreover, the state space for the flight phase is taken as the tangent bundle of $\mathcal{Q}_f$, that is,*

$$\mathcal{X}_f := T\mathcal{Q}_f := \{(q'_f, \dot{q}'_f)' | q_f \in \mathcal{Q}_f, \dot{q}_f \in \mathbb{R}^9\}.$$

6.2.2 Dynamics of the Stance Phase

During the stance phase, we assume that the position of the leg end is on the origin of the world frame (i.e., $p_{cm} = \Gamma_l(q)$). Let $F := (F_x, F_y, F_z)' \in \mathbb{R}^{3\times 1}$ represent the ground reaction force at the leg end. Then, by applying the principle of virtual work,[1] equation (6.5) can be reduced as follows:

$$D_s(q_s)\ddot{q}_s + C_s(q_s, \dot{q}_s)\dot{q}_s + G_s(q_s) = Bu$$

$$m_{tot}\frac{\partial \Gamma_l}{\partial q_s}(q_s)\ddot{q}_s + m_{tot}\frac{\partial}{\partial q_s}\left(\frac{\partial \Gamma_l}{\partial q_s}(q_s)\dot{q}_s\right)\dot{q}_s + m_{tot}\begin{bmatrix} 0 \\ 0 \\ g_0 \end{bmatrix} = F, \tag{6.6}$$

---

[1] Applying the principle of the virtual work in equation (6.5) results in $A\ddot{q} + C\dot{q} = Bu - \partial\Gamma_l/\partial q\, F$ and $m_{tot}\ddot{p}_{cm} = m_{tot}[0\ \ 0\ \ -g_0]' + F$, which together with $p_{cm} = \Gamma_l(q)$ yield equation (6.6).

where

$$D_s := A + m_{\text{tot}} \frac{\partial \Gamma_l}{\partial q_s}' \frac{\partial \Gamma_l}{\partial q_s} \in \mathbb{R}^{6 \times 6}$$

$$C_s := C + m_{\text{tot}} \frac{\partial \Gamma_l}{\partial q_s}' \frac{\partial}{\partial q_s} \left(\frac{\partial \Gamma_l}{\partial q_s} \dot{q}_s \right) \in \mathbb{R}^{6 \times 6}$$

$$G_s := m_{\text{tot}} \frac{\partial \Gamma_l}{\partial q_s}' \begin{bmatrix} 0 \\ 0 \\ g_0 \end{bmatrix} \in \mathbb{R}^{6 \times 1}.$$

Remark 6.2 (The Cyclic Variable of the Stance Variable) *The configuration manifold of the stance phase can be expressed as the submanifold $\mathcal{Q}_s := \{q_f \in \mathcal{Q}_f | p_l(q_f) = 0_{3 \times 1}\}$ and the stance phase Lagrangian can be given by $\mathcal{L}_s := \mathcal{L}_f | T\mathcal{Q}_s$. Moreover, θ_3 is the cyclic variable during the stance phase in the sense that $\frac{\partial \mathcal{K}_s}{\partial \theta_3} = 0$. Thus, D_s and C_s in equation (6.6) are independent of θ_3.*

Remark 6.3 (Validity of the Stance Phase) *The stance phase model is valid if the ground reaction force, F, satisfies the constrains (i) $F_z > 0$ (to avoid takeoff) and (ii) $|\sqrt{F_x^2 + F_y^2}/F_z| < \mu_s$ (to avoid slipping), where μ_s denotes the static friction coefficient between the leg end and the ground. By defining the state vector $x_s := (q_s', \dot{q}_s')' \in \mathcal{X}_s$, where*

$$\mathcal{X}_s := T\mathcal{Q}_s := \{(q_s', \dot{q}_s')' | q_s \in \mathcal{Q}_s, \dot{q}_s \in \mathbb{R}^6\},$$

a state equation for describing the evolution of the mechanical system during the stance phase can be expressed as $\dot{x}_s = f_s(x_s) + g_s(x_s)u$.

6.2.3 Transition Maps

The transition from flight to stance (impact) takes place when the height of the leg end becomes zero. Thus, define the impact switching hypersurface as

$$\mathcal{S}_f^s := \{x_f = (q_f', \dot{q}_f')' \in \mathcal{X}_f | p_l^v(q_f) := z_{\text{cm}} - \Gamma_l^v(q) = 0\},$$

where the superscript "v" denotes the vertical component of the position vector. This transition can be modeled as $x_s^+ = \Delta_f^s(x_f^-)$, where $\Delta_f^s : \mathcal{S}_f^s \to \mathcal{X}_s$ denotes the impact map and the superscripts "$-$" and "$+$" represent the state of the mechanical system just before and after the transition, respectively. By extending the planar impact model presented in Ref. [18, pp. 74–75] for the mechanical system of Fig. 6.1, Δ_f^s can be expressed as

$$\Delta_f^s(x_f^-) := \begin{bmatrix} [I_{6 \times 6} \ 0_{6 \times 3}] q_f^- \\ \tilde{\Delta}_f^s(q_f^-) \dot{q}_f^- \end{bmatrix}, \tag{6.7}$$

in which

$$\tilde{\Delta}_f^s(q_f^-) := \left(A + m_{\text{tot}} \frac{\partial \Gamma_l'}{\partial q} \frac{\partial \Gamma_l}{\partial q} \right)^{-1} \left[A \; m_{\text{tot}} \frac{\partial \Gamma_l'}{\partial q} \right].$$

Remark 6.4 (Validity of the Impact Model) *The impact model is valid if the intensity of the impulsive ground reaction force, $F := (F_x, F_y, F_z)'$, satisfies the constraints (i) $F_z > 0$ and (ii) $|\sqrt{F_x^2 + F_y^2}/F_z| < \mu_s$. In addition, F can be given by*

$$F = m_{\text{tot}}(\dot{p}_{\text{cm}}^+ - \dot{p}_{\text{cm}}^-) = m_{\text{tot}} \left(\frac{\partial \Gamma_l}{\partial q}(q^-) \tilde{\Delta}_f^s(q_f^-) \dot{q}_f^- - \dot{p}_{\text{cm}}^- \right). \tag{6.8}$$

The transition from stance to flight (takeoff) occurs when the vertical component of the leg end acceleration becomes positive. However for simplicity, it is assumed that this transition takes place when the angle of the virtual leg in the sagittal plane that is denoted by γ (see Fig. 6.1) passes through the threshold value γ^-. We remark that for the mechanical system of Fig. 6.1,[2] γ can be expressed as

$$\gamma(q_s) = -\varphi_2 + \frac{\varphi_3}{2} + \theta_3 + \frac{\pi}{2}.$$

Consequently, the takeoff switching hypersurface can be defined as

$$S_s^f := \{ x_s = (q_s', \dot{q}_s')' \in \mathcal{X}_s | \gamma(q_s) - \gamma^- = 0 \}.$$

In addition, by assuming that the position and velocity remain continuous during this transition, the takeoff map $\Delta_s^f : S_s^f \to \mathcal{X}_f$ can be expressed as

$$\Delta_s^f(x_s^-) := \begin{bmatrix} \begin{bmatrix} q_s^- \\ \Gamma_l(q_s^-) \end{bmatrix} \\ \begin{bmatrix} \dot{q}_s^- \\ \frac{\partial \Gamma_l}{\partial q}(q_s^-) \dot{q}_s^- \end{bmatrix} \end{bmatrix}. \tag{6.9}$$

Remark 6.5 (Validity of the Takeoff Model) *The transition model from stance to flight is valid if the feedback law for the flight phase is designed such that the vertical component of the leg end acceleration is positive at the beginning of the flight phase (i.e., $\ddot{p}_l^v > 0$). To achieve this objective, it is assumed that control inputs may have discontinuities during transitions between continuous phases.*

---

[2] It is assumed that the lengths of the tibia and shin links are equal, that is, $L_1 = L_2$.

6.2.4 Hybrid Model

A symmetric monopedal running motion along the x-axis of the world frame can be considered as a periodic orbit composed of two consecutive steps

$$\mathcal{O} = \underbrace{\mathcal{O}_s^1 \cup \mathcal{O}_f^1}_{\text{step 1}} \cup \underbrace{\mathcal{O}_s^2 \cup \mathcal{O}_f^2}_{\text{step 2}}$$

in the overall state space of the mechanical system, in which $\mathcal{O}_i^j$ for $i \in \{s, f\}$ and $j \in \{1, 2\}$ denotes the intersection of $\mathcal{O}$ and the state space of the jth continuous phase of type i. This is resulted from the property that the leg would be in the left-hand side of the x-axis during a step if it was on the right-hand side of the same axis in the previous step. In other words, if we assume that T_s^* and T_f^* are the time durations of the stance and flight phases on a step of $\mathcal{O}$ and define $T^* := T_s^* + T_f^*$ as the time duration of a step, then symmetry implies that for every $t \geq 0$ on the periodic orbit

$$\begin{aligned} \varphi(t + T^*) &= S_1\, \varphi(t) \\ \theta(t + T^*) &= S_2\, \theta(t) \\ p_{\text{cm}}(t + T^*) &= S_3\, p_{\text{cm}}(t), \end{aligned} \tag{6.10}$$

where $S_1 := \text{diag}\{-1, 1, 1\}$, $S_2 := \text{diag}\{-1, 1, -1\}$, and $S_3 := \text{diag}\{1, -1, 1\}$ (see Fig. 6.2 as a typical periodic motion with $T^* = 0.5730(s)$). This fact motivates us to study the period-one solutions and stabilization for the following open-loop hybrid system composed of four continuous phases:

$$\Sigma_s^1 : \begin{cases} \dot{x}_s^1 = f_s\left(x_s^1\right) + g_s\left(x_s^1\right) u & x_s^{1-} \notin \mathcal{S}_s^f \\ x_f^{1+} = \Delta_s^f\left(x_s^{1-}\right) & x_s^{1-} \in \mathcal{S}_s^f \end{cases}$$

$$\Sigma_f^1 : \begin{cases} \dot{x}_f^1 = f_f\left(x_f^1\right) + g_f\left(x_f^1\right) u & x_f^{1-} \notin \mathcal{S}_f^s \\ x_s^{2+} = \Delta_f^s\left(x_f^{1-}\right) & x_f^{1-} \in \mathcal{S}_f^s \end{cases}$$

$$\Sigma_s^2 : \begin{cases} \dot{x}_s^2 = f_s\left(x_s^2\right) + g_s\left(x_s^2\right) u & x_s^{2-} \notin \mathcal{S}_s^f \\ x_f^{2+} = \Delta_s^f\left(x_s^{2-}\right) & x_s^{2-} \in \mathcal{S}_s^f \end{cases} \tag{6.11}$$

$$\Sigma_f^2 : \begin{cases} \dot{x}_f^2 = f_f\left(x_f^2\right) + g_f\left(x_f^2\right) u & x_f^{2-} \notin \mathcal{S}_f^s \\ x_s^{1+} = \Delta_f^s\left(x_f^{2-}\right) & x_f^{2-} \in \mathcal{S}_f^s, \end{cases}$$

where the notation $(.)_i^j$ for $i \in \{s, f\}$ and $j \in \{1, 2\}$ corresponds to the jth continuous phase of type i. In addition, the superscripts "$-$" and "$+$" denote the state just before and after the discrete transitions.

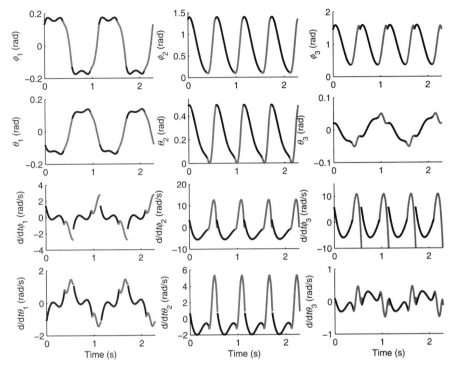

Figure 6.2 Angular positions and velocities during four consecutive steps of the optimal motion ($T^* = 0.5730(s)$). Bold and light curves correspond to stance and flight phases, respectively. Discontinuities are due to impacts. (See the color version of this figure in the color plates section.)

6.3 DESIGN OF A PERIOD-ONE SOLUTION FOR THE OPEN-LOOP MODEL OF RUNNING

The objective of this section is to present a motion planning algorithm based on a finite-dimensional nonlinear optimization problem with equality and inequality constraints to generate a feasible period-one solution for the open-loop hybrid model of running in equation (6.11). We first present the following definition.

Definition 6.1 (Feasible Periodic Orbit) *The periodic orbit $\mathcal{O}$ of the open-loop hybrid model in equation (6.11) is said to be feasible if*

1. *the constraints on the joint angles representing an anthropomorphic gait and the constraints on the angular velocities representing the actuation limits are satisfied on $\mathcal{O}$;*

2. *the open-loop control input corresponding to the trajectory $\mathcal{O}$ is admissible in the sense that $\|u\|_{\mathcal{L}_\infty} := \sup_{t \geq 0} \|u(t)\|$ is less than a physically realizable value $u_{\max} > 0$;*

3. *the ground reaction forces during stance phases of the orbit $\mathcal{O}$ are admissible as stated in Remark 6.3;*

4. *the intensity of impulsive ground reaction forces during impacts are admissible as stated in Remark 6.4;*

5. *the vertical component of the leg end acceleration at the beginning of flight phases is positive;*

6. *the height of the leg end during flight phases is positive.*

Now let $N \geq 1$ and $M \geq 0$ be two integer numbers. For any matrix $\alpha = [\alpha_0 \cdots \alpha_M] = \text{col}\{\alpha_i\}_{i=0}^M \in \mathbb{R}^{N \times (M+1)}$, the Bézier polynomial with the parameter α denoted by $\mathfrak{B}(., \alpha) : \mathbb{R} \to \mathbb{R}^N$ is defined as follows:

$$\mathfrak{B}(s, \alpha) := \sum_{i=0}^{M} \frac{M!}{i!(M-i)!} \alpha_i s^i (1-s)^{M-i}. \tag{6.12}$$

To develop the motion planning algorithm, assume that

$$x_s^{1-*} := \begin{bmatrix} q_s^{1-*} \\ \dot{q}_s^{1-*} \end{bmatrix} \in \mathcal{S}_s^f, \quad x_f^{1-*} := \begin{bmatrix} q_f^{1-*} \\ \dot{q}_f^{1-*} \end{bmatrix} \in \mathcal{S}_f^s \tag{6.13}$$

represent the states of the mechanical system at the end of the first stance and flight phases of the periodic orbit $\mathcal{O}$ (just before the corresponding transitions), respectively, where the superscript "$*$" denotes the quantities corresponding to the periodic orbit. By using equation (6.10) to generate a symmetric periodic gait along the x-axis, the final states of the second stance and flight phases (i.e., x_s^{2-*} and x_f^{2-*}) are defined as follows:

$$\begin{aligned} x_s^{2-*} &:= \mathbf{S}_s \, x_s^{1-*} \\ x_f^{2-*} &:= \mathbf{S}_f \, x_f^{1-*}, \end{aligned} \tag{6.14}$$

where

$$\mathbf{S}_s := \text{block diag}\{S_1, S_2, S_1, S_2\} \in \mathbb{R}^{12 \times 12}$$
$$\mathbf{S}_f := \text{block diag}\{S_1, S_2, S_3, S_1, S_2, S_3\} \in \mathbb{R}^{18 \times 18}.$$

We will assume that phases are executed in a fixed order

$$\text{stance}_1 \to \text{flight}_1 \to \text{stance}_2 \to \text{flight}_2 \to \text{stance}_1. \tag{6.15}$$

According to the order in equation (6.15) and transition maps in equations (6.7) and (6.9), the states just after the transitions on each of the continuous phases of $\mathcal{O}$ can be given by

$$x_s^{1+*} = \Delta_f^s \left(x_f^{2-*} \right) \qquad x_s^{2+*} = \Delta_f^s \left(x_f^{1-*} \right)$$
$$x_f^{1+*} = \Delta_s^f \left(x_s^{1-*} \right) \qquad x_f^{2+*} = \Delta_s^f \left(x_s^{2-*} \right). \tag{6.16}$$

Since the body joints are independently actuated, we choose a Bézier polynomial evolution of time for the body angles φ during the stance and flight phases. For this purpose, define the scaled times

$$s_s := \frac{t}{T_s^*}, \qquad s_f := \frac{t - T_s^*}{T_f^*}.$$

Suppose that $M_s \geq 3$ and $M_f \geq 3$ are the degrees of the Bézier polynomials during the stance and flight phases, respectively. In addition, let the matrices $\alpha^{1*}, \alpha^{2*} \in \mathbb{R}^{3 \times (M_s+1)}$ and $\beta^{1*}, \beta^{2*} \in \mathbb{R}^{3 \times (M_f+1)}$ be the parameters of the Bézier polynomials during the first and second stance and flight phases of the periodic orbit $\mathcal{O}$. We remark that symmetric gait conditions in equation (6.10) imply that $\alpha^{2*} = S_1 \alpha^{1*}$ and $\beta^{2*} = S_1 \beta^{1*}$. Next, let

$$\varphi(t) = \mathcal{B} \left(s_s, \alpha^{1*} \right), \qquad 0 \leq t \leq T_s^*$$
$$\varphi(t) = \mathcal{B} \left(s_f, \beta^{1*} \right), \qquad T_s^* \leq t \leq T_s^* + T_f^* = T^*. \tag{6.17}$$

Following properties of Bézier polynomials given in Remark 3.16, the initial and final conditions on the angular position and velocity of the body joints can be taken into account in the following manner:

$$\alpha_0^{1*} = \left(q_s^{1+*} \right)_\varphi \qquad\qquad \beta_0^{1*} = \left(q_f^{1+*} \right)_\varphi$$
$$\alpha_1^{1*} = \alpha_0^{1*} + \frac{T_s^*}{M_s} \left(\dot{q}_s^{1+*} \right)_\varphi \qquad \beta_1^{1*} = \beta_0^{1*} + \frac{T_f^*}{M_f} \left(\dot{q}_f^{1+*} \right)_\varphi$$
$$\alpha_{M_s}^{1*} = \left(q_s^{1-*} \right)_\varphi \qquad\qquad \beta_{M_f}^{1*} = \left(q_f^{1-*} \right)_\varphi \tag{6.18}$$
$$\alpha_{M_s-1}^{1*} = \alpha_{M_s}^{1*} - \frac{T_s^*}{M_s} \left(\dot{q}_s^{1-*} \right)_\varphi \quad \beta_{M_f-1}^{1*} = \beta_{M_f}^{1*} - \frac{T_f^*}{M_f} \left(\dot{q}_f^{1-*} \right)_\varphi,$$

where the subscript "φ" denotes the components corresponding to the body joints. Moreover, if we define $H(q, \dot{q}) := C_s(q, \dot{q})\dot{q} + G_s(q)$ and decompose dynamical equation (6.6) into φ and θ components, the evolution of θ during the stance phase is given by

$$\ddot{\theta} = -D_{s,\theta\theta}^{-1}(q) \, D_{s,\theta\varphi}(q) \, \ddot{\varphi} - D_{s,\theta\theta}^{-1}(q) \, H_\theta(q, \dot{q}), \tag{6.19}$$

where $D_{s,\theta\theta}$ and $D_{s,\theta\varphi}$ are the (3×3) lower right and left submatrices of D_s, respectively. Furthermore, H_θ represents the last three rows of H. We remark that in this latter equation, φ can be considered as an input. A similar procedure during the flight phase (see equation (6.5)) yields

$$\ddot{\theta} = -A_{\theta\theta}^{-1}(q) A_{\theta\varphi}(q) \ddot{\varphi} - A_{\theta\theta}^{-1}(q) J_\theta(q, \dot{q})$$

$$m_{\text{tot}} \ddot{p}_{\text{cm}} = m_{\text{tot}} \begin{bmatrix} 0 \\ 0 \\ -g_0 \end{bmatrix}, \tag{6.20}$$

in which $J(q, \dot{q}) := C(q, \dot{q})\dot{q}$. Now we are in a position to present the following lemma that states that for the motion planning algorithm, it is sufficient to generate and study the first step of $\mathcal{O}$.

Lemma 6.1 (Symmetric Gait) *For the mechanical system of Fig. 6.1, the following statements are true.*

1. *Assume that $\theta^*(t), 0 \leq t \leq T_s^*$ is the unique solution of equation (6.19) when $\varphi(t) = \varphi^*(t)$. Then, for the initial condition*

$$\theta(0) = S_2 \, \theta^*(0)$$
$$\dot{\theta}(0) = S_2 \, \dot{\theta}^*(0)$$

   and the input

$$\varphi(t) = S_1 \, \varphi^*(t), \quad 0 \leq t \leq T_s^*,$$

   the trajectory

$$\theta(t) = S_2 \, \theta^*(t)$$

   is the unique solution for (6.19).

2. *Assume that $(\theta^*(t), p_{\text{cm}}^*(t))$ for $T_s^* \leq t \leq T^*$ is the unique solution of equation (6.20) when $\varphi(t) = \varphi^*(t)$. Then, for the initial condition*

$$(\theta(T_s^*), p_{\text{cm}}(T_s^*)) = (S_2 \, \theta^*(T_s^*), S_3 \, p_{\text{cm}}^*(T_s^*))$$
$$(\dot{\theta}(T_s^*), \dot{p}_{\text{cm}}(T_s^*)) = (S_2 \, \dot{\theta}^*(T_s^*), S_3 \, \dot{p}_{\text{cm}}^*(T_s^*))$$

   and the input

$$\varphi(t) = S_1 \, \varphi^*(t), \quad T_s^* \leq t \leq T^*,$$

   the trajectory

$$(\theta(t), p_{\text{cm}}(t)) = (S_2 \, \theta^*(t), S_3 \, p_{\text{cm}}^*(t))$$

   is the unique solution of (6.20).

The proof is given in Appendix C.1. On the basis of Lemma 6.1, we consider the motion planning algorithm only during the first stance and flight phases ($\mathcal{O}_s^1$ and $\mathcal{O}_f^1$). By this approach, $\mathcal{O}_s^2$ and $\mathcal{O}_f^2$ can be obtained by applying the linear maps $\mathbf{S}_s$ and $\mathbf{S}_f$ on $\mathcal{O}_s^1$ and $\mathcal{O}_f^1$, respectively, that is,

$$\mathcal{O}_s^2 = \mathbf{S}_s \, \mathcal{O}_s^1 , \qquad \mathcal{O}_f^2 = \mathbf{S}_f \, \mathcal{O}_f^1.$$

Consequently, the evolution of the mechanical system on the periodic orbit $\mathcal{O}$ can be completely determined by the following vector of parameters:

$$\Theta^* := \left(q_s^{1-*\prime}, \dot{q}_s^{1-*\prime}, q_f^{1-*\prime}, \dot{q}_f^{1-*\prime}, T_s^*, T_f^*, \alpha_2^{1*\prime}, \cdots \alpha_{M_s-2}^{1*\prime}, \beta_2^{1*\prime}, \ldots, \beta_{M_f-2}^{1*\prime} \right)',$$
(6.21)

which are utilized in the first step. Next, we present the following motion planning algorithm.

Algorithm 6.1 *Motion Planning Algorithm for Generating a Periodic Orbit*

1. Choose the degrees of the Bézier polynomials as $M_s \geq 3$ and $M_f \geq 3$.
2. Select Θ^* and using equations (6.13), (6.14), and (6.16) calculate the initial conditions for the first stance and flight phases. From equation (6.18), calculate the first and last two columns of the parameter matrices $\alpha^{1*} = [\alpha_0^{1*} \cdots \alpha_{M_s}^{1*}] = \mathrm{col}\{\alpha_i^{1*}\}_{i=0}^{M_s}$ and $\beta^{1*} = [\beta_0^{1*} \cdots \beta_{M_f}^{1*}] = \mathrm{col}\{\beta_i^{1*}\}_{i=0}^{M_f}$.
3. Integrate equation (6.19) on the interval $[0, T_s^*]$ with the initial condition obtained in Step 2 for the first stance phase. Calculate the open-loop control input and the ground reaction force by applying equation (6.6).
4. Integrate equation (6.20) on the interval $[T_s^*, T^*]$ with the initial condition obtained in Step 2 for the first flight phase. Calculate the open-loop control input by applying equation (6.5). Obtain the vertical acceleration of the leg end at the beginning of the flight phase as follows:

$$\ddot{p}_l^v = -g_0 - \frac{\partial \Gamma_l^v}{\partial q} \ddot{q} - \frac{\partial}{\partial q}\left(\frac{\partial \Gamma_l^v}{\partial q} \dot{q} \right) \dot{q}.$$

5. Since θ and (θ, p_{cm}) are unactuated degrees of freedom during the stance and flight phases, to have a periodic orbit for the open-loop hybrid model of running (6.11), Θ^* should be designed such that the final values of the position and velocity vectors corresponding to these DOF are equal to their predetermined values that are given in Θ^*. To achieve this goal, introduce the equality constraint vector $c_e(\Theta^*)$ and evaluate its components as follows:

$$\begin{aligned}
c_{e,1}(\Theta^*) &:= \theta(T_s^*) - \left(q_s^{1-*}\right)_\theta & c_{e,5}(\Theta^*) &:= p_{cm}(T^*) - \left(q_f^{1-*}\right)_{p_{cm}} \\
c_{e,2}(\Theta^*) &:= \dot{\theta}(T_s^*) - \left(\dot{q}_s^{1-*}\right)_\theta & c_{e,6}(\Theta^*) &:= \dot{p}_{cm}(T^*) - \left(\dot{q}_f^{1-*}\right)_{p_{cm}} \\
c_{e,3}(\Theta^*) &:= \theta(T^*) - \left(q_f^{1-*}\right)_\theta & c_{e,7}(\Theta^*) &:= p_l^v\left(q_f^{1-*}\right), \\
c_{e,4}(\Theta^*) &:= \dot{\theta}(T^*) - \left(\dot{q}_f^{1-*}\right)_\theta &
\end{aligned}$$
(6.22)

where the subscripts "θ" and "p_{cm}" denote those components corresponding to θ and p_{cm}, respectively.

6. Evaluate a cost function and an inequality constraint vector $c_{ie}(\Theta^*)$ such that $c_{ie}(\Theta^*) \leq 0$ guarantees the feasibility of the optimal trajectory (as stated in Definition 6.1) and invertibility of the decoupling matrix during the stance phase (to be stated in Remark 6.7).

7. Repeat Steps 2–6 until $c_e(\Theta^*) = 0$, $c_{ie}(\Theta^*) \leq 0$ and the cost function (6.23) is minimized.

Remark 6.6 (Interpretation of the Equality Constraints) *The equality constraints $c_{e,j}(\Theta^*) = 0$, $j = 1, \ldots, 6$ are necessary and sufficient conditions by which the open-loop hybrid model of running (6.11) has a period-one solution. Moreover, $c_{e,7}(\Theta^*) = p_l^v(q_f^{1-*}) = 0$ in equation (6.22) implies that the height of the leg end at the end of the flight phase is zero, that is, $x_f^{1-*} \in S_f^s$. We also observe that the threshold γ^- and thereby S_s^f are determined on the basis of the periodic orbit $\mathcal{O}$.*

6.4 NUMERICAL EXAMPLE

This section presents a numerical example for the motion planning algorithm developed in Section 6.3. The physical parameters of the monopedal robot in Fig. 6.1 are given in Table 6.1.[3] We assume that the inertia tensor in the torso frame can be given by $I_4 = \text{diag}\{I_{4,xx}, I_{4,yy}, I_{4,zz}\}$. In the optimization problem, it is assumed that the body angles φ_1, φ_2, and φ_3 lie in the intervals $[-10°, 10°]$, $[-80°, 80°]$, and $[1°, 120°]$, respectively. Also, the orientations θ_1, θ_2, and θ_3 lie in $[-10°, 10°]$, $[0°, 60°]$, and $[-30°, 30°]$, respectively. Moreover, suppose that T_s^*, $T_f^* \in [0.1, 1](s)$ and α_i^{l*}, $\beta_j^{l*} \in [-1, 1]^3$ for $i = 2, \ldots, M_s - 2$, $j = 2, \ldots$, and $M_f - 2$. The static friction coefficient between the end of the leg and the ground is equal to $\mu = \frac{2}{3}$. Furthermore, according to the actuation limits and considering the gear reduction at the body joints, we choose $u_{\text{max}} = 300$ (Nm) and the maximum absolute value of the joint angular velocities as 20 (rad/s). A two-stage strategy is used to solve the motion planning algorithm. In the first stage, the cost function is chosen as 1 and by using the `fmincon` function of MATLAB's Optimization Toolbox, we search for a feasible periodic solution of the open-loop hybrid model of equation (6.11) that will be used in the next stage as an initial guess. By using the `fmincon` function, the motion planning algorithm during the second stage is continued to minimize the following desired cost function:

$$\mathcal{I}(\Theta^*) := \frac{1}{L_s} \int_0^{T^*} \|u(t)\|_2^2 \, dt, \tag{6.23}$$

---

[3] In Table 6.1, d_1 and d_2 represent the distances between the lumped masses m_1 and m_2 and the hip and knee joints, respectively.

TABLE 6.1 Physical Parameters of the Monopedal Robot

m_1(kg)	m_2(kg)	m_3(kg)	m_4(kg)	d_1(m)	d_2(m)
1	1	5	0.5	0.25	0.25

L_1(m)	L_2(m)	L_3(m)	$I_{4,xx}$(kgm^2)	$I_{4,yy}$(kgm^2)	$I_{4,zz}$(kgm^2)
0.5	0.5	0.5	0.2	0.2	0.2

TABLE 6.2 Components of q_s^{1-*} and $\dot{q}_s^{1-*}$

φ_1(rad)	0.1534	$\dot{\varphi}_1(\frac{rad}{s})$	−0.8932
φ_2(rad)	0.1142	$\dot{\varphi}_2(\frac{rad}{s})$	−0.9604
φ_3(rad)	0.3738	$\dot{\varphi}_3(\frac{rad}{s})$	0.9428
θ_1(rad)	−0.1169	$\dot{\theta}_1(\frac{rad}{s})$	0.8591
θ_2(rad)	0.0427	$\dot{\theta}_2(\frac{rad}{s})$	−0.8987
θ_3(rad)	−0.0404	$\dot{\theta}_3(\frac{rad}{s})$	−0.2346

where L_s represents the step length. By choosing $M_s = 4$ and $M_f = 4$, a local minimum is obtained with the components given in Tables 6.2–6.5. The value of the cost function evaluated in the optimal point is also equal to $\mathcal{I}(\Theta^*) = 3.9150 \times 10^3 (N^2ms)$. The optimal running motion has a period of $2T^* = 2(T_s^* + T_f^*) = 2(0.3846 + 0.1884) = 1.1459(s)$, a step length of $L_s = 0.2292(m)$, and an average running speed

TABLE 6.3 Components of q_f^{1-*} and $\dot{q}_f^{1-*}$

φ_1(rad)	−0.1313	$\dot{\varphi}_1(\frac{rad}{s})$	−2.6961
φ_2(rad)	1.3450	$\dot{\varphi}_2(\frac{rad}{s})$	−0.2557
φ_3(rad)	1.4494	$\dot{\varphi}_3(\frac{rad}{s})$	−9.9276
θ_1(rad)	0.0808	$\dot{\theta}_1(\frac{rad}{s})$	1.2216
θ_2(rad)	0.4886	$\dot{\theta}_2(\frac{rad}{s})$	0.8391
θ_3(rad)	−0.0192	$\dot{\theta}_3(\frac{rad}{s})$	0.1276
x_{cm}(m)	0.1946	$\dot{x}_{cm}(\frac{m}{s})$	0.4489
y_{cm}(m)	−0.0374	$\dot{y}_{cm}(\frac{m}{s})$	−0.0282
z_{cm}(m)	0.6779	$\dot{z}_{cm}(\frac{m}{s})$	−1.9809

TABLE 6.4 Stance and Flight Phase Times

T_s^*(s)	0.3846
T_f^*(s)	0.1884

TABLE 6.5   Third Columns of the Parameter Matrices α^{1*} and β^{1*}

$\alpha_3^{1*'}$	0.0247	0.4540	0.9027
$\beta_3^{1*'}$	0.1473	−0.0547	1.0000

of $0.4(\frac{m}{s})$. In Figs. 6.2 and 6.3, the bold and light curves correspond to stance and flight phases, respectively. Figure 6.2 depicts the angular position and velocity of the robot during four consecutive steps (i.e., two periods) of the optimal motion, respectively. Discontinuities in Fig. 6.2 are due to impacts. Figure 6.3 shows open-loop control inputs and horizontal and vertical components of the ground reaction force at the end of the leg during four consecutive steps of the optimal motion. Two types of discontinuity due to transitions between continuous phases are shown. On the optimal periodic orbit, $\ddot{p}_l^v = 13.8115(\frac{m}{s^2}) > 0$ at the beginning of the flight phases and, hence, the takeoff model is valid.

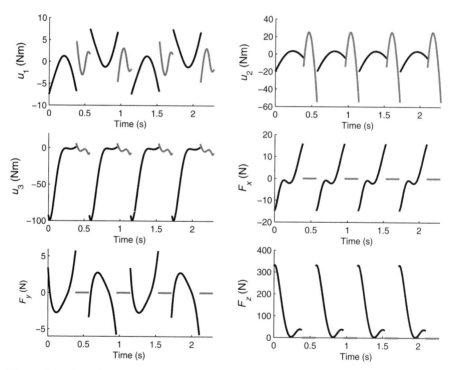

Figure 6.3 Open-loop control inputs and horizontal and vertical components of the ground reaction force at the leg end during four consecutive steps of the optimal motion. Bold and light curves correspond to stance and flight phases, respectively. Discontinuities are due to transitions between continuous phases. Impulsive ground reaction forces are not presented. (See the color version of this figure in the color plates section.)

6.5 WITHIN-STRIDE CONTROLLERS

This section presents a design method to obtain time-invariant control laws during the stance and flight phases to realize the desired period-one orbit $\mathcal{O}$ as an exponentially stable orbit. Assume that $\mathcal{O}$ is a periodic orbit for the time-invariant closed-loop hybrid model of running. On the basis of extended method of Poincaré sections for hybrid systems (Theorem 2.2 of Chapter 2), an equivalence can be established between the stabilization problem of $\mathcal{O}$ for the closed-loop hybrid model and that of the corresponding equilibrium point for a discrete-time system defined on the basis of a two-step Poincaré return map. The stabilization issue will be studied in Section 6.7. In order to reduce the dimension of the stabilization problem, by extending the ideas developed in Refs. [56, 59], a two-level control action is proposed. At the first level of control action, *within-stride controllers* including stance and flight phase controllers are employed to create a family of attractive forward invariant manifolds in each of the continuous phases. At the second level, the parameters of within-stride controllers are updated by *event-based update laws* during discrete transitions between continuous phases to achieve hybrid invariance and stabilization.

6.5.1 Stance Phase Control Law

Based on Refs. [52, 56], by assuming that the angle γ of the virtual leg is a strictly increasing function of time during the stance phases of the desired periodic orbit $\mathcal{O}$, we choose the nominal holonomic output functions

$$h^j_{s,\mathcal{O}} : \mathcal{Q}_s \rightarrow \mathbb{R}^3, j = 1, 2$$

such that they vanish on the orbits $\mathcal{O}^j_s$ for $j = 1, 2$. In our notation, "s" and "j" denote the jth stance phase. Moreover, "h" and the subscript "$\mathcal{O}$" correspond to the holonomic function vanishing on the periodic orbit. To make this notion precise, according to the symmetry, let $\Gamma(t), 0 \leq t \leq T^*_s$ be the time evolution of γ during the first and second stance phases of the periodic orbit $\mathcal{O}$. Then, we define $h^j_{s,\mathcal{O}}(q_s) := \varphi - \varphi^j_{s,d}(\gamma)$, where

$$\varphi^j_{s,d}(\gamma) := \mathcal{B}\left(\frac{\Gamma^{-1}(\gamma)}{T^*_s}, \alpha^{j*}\right) \qquad j = 1, 2 \qquad (6.24)$$

is the desired evolution of body angles φ on $\mathcal{O}^j_s$ in terms of γ and $t = \Gamma^{-1}(\gamma)$ represents the inverse of function $\gamma = \Gamma(t)$. We remark that α^{j*} and T^*_s are obtained from the motion planning algorithm in Section 6.3. Since γ is a strictly increasing function of time on $\mathcal{O}^j_s$, the desired evolutions of φ can be expressed as a function of γ instead of the time. We observe that by zeroing the nominal output function $h^j_{s,\mathcal{O}}(q_s)$, the evolution of the body angles can be expressed in terms of γ. To define a modified holonomic output function for the system during the stance phases, a corrective term as a Bézier polynomial is also added to $h^j_{s,O}$ in the following manner:

$$
\begin{aligned}
y^j_s \left(x^j_s; \xi^j \right) &:= h^j_s \left(q_s; \xi^j \right) \\
&:= h^j_{s,O}(q_s) + \mathcal{B} \left(\frac{\gamma - \gamma^{j+}}{\gamma^- - \gamma^{j+}}, a^j \right) \\
&= \varphi - \left(\varphi^j_{s,d}(\gamma) - \mathcal{B} \left(\frac{\gamma - \gamma^{j+}}{\gamma^- - \gamma^{j+}}, a^j \right) \right) \\
&=: \varphi - \Phi^j_{s,d} \left(\gamma; \xi^j \right), \qquad j = 1, 2.
\end{aligned}
\tag{6.25}
$$

The additive term is introduced to create hybrid invariance and stabilization and it vanishes as $t \to \infty$. In equation (6.25), the additive Bézier polynomial is assumed to be degree $N_s \geq 3$. In addition, $a^j := \mathrm{col}\{a^j_i\}^{N_s}_{i=0} \in \mathbb{R}^{3 \times (N_s+1)}$ represents the coefficients of the additive Bézier polynomial. Note that γ^- is the fixed threshold value used in Remark 6.6 to define the takeoff switching hypersurface S^f_s. Also, γ^{j+}, $j = 1, 2$ denotes the initial value of γ during the jth stance phase. The values of γ^{j+} and a^j for $j = 1, 2$ may differ during consecutive steps. Due to this fact, we form the following vector:

$$
\xi^j := \left(a^{j'}_0, a^{j'}_1, \ldots, a^{j'}_{N_s-1}, a^{j'}_{N_s}, \gamma^{j+} \right)', \qquad j = 1, 2
\tag{6.26}
$$

as the *parameter vector of the jth stance phase controller*. The parameter vector $\xi^j \in \mathbb{R}^{3(N_s+1)+1}$ is held constant during stance phases, that is, $\dot{\xi}^j = 0$, and updated during transitions from flight to stance by an event-based control law. In addition, $\Phi^j_{s,d}$ is the modified desired evolution of the body angles φ in terms of γ.

By using the input–output linearization [103], it can be formally shown that

$$
\ddot{y}^j_s \left(x^j_s; \xi^j \right) = L_{g_s} L_{f_s} y^j_s \left(x^j_s; \xi^j \right) u + L^2_{f_s} y^j_s \left(x^j_s; \xi^j \right), \qquad j = 1, 2,
$$

where

$$
L_{g_s} L_{f_s} y^j_s \left(x^j_s; \xi^j \right) = \frac{\partial h^j_s}{\partial q_s} D^{-1}_s B \in \mathbb{R}^{3 \times 3}
$$

$$
L^2_{f_s} y^j_s \left(x^j_s; \xi^j \right) = \frac{\partial}{\partial q_s} \left(\frac{\partial h^j_s}{\partial q_s} \dot{q}_s \right) \dot{q}_s - \frac{\partial h^j_s}{\partial q_s} D^{-1}_s (C_s \dot{q}_s + G_s) \in \mathbb{R}^{3 \times 1}.
$$

Let ξ^{j*} be the nominal value of ξ^j on the periodic orbit $\mathcal{O}^j_s$, that is,

$$
\xi^{j*} := \begin{bmatrix} 0_{3(N_s+1) \times 1} \\ \gamma^{j+*} \end{bmatrix}.
$$

If the decoupling matrix $L_{g_s} L_{f_s} y_s^j(x_s^j; \xi^{j*})$ is invertible on the orbit $\mathcal{O}_s^j$, there exists an open neighborhood $\mathcal{N}^j$ of $\mathcal{O}_s^j \times \xi^{j*}$ such that for every $(x_s^j, \xi^j) \in \mathcal{N}^j(\mathcal{O}_s^j \times \xi^{j*})$, the feedback law

$$u\left(x_s^j; \xi^j\right) = -\left(L_{g_s} L_{f_s} y_s^j\left(x_s^j; \xi^j\right)\right)^{-1} \left(L_{f_s}^2 y_s^j\left(x_s^j; \xi^j\right) - v_s^j\left(y_s^j, \dot{y}_s^j\right)\right) \quad (6.27)$$

is well defined and results in $\ddot{y}_s^j = v_s^j(y_s^j, \dot{y}_s^j)$. We will assume that the continuous laws $v_s^j(y_s^j, \dot{y}_s^j)$ are designed such that the origin of the closed-loop system $\ddot{y}_s^j = v_s^j(y_s^j, \dot{y}_s^j)$ is globally finite-time stable for $j = 1$ and globally exponentially quickly stable for $j = 2$. To achieve this result, the methods developed in Refs. [46, 93] can be used for the design of v_s^1, and v_s^2 is also defined as follows:

$$v_s^2\left(y_s^2, \dot{y}_s^2\right) := -\frac{1}{\varepsilon} K_D \, \dot{y}_s^2 - \frac{1}{\varepsilon^2} K_P \, y_s^2,$$

where $K_P, K_D \in \mathbb{R}^{3\times 3}$ are positive definite diagonal matrices and ε is a sufficiently small positive scalar. For the later purposes, define $\mathcal{Z}_{s,\xi^j}^j$ as the *parametric stance phase zero dynamics manifold* associated with the output y_s^j,

$$\mathcal{Z}_{s,\xi^j}^j := \left\{x_s^j \in \mathcal{X}_s \middle| y_s^j\left(x_s^j; \xi^j\right) = 0_{3\times 1}, L_{f_s} y_s^j\left(x_s^j; \xi^j\right) = 0_{3\times 1}\right\}.$$

Furthermore, let

$$\dot{z}_s^j = f_{\text{zero},s}^j\left(z_s^j; \xi^j\right)$$

denote the corresponding *parametric stance phase zero dynamics*, where $f_{\text{zero},s}^j$ is the restriction of the stance phase closed-loop vector field to the zero dynamics manifold $\mathcal{Z}_{s,\xi^j}^j$. The following lemma presents proper local coordinates for the zero dynamics manifold and a closed-form expression for the stance phase zero dynamics.

Lemma 6.2 (Stance Phase Zero Dynamics) *Let $\tilde{\theta} := (\theta_1, \gamma, \theta_3)'$ and $\tilde{q} := (\varphi', \tilde{\theta}')'$. Then,*

$$\tilde{q} = T(q) := T_0 \, q + T_1.$$

Moreover, define the mass-inertia matrix in the new coordinates as

$$\mathcal{D}(\tilde{q}) := (T_0^{-1})' \, D_s(T^{-1}(\tilde{q})) \, T_0^{-1}$$

and the conjugate momenta as

$$\sigma_s := (\sigma_{s,1}, \sigma_{s,2}, \sigma_{s,3})' := \frac{\partial \mathcal{L}_s}{\partial \dot{\tilde{\theta}}}' = E_2 \, \mathcal{D}(\tilde{q}) \, \dot{\tilde{q}},$$

where $E_2 := [0_{3\times3} \ I_{3\times3}]$. *Then, in the local coordinates* $(\tilde{\theta}, \sigma_s)$ *for the manifold* $\mathcal{Z}^j_{s,\xi^j}$, *the parametric stance phase zero dynamics can be given by*

$$\dot{\tilde{\theta}} = \left(\mathcal{D}_{\tilde{\theta}\tilde{\theta}}(\tilde{q}) + \mathcal{D}_{\tilde{\theta}\varphi}(\tilde{q}) \frac{\partial \Phi^j_{s,d}}{\partial \gamma}(\gamma; \xi^j) e'_2 \right)^{-1} \sigma_s$$

$$\dot{\sigma}_{s,1} = \frac{1}{2} \sigma'_s \lambda^{j'}(\tilde{\theta}; \xi^j) \frac{\partial \mathcal{D}}{\partial \theta_1}(\tilde{q}) \lambda^j(\tilde{\theta}; \xi^j) \sigma_s + m_{\text{tot}} g_0 \sin(\theta_3) x_{\text{cm}}(\tilde{q})$$
$$- m_{\text{tot}} g_0 \cos(\theta_3) y_{\text{cm}}(\tilde{q})$$

$$\dot{\sigma}_{s,2} = \frac{1}{2} \sigma'_s \lambda^{j'}(\tilde{\theta}; \xi^j) \frac{\partial \mathcal{D}}{\partial \gamma}(\tilde{q}) \lambda^j(\tilde{\theta}; \xi^j) \sigma_s + m_{\text{tot}} g_0 \cos(\theta_3) \cos(\theta_1) x_{\text{cm}}(\tilde{q})$$
$$+ m_{\text{tot}} g_0 \sin(\theta_3) \cos(\theta_1) y_{\text{cm}}(\tilde{q})$$

$$\dot{\sigma}_{s,3} = 0,$$

where at the right-hand side $\tilde{q}$ *can be expressed in terms of* $\tilde{\theta}$ *(i.e.,* $\tilde{q} = \tilde{q}(\tilde{\theta})$*),* $e_2 :=$ $[0 \ 1 \ 0]'$ *and*

$$\lambda^j(\tilde{\theta}; \xi^j) := \begin{bmatrix} \frac{\partial}{\partial q_s} h^j_s(T^{-1}(\tilde{q}(\tilde{\theta})); \xi^j) T_0^{-1} \\ E_2 \mathcal{D}(\tilde{q}(\tilde{\theta})) \end{bmatrix}^{-1} \begin{bmatrix} 0_{3\times3} \\ I_{3\times3} \end{bmatrix}.$$

The proof is given in Appendix C.2.

Remark 6.7 (Invertibility of the Decoupling Matrix on $\mathcal{O}$**)** *Appendix C.3 shows that* $L_{g_s} L_{f_s} y^j_s(x^j_s; \xi^{j*})$ *is invertible on the orbit* $\mathcal{O}^j_s$ *if and only if the scalar function*

$$\kappa^j(\tilde{\theta}) := 1 + e'_2 \mathcal{D}^{-1}_{\tilde{\theta}\tilde{\theta}} \mathcal{D}_{\tilde{\theta}\varphi} \frac{\partial \varphi^j_{s,d}}{\partial \gamma}$$

is nonzero on $\mathcal{O}^j_s$.

6.5.2 Flight Phase Control Law

Analogous to the development for the stance phase, assume that x_{cm} is a strictly increasing function of time on the orbits $\mathcal{O}^j_f$, $j = 1, 2$, and, by considering symmetry, denote its time evolution by $X_{\text{cm}}(t)$, $T^*_s \leq t \leq T^*$. Then, we define the nominal holonomic output functions

$$h^j_{f,O} : \mathcal{Q}_f \to \mathbb{R}^3, \quad j = 1, 2$$

such that they vanish on the the orbits $\mathcal{O}_f^j$. In our notation, "f" and "j" represent the jth flight phase. The nominal holonomic function can be expressed as $h_{f,O}^j(q_f) := \varphi - \varphi_{f,d}^j(x_{cm})$, in which

$$\varphi_{f,d}^j(x_{cm}) := \mathcal{B}\left(\frac{X_{cm}^{-1}(x_{cm}) - T_s^*}{T_f^*}, \beta^{j*} \right), \qquad j = 1, 2 \qquad (6.28)$$

denotes the desired evolution of body angles φ in terms of x_{cm} and $t = X_{cm}^{-1}(x_{cm})$ represents the inverse of function $x_{cm} = X_{cm}(t)$. We remark that T_s^*, T_f^*, and β^{j*} are obtained from the motion planning algorithm in Section 6.3. Zeroing the output function (6.28) forces the desired evolution of the body angles to be constrained to x_{cm}. By adding a corrective Bézier polynomial to the output function (6.28) for hybrid invariance and stabilization, define the following modified holonomic output for the system during flight phases:

$$
\begin{aligned}
y_f^j(x_f^j; \psi^j) &:= h_f^j(q_f; \psi^j) \\
&:= h_{f,O}^j(q_f) + \mathcal{B}\left(\frac{x_{cm} - x_{cm}^{j+}}{\dot{x}_{cm}^{j+} T_f^j}, b^j \right) \\
&= \varphi - \left(\varphi_{f,d}^j(x_{cm}) - \mathcal{B}\left(\frac{x_{cm} - x_{cm}^{j+}}{\dot{x}_{cm}^{j+} T_f^j}, b^j \right) \right) \\
&=: \varphi - \Phi_{f,d}^j(x_{cm}; \psi^j), \qquad j = 1, 2,
\end{aligned}
\qquad (6.29)
$$

where, $N_f \geq 1$ and $b^j := \mathrm{col}\{b_i^j\}_{i=0}^{N_f} \in \mathbb{R}^{3 \times (N_f+1)}$ are the degree and coefficients of the additive Bézier polynomial, respectively. Moreover, x_{cm}^{j+} and $\dot{x}_{cm}^{j+}$ represent the first components of the position and velocity of the COM at the beginning of the jth flight phase. T_f^j is also an estimate of the flight phase time duration. Consequently, the *parameter vector of the jth flight phase controller* can be defined as

$$\psi^j := \left(b_0^{j'}, b_1^{j'}, \ldots, b_{N_f-1}^{j'}, b_{N_f}^{j'}, x_{cm}^{j+}, \dot{x}_{cm}^{j+}, T_f^j \right)', \qquad j = 1, 2. \qquad (6.30)$$

In addition, $\Phi_{f,d}^j$ is the modified desired evolution of the body angles φ in terms of x_{cm}. The parameter vector $\psi^j \in \mathbb{R}^{3(N_f+1)+3}$ is held constant during flight phases, that is, $\dot{\psi}^j = 0$, and updated during transitions from stance to flight by an event-based control law. Using the input–output linearization, it can be shown that

$$\ddot{y}_f^j = L_{f_f} L_{g_f} y_f^j\left(x_f^j; \psi^j \right) u + L_{f_f}^2 y_f\left(x_f^j; \psi^j \right), \qquad j = 1, 2, \qquad (6.31)$$

where

$$L_{g_f} L_{f_f} y_f^j \left(x_f^j; \psi^j \right) = \frac{\partial h_f^j}{\partial q_f} D_f^{-1} B_f \in \mathbb{R}^{3 \times 3}$$

$$L_{f_f}^2 y_f^j \left(x_f^j; \psi^j \right) = \frac{\partial}{\partial q_f} \left(\frac{\partial h_f^j}{\partial q_f} \dot{q}_f \right) \dot{q}_f - \frac{\partial h_f^j}{\partial q_f} D_f^{-1} (C_f \dot{q}_f + G_f) \in \mathbb{R}^{3 \times 1}.$$

(6.32)

The decoupling matrix, associated with the output y_f^j, is invertible on $\mathcal{X}_f \times \mathbb{R}^{3(N_f+1)+3}$ because

$$L_{g_f} L_{f_f} y_f^j \left(x_f^j; \psi^j \right) = \frac{\partial h_f^j}{\partial q_f} D_f^{-1} B_f = E_1 A^{-1}$$

and A is positive definite, where $E_1 := [I_{3 \times 3} \ 0_{3 \times 3}]$. Thus, the feedback law

$$u \left(x_f^j; \psi^j \right) = - \left(L_{g_f} L_{f_f} y_f^j \left(x_f^j; \psi^j \right) \right)^{-1} \left(L_{f_f}^2 y_f^j \left(x_f^j; \psi^j \right) - v_f^j \left(y_f^j, \dot{y}_f^j \right) \right) \quad (6.33)$$

is well defined for every $(x_f^j, \psi^j) \in \mathcal{X}_f \times \mathbb{R}^{3(N_f+1)+3}$ and results in $\ddot{y}_f^j = v_f^j(y_f^j, \dot{y}_f^j)$. Furthermore, by choosing

$$v_f^j \left(y_f^j, \dot{y}_f^j \right) := -\frac{1}{\varepsilon} K_D \dot{y}_f^j - \frac{1}{\varepsilon^2} K_P y_f^j, \qquad j = 1, 2,$$

the origin is globally exponentially quickly stable for the closed-loop system $\ddot{y}_f^j = v_f^j(y_f^j, \dot{y}_f^j)$. The *parametric flight phase zero dynamics manifold* associated with the output y_f^j is defined as follows:

$$\mathcal{Z}_{f,\psi^j}^j := \left\{ x_f^j \in \mathcal{X}_f \mid y_f^j \left(x_f^j; \psi^j \right) = 0_{3 \times 1}, \ L_{f_f} y_f^j \left(x_f^j; \psi^j \right) = 0_{3 \times 1} \right\}.$$

The corresponding *parametric flight phase zero dynamics* can also be expressed as

$$\dot{z}_f^j = f_{\text{zero}, f}^j \left(z_f^j; \psi^j \right),$$

in which $f_{\text{zero}, f}^j$ is the restriction of the flight phase closed-loop vector field to the zero dynamics manifold $\mathcal{Z}_{f,\psi^j}^j$.

Lemma 6.3 (Flight Phase Zero Dynamics) *Define the conjugate momenta*

$$\sigma_f := (\sigma_{f,1}, \sigma_{f,2}, \sigma_{f,3})' := \frac{\partial \mathcal{L}_f}{\partial \dot{\theta}}' = E_2 A(q) \dot{q},$$

where $E_2 := [0_{3\times3} \ I_{3\times3}]$. Then, in the global coordinates $(\theta, p_{cm}, \sigma_f, \dot{p}_{cm})$ for the manifold $\mathcal{Z}^j_{f,\psi^j}$, the parametric flight phase zero dynamics can be given by

$$\dot{\theta} = A_{\theta\theta}^{-1}(q)\,\sigma_f - A_{\theta\theta}^{-1}(q)\,A_{\theta\varphi}(q)\frac{\partial \Phi^j_{f,d}}{\partial x_{cm}}(x_{cm}; \psi^j)\,\dot{x}_{cm}$$

$$\dot{\sigma}_{f,1} = \frac{1}{2}\left[\sigma'_f \ \dot{x}_{cm}\right]\mu^{j'}(\theta, x_{cm}; \psi^j)\frac{\partial A}{\partial \theta_1}(q)\,\mu^j(\theta, x_{cm}; \psi^j)\begin{bmatrix}\sigma_f \\ \dot{x}_{cm}\end{bmatrix}$$

$$\dot{\sigma}_{f,2} = \frac{1}{2}\left[\sigma'_f \ \dot{x}_{cm}\right]\mu^{j'}(\theta, x_{cm}; \psi^j)\frac{\partial A}{\partial \theta_2}(q)\,\mu^j(\theta, x_{cm}; \psi^j)\begin{bmatrix}\sigma_f \\ \dot{x}_{cm}\end{bmatrix}$$

$$\dot{\sigma}_{f,3} = 0$$

$$m_{tot}\,\ddot{p}_{cm} = m_{tot}\begin{bmatrix}0 \\ 0 \\ -g_0\end{bmatrix},$$

where q can be expressed as a function of θ and x_{cm} (i.e., $q = q(\theta, x_{cm})$) and

$$\mu^j(\theta, x_{cm}; \psi^j) := \begin{bmatrix} 0_{3\times3} & \frac{\partial \Phi^j_{f,d}}{\partial x_{cm}}(x_{cm}; \psi^j) \\ A_{\theta\theta}^{-1}(q) & -A_{\theta\theta}^{-1}(q)\,A_{\theta\varphi}(q)\frac{\partial \Phi^j_{f,d}}{\partial x_{cm}}(x_{cm}; \psi^j) \end{bmatrix}.$$

The proof is similar to that presented for Lemma 6.2.

6.6 EVENT-BASED UPDATE LAWS FOR HYBRID INVARIANCE

In order to render stance and flight phase zero dynamics manifolds hybrid invariant for the closed-loop hybrid model of running, this section presents a policy for takeoff and impact update laws. These update laws will result in a reduced-order hybrid model for which stabilization will be studied in Section 6.7.

Definition 6.2 (Regular Parameter Vector of the Stance Phase) *The parameter vector of the jth stance phase controller ξ^j, $j = 1, 2$ is said to be regular if*

1. *there exists an open neighborhood $\mathcal{V}^j$ of $\mathcal{O}^j_s$ such that for every $x^j_s \in \mathcal{V}^j(\mathcal{O}^j_s) \cap \mathcal{Z}^j_{s,\xi^j}$, the decoupling matrix $L_{g_s}L_{f_s}y^j_s(x^j_s; \xi^j)$ is invertible,*

2. *$\gamma^{j+} \neq \gamma^-$ (see equation (6.25)), and*

3. *$a^j_{N_s-1} = a^j_{N_s} = 0_{3\times1}$.*

Definition 6.3 (Regular Parameter Vector of the Flight Phase)   *The param-eter vector of the jth flight phase controller ψ^j, $j = 1, 2$ is said to be regular if $\ddot{x}_{cm}^{j+}$, $T_f^j \neq 0$ (see equation (6.29)).*

Next, let ξ^j and ψ^j for $j = 1, 2$ be regular vector of parameters. Then, by the continuous within-stride feedback laws developed in Section 6.3, $\mathcal{Z}^j_{s,\xi^j}$ and $\mathcal{Z}^j_{f,\psi^j}$ are forward invariant under the stance and flight phases closed-loop dynamics, respectively. It can be shown that for a regular vector of parameters ξ^j, $\mathcal{Z}^j_{s,\xi^j}$ is a six-dimensional embedded submanifold of $\mathcal{X}_s$. This fact in combination with the definition of the switching hypersurface $\mathcal{S}_s^f$ as a level set of the virtual leg angle γ, that is,

$$\mathcal{S}_s^f = \{x_s \in \mathcal{X}_s | \gamma(q_s) - \gamma^- = 0\}$$

implies that $\mathcal{S}_s^f \cap \mathcal{Z}^j_{s,\xi^j}$ is a five-dimensional submanifold of $\mathcal{X}_s$. In addition, by Remark 3.16, $a^j_{N_s-1} = a^j_{N_s} = 0_{3\times1}$ implies that

$$\mathcal{B}(1, a^j) = a^j_{N_s} = 0_{3\times1}$$

$$\frac{\partial}{\partial s}\mathcal{B}(1, a^j) = N_s\left(a^j_{N_s} - a^j_{N_s-1}\right) = 0_{3\times1},$$

and consequently $\mathcal{S}_s^f \cap \mathcal{Z}^j_{s,\xi^j}$ is independent of the coefficient matrix of the additive Bézier polynomial during the jth stance phase (i.e., a^j) and, for simplicity, this common intersection is denoted by $\mathcal{S}_s^f \cap \mathcal{Z}_s^j$. It can also be easily shown that for every regular ψ^j, $\mathcal{Z}^j_{f,\psi^j}$ is a 12-dimensional embedded submanifold of $\mathcal{X}_f$.

According to the order stance$_1$ → flight$_1$ → stance$_2$ → flight$_2$ → stance$_1$, in which the phases are executed, we will denote the event-based update laws by the 4-tuple

$$\left(\pi^{1\to1}_{s\to f},\ \pi^{1\to2}_{f\to s},\ \pi^{2\to2}_{s\to f},\ \pi^{2\to1}_{f\to s}\right).$$

Let $\dot{x}_s^j = f_{cl,s}^j(x_s^j; \xi^j)$ and $\dot{x}_f^j = f_{cl,f}^j(x_f^j; \psi^j)$ be the closed-loop dynamics of the jth stance and flight phases, respectively, where $j \in \{1, 2\}$. For every initial condition $x_{s,0}^j \in \mathcal{X}_s$ and parameter vector $\xi^j \in \mathbb{R}^{3(N_s+1)+1}$, the flow of the jth stance phase is denoted by $\mathcal{F}_s^j(x_{s,0}^j; \xi^j)$ and defined as the solution of the initial-value problem $\dot{x}_s^j = f_{cl,s}^j(x_s^j; \xi^j)$, $x_s^j(0) = x_{s,0}^j$ evaluated at the takeoff time. In an analogous manner, for every initial condition $x_{f,0}^j \in \mathcal{X}_f$ and parameter vector $\psi^j \in \mathbb{R}^{3(N_f+1)+3}$, the flow of the jth flight phase is denoted by $\mathcal{F}_f^j(x_{f,0}^j; \psi^j)$ and defined as the solution of $\dot{x}_f^j = f_{cl,f}^j(x_f^j; \psi^j)$, $x_f^j(0) = x_{f,0}^j$ evaluated at the impact time.

Next, we present the following definition.

Definition 6.4 (Hybrid Invariance) *Let $\mathcal{F}_i^j$ for $i \in \{s, f\}$ and $j \in \{1, 2\}$ represent the flow of the jth closed-loop phase of type i. Define the family of the zero dynamics manifolds for the first stance phase as*

$$\mathbf{Z}_s^1 := \left\{ \mathcal{Z}_{s, \xi^1}^1 : \xi^1 \text{ is regular} \right\}. \tag{6.34}$$

$\mathbf{Z}_s^1$ *is said to be hybrid invariant for the closed-loop hybrid model of running under the 4-tuple event-based update law $(\pi_{s \to f}^{1 \to 1}, \pi_{f \to s}^{1 \to 2}, \pi_{s \to f}^{2 \to 2}, \pi_{f \to s}^{2 \to 1})$ if there exists an open neighborhood $\mathcal{N}$ of x_s^{1-*} such that for every $x_s^{1-} \in \mathcal{N}(x_s^{1-*}) \cap \mathcal{S}_s^f \cap \mathcal{Z}_s^1$, the following update sequence (see Fig. 6.4):*

$$\psi^1 := \pi_{s \to f}^{1 \to 1}\left(x_s^{1-} \right) \qquad x_f^{1-} := \mathcal{F}_f^1\left(\Delta_s^f\left(x_s^{1-} \right); \psi^1 \right)$$

$$\xi^2 := \pi_{f \to s}^{1 \to 2}\left(x_f^{1-} \right) \qquad x_s^{2-} := \mathcal{F}_s^2\left(\Delta_f^s\left(x_f^{1-} \right); \xi^2 \right)$$

$$\psi^2 := \pi_{s \to f}^{2 \to 2}\left(x_s^{2-} \right) \qquad x_f^{2-} := \mathcal{F}_f^2\left(\Delta_s^f\left(x_s^{2-} \right); \psi^2 \right)$$

$$\xi^1 := \pi_{f \to s}^{2 \to 1}\left(x_f^{2-} \right)$$

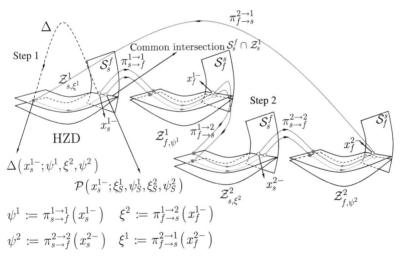

Figure 6.4 Geometric description of hybrid invariance. The plot depicts that under the 4-tuple event-based update law $(\pi_{s \to f}^{1 \to 1}, \pi_{f \to s}^{1 \to 2}, \pi_{s \to f}^{2 \to 2}, \pi_{f \to s}^{2 \to 1})$, the family of the zero dynamics manifolds for the first stance phase $\mathbf{Z}_s^1$ is hybrid invariant, that is, $\Delta(x_s^{1-}; \psi^1, \xi^2, \psi^2) \in \mathcal{Z}_{s, \xi^1}^1$, where $\Delta(x_s^{1-}; \psi^1, \xi^2, \psi^2) := \Delta_f^s(x_f^{2-})$ is the two-step reset map. In addition, $a_{N_s-1}^j = a_{N_s}^j = 0_{3 \times 1}$ results in the common intersection $\mathcal{S}_s^f \cap \mathcal{Z}_s^j$. Plot also illustrates the five-dimensional restricted Poincaré return map $\mathcal{P}(x_s^{1-}; \xi_\mathcal{S}^1, \psi_\mathcal{S}^1, \xi_\mathcal{S}^2, \psi_\mathcal{S}^2)$ and the HZD. (See the color version of this figure in the color plates section.)

results in (i) regular ξ^j, ψ^j for $j = 1, 2$, and (ii) two-step reset invariance, that is,

$$\Delta\left(x_s^{1-}; \psi^1, \xi^2, \psi^2\right) \in \mathcal{Z}_{s,\xi^1}^1, \tag{6.35}$$

where

$$\Delta : \left(\mathcal{S}_s^f \cap \mathcal{Z}_s^1\right) \times \mathbb{R}^{3(N_f+1)+3} \times \mathbb{R}^{3(N_s+1)+1} \times \mathbb{R}^{3(N_f+1)+3} \to \mathcal{X}_s$$

is the two-step reset map given by

$$\Delta\left(x_s^{1-}; \psi^1, \xi^2, \psi^2\right) := \Delta_f^s\left(x_f^{2-}\right). \tag{6.36}$$

Figure 6.4 represents a geometric description for hybrid invariance. In other words, hybrid invariance means that for every final state at the first stance phase of the current step $x_s^{1-} \in \mathcal{S}_s^f \cap \mathcal{Z}_s^1$, the initial state at the first stance phase of two-step ahead belongs to the family $\mathbf{Z}_s^1$. By definition, the static event-based laws

$$\psi^1 = \pi_{s \to f}^{1 \to 1}\left(x_s^{1-}\right), \quad \psi^2 = \pi_{s \to f}^{2 \to 2}\left(x_s^{2-}\right)$$

are called *takeoff update laws* and the static event-based laws

$$\xi^2 = \pi_{f \to s}^{1 \to 2}\left(x_f^{1-}\right), \quad \xi^1 = \pi_{f \to s}^{2 \to 1}\left(x_f^{2-}\right)$$

are called *impact update laws*.

6.6.1 Takeoff Update Laws

According to the the takeoff map Δ_s^f, at the end of the jth stance phase, calculate the initial state of the jth flight phase as $x_f^{j+} = \Delta_s^f(x_s^{j-})$. On the basis of x_f^{j+}, obtain φ^+, $\dot{\varphi}^+$, x_{cm}^+, and $\dot{x}_{cm}^+$ and then, update the parameters of the modified output function in equation (6.29) as

$$x_{cm}^{j+} = x_{cm}^+, \quad \dot{x}_{cm}^{j+} = \dot{x}_{cm}^+.$$

The other components of the vector of parameters ψ^j are updated by the following policy:

$$\begin{aligned}
T_f^j &= T_f^* \\
b_0^j &= -\varphi^+ + \varphi_{f,d}^j\left(x_{cm}^{j+}\right) \\
b_1^j &= b_0^j - \frac{T_f^j}{N_f}\left(\dot{\varphi}^+ - \frac{\partial \varphi_{f,d}^j}{\partial x_{cm}}\left(x_{cm}^{j+}\right)\dot{x}_{cm}^{j+}\right).
\end{aligned} \tag{6.37}$$

By Remark 3.16, equation (6.37) results in

$$y_f^j\left(x_f^{j+};\psi^j\right) = 0_{3\times 1}$$
$$L_{f_f}y_f^j\left(x_f^{j+};\psi^j\right) = 0_{3\times 1},$$

and consequently $x_f^{j+} = \Delta_s^f(x_s^{j-}) \in \mathcal{Z}_{f,\psi^j}^j$ (see Fig. 6.4), where $\psi^j := \pi_{s\to f}^{j\to j}(x_s^{j-})$.
Therefore, the vector of parameters ψ^j can be split into two vectors $\psi_{HI}^j \in \mathbb{R}^9$ and
$\psi_S^j \in \mathbb{R}^{3(N_f-1)}$ that are employed for hybrid invariance and stabilization, respectively.
In particular,

$$\psi_{HI}^j := \left(b_0^{j'}, b_1^{j'}, x_{cm}^{j+}, \dot{x}_{cm}^{j+}, T_f^j\right)'$$
$$\psi_S^j := \left(b_2^{j'}, \ldots, b_{N_f}^{j'}\right)'.$$

The update policy for ψ_S^j will be presented in Section 6.7.

6.6.2 Impact Update Laws

By considering the impact map Δ_f^s and using the state of the system at the end of kth
flight phase, calculate the state of the mechanical system at the beginning of the jth
stance phase as $x_s^{j+} = \Delta_f^s(x_f^{k-})$, where for $k = 1$, $j(k) = 2$ and for $k = 2$, $j(k) = 1$.
On the basis of x_s^{j+}, calculate φ^+, $\dot{\varphi}^+$, γ^+, and $\dot{\gamma}^+$, and update the parameters of the
modified output function in equation (6.25) as $\gamma^{j+} = \gamma^+$ and

$$a_0^j = -\varphi^+ + \varphi_{s,d}^j(\gamma^{j+})$$
$$a_1^j = a_0^j - \frac{\gamma^- - \gamma^{j+}}{N_s\dot{\gamma}^+}\left(\dot{\varphi}^+ - \frac{\partial\varphi_{s,d}^j}{\partial\gamma}(\gamma^{j+})\dot{\gamma}^+\right)$$
$$a_{N_s-1}^j = 0_{3\times 1}$$
$$a_{N_s}^j = 0_{3\times 1}.$$
(6.38)

By Remark 3.16, equation (6.38) results in

$$y_s^j\left(x_s^{j+};\xi^j\right) = 0_{3\times 1}$$
$$L_{f_s}y_s^j\left(x_s^{j+};\xi^j\right) = 0_{3\times 1},$$

and consequently $x_s^{j+} = \Delta_f^s(x_f^{k-}) \in \mathcal{Z}_{s,\xi^j}^j$ (see Fig. 6.4), where $\xi^j := \pi_{f\to s}^{k\to j}(x_f^{k-})$.
Analogous to the development for the takeoff update law, the vector of parameters

ξ^j can also be split into $\xi^j_{HI} \in \mathbb{R}^{13}$ and $\xi^j_S \in \mathbb{R}^{3(N_s-3)}$ according to hybrid invariance and stabilization, respectively. In particular,

$$\xi^j_{HI} := \left(a^{j'}_0, a^{j'}_1, a^{j'}_{N_s-1}, a^{j'}_{N_s}, \gamma^{j+} \right)'$$

$$\xi^j_S := \left(a^{j'}_2, \ldots, a^{j'}_{N_s-2} \right)'.$$

The update policy for ξ^j_S will be presented in Section 6.7. The following lemma presents the main result of this section.

Lemma 6.4 (Hybrid Invariance) *Assume that the stabilizing update laws for ψ^j_S and ξ^j_S, $j = 1, 2$ are continuous functions with respect to x^{1-}_s and vanish on the periodic orbit $\mathcal{O}$. Then, the family of the zero dynamics manifolds $\mathbf{Z}^1_s$ in equation (6.34) is hybrid invariant under the 4-tuple event-based update law $(\pi^{1\rightarrow1}_{s\rightarrow f}, \pi^{1\rightarrow2}_{f\rightarrow s}, \pi^{2\rightarrow2}_{s\rightarrow f}, \pi^{2\rightarrow1}_{f\rightarrow s})$ for which ψ^j_{HI} and ξ^j_{HI} are given in equations (6.37) and (6.38), respectively.*

Proof. Continuity of the stabilizing update laws $\psi^j_S(x^{1-}_s)$ and $\xi^j_S(x^{1-}_s)$ for $j = 1, 2$ and the fact that on the periodic orbit $\mathcal{O}$, $a^j = 0_{3\times(N_s+1)}$, $b^j = 0_{3\times(N_f+1)}$ (the additive Bézier polynomials of the modified outputs in equations (6.25) and (6.29) are zero on $\mathcal{O}$), $\kappa^j(\bar{\theta}) \neq 0$ for $j = 1, 2$ (the decoupling matrices corresponding to the outputs (6.25) on the stance phases of the periodic orbit are invertible as stated in Remark 6.7), $\gamma^{+*} \neq \gamma^-$, $\dot{\gamma}^{+*} \neq 0$, $\dot{x}^{+*}_{cm} \neq 0$, and $T^*_f \neq 0$ imply that there exists an open neighborhood $\mathcal{N}$ of x^{1-*}_s such that for every $x^{1-}_s \in \mathcal{N}(x^{1-*}_s) \cap \mathcal{S}^f_s \cap \mathcal{Z}^1_s$ the update laws of ψ^j_{HI} and ξ^j_{HI} in equations (6.37) and (6.38) are well defined and result in hybrid invariance, that is, $\Delta^s_f(x^{2-}_f) \in \mathbf{Z}^1_s$. ∎

6.7   STABILIZATION PROBLEM

In this section, static event-based update laws for stabilizing parameters $\psi^j_S \in \mathbb{R}^{3(N_f-1)}$ and $\xi^j_S \in \mathbb{R}^{3(N_s-3)}$, $j = 1, 2$ in terms of x^{1-}_s are presented such that the periodic orbit $\mathcal{O}$ is an exponentially stable limit cycle for the closed-loop hybrid model of running. Under the assumptions of Lemma 6.4, by applying the 4-tuple event-based update law $(\pi^{1\rightarrow1}_{s\rightarrow f}, \pi^{1\rightarrow2}_{f\rightarrow s}, \pi^{2\rightarrow2}_{s\rightarrow f}, \pi^{2\rightarrow1}_{f\rightarrow s})$, the family of the manifolds $\mathbf{Z}^1_s$ is hybrid invariant. In addition, as mentioned previously, $a^j_{N_s-1} = a^j_{N_s} = 0_{3\times1}$ results in the common intersection $\mathcal{S}^f_s \cap \mathcal{Z}^j_s$. For simplicity, the switching map Δ can be denoted by

$$\Delta \left(x^{1-}_s; \psi^1_S, \xi^2_S, \psi^2_S \right)$$

(see Fig. 6.4). Thus, to study the stabilization problem, we consider the following reduced-order system with impulse effects:

$$
\Sigma_{s,\text{cl}} : \begin{cases} \dot{z}_s^1 = f_{\text{zero},s}^1 \left(z_s^1; \xi_S^1 \right) & z_s^{1-} \notin S_s^f \cap Z_s^1 \\ z_s^{1+} = \Delta \left(z_s^{1-}; \psi_S^1, \xi_S^2, \psi_S^2 \right) & z_s^{1-} \in S_s^f \cap Z_s^1, \end{cases}
\tag{6.39}
$$

which is called HZD. From Lemma 6.2,

$$
z_s^1 := (\theta_1, \gamma, \theta_3, \sigma_{s,1}, \sigma_{s,2}, \sigma_{s,3})'
$$

is a valid coordinates transformation for the parametric zero dynamics manifold Z_{s,ξ^1}^1. In addition by considering the fact that on S_s^f, $\gamma = \gamma^-$, then

$$
z_s^{1-} := (\theta_1, \theta_3, \sigma_{s,1}, \sigma_{s,2}, \sigma_{s,3})'
$$

is a valid coordinates transformation for $S_s^f \cap Z_{s,\xi^1}^1 = S_s^f \cap Z_s^1$. By defining the five-dimensional Poincaré return map for the HZD in equation (6.39) as

$$
\mathcal{P}: \left(S_s^f \cap Z_s^1 \right) \times \mathbb{R}^{3(N_s-3)} \times \mathbb{R}^{3(N_f-1)} \times \mathbb{R}^{3(N_s-3)} \times \mathbb{R}^{3(N_f-1)} \to S_s^f \cap Z_s^1
$$

by

$$
\mathcal{P} \left(z_s^{1-}; \xi_S^1, \psi_S^1, \xi_S^2, \psi_S^2 \right) := \mathcal{F}_s^1 \left(\Delta(z_s^{1-}; \psi_S^1, \xi_S^2, \psi_S^2); \xi_S^1 \right),
\tag{6.40}
$$

the following discrete-time system with the state space $S_s^f \cap Z_s^1$ can be introduced to study the stabilization problem

$$
z_s^{1-}[k+1] = \mathcal{P} \left(z_s^{1-}[k]; \xi_S^1[k], \psi_S^1[k], \xi_S^2[k], \psi_S^2[k] \right), \quad k = 1, 2, \dots.
\tag{6.41}
$$

Figure 6.4 presents a geometric description for the Poincaré return map and stabilization problem.

Theorem 6.1 (Exponential Stability) *Consider the open-loop hybrid model of running (6.11) by the within-stride controllers (6.27) and (6.33), and the event-based update laws with ψ_{HI}^j and ξ_{HI}^j given in equations (6.37) and (6.38). Let*

$$
A_{ol} := \frac{\partial \mathcal{P}}{\partial z_s^{1-}} \left(z_s^{1-}; \xi_S^1, \psi_S^1, \xi_S^2, \psi_S^2 \right) \Big|_{z_s^{1-}=z_s^{1-*}, \xi_S^j=\xi_S^{j*}, \psi_S^j=\psi_S^{j*}}
$$

$$
B_s^j := \frac{\partial \mathcal{P}}{\partial \xi_S^j} \left(z_s^{1-}; \xi_S^1, \psi_S^1, \xi_S^2, \psi_S^2 \right) \Big|_{z_s^{1-}=z_s^{1-*}, \xi_S^j=\xi_S^{j*}, \psi_S^j=\psi_S^{j*}}
\tag{6.42}
$$

$$
B_f^j := \frac{\partial \mathcal{P}}{\partial \psi_S^j} \left(z_s^{1-}; \xi_S^1, \psi_S^1, \xi_S^2, \psi_S^2 \right) \Big|_{z_s^{1-}=z_s^{1-*}, \xi_S^j=\xi_S^{j*}, \psi_S^j=\psi_S^{j*}}
$$

and

$$B_{ol} := \begin{bmatrix} B_s^1 & B_f^1 & B_s^2 & B_f^2 \end{bmatrix},$$

where z_s^{1-*} is the intersection of the orbit $\mathcal{O}_s^1$ and $\mathcal{S}_s^f$, and $\xi_S^{j*} := 0_{3(N_s-3)\times 1}$ and $\psi_S^{j*} := 0_{3(N_f-1)\times 1}$ for $j = 1, 2$ are the nominal stabilizing parameters. Then, if (A_{ol}, B_{ol}) is controllable, there exists the gain matrix

$$K := \begin{bmatrix} K_s^{1'} & K_f^{1'} & K_s^{2'} & K_f^{2'} \end{bmatrix}'$$

such that by using the static update laws

$$\xi_S^j = -K_s^j \left(z_s^{1-} - z_s^{1-*} \right)$$

$$\psi_S^j = -K_f^j \left(z_s^{1-} - z_s^{1-*} \right)$$

for $j = 1, 2$, the periodic orbit $\mathcal{O}$ is exponentially stable for the closed-loop hybrid model of running.

Proof. Since (i) the event-based update laws ψ_{HI}^j and ξ_{HI}^j for $j = 1, 2$ are continuously differentiable (i.e., C^1) and (ii) v_f^j, $j = 1, 2$ and v_s^2 are C^1, $\Delta(\cdot; \psi_S^1, \xi_S^2, \psi_S^2)$ is C^1, which, in combination with the fact that the continuous law v_s^1 [46, 93] vanishes on $\mathcal{Z}_{s,\xi^1}^1$, follows that $\mathcal{P}(\cdot, ; \xi_S^1, \psi_S^1, \xi_S^2, \psi_S^2)$ is C^1. Moreover, similar to the proof of Lemma C.6 of Ref. [18, p. 450], it can be shown that $\mathcal{P}$ is also C^1 with respect to $(\xi_S^1, \psi_S^1, \xi_S^2, \psi_S^2)$ in an open neighborhood of $(\xi_S^{1*}, \psi_S^{1*}, \xi_S^{2*}, \psi_S^{2*})$. Hence, A_{ol} and B_{ol} are well defined. In addition, controllability of (A_{ol}, B_{ol}) implies the existence of the gain matrix K such that $|\mathrm{eig}(A_{ol} - B_{ol}K)| < 1$, where $\mathrm{eig}(\cdot)$ denotes the eigenvalues, which, in turn, implies that z_s^{1-*} is locally exponentially stable for the closed-loop system $z_s^{1-}[k + 1] = \mathcal{P}_{cl}(z_s^{1-}[k])$, where

$$\mathcal{P}_{cl} \left(z_s^{1-} \right) := \mathcal{P} \left(z_s^{1-}; \xi_S^1 \left(z_s^{1-} \right), \psi_S^1 \left(z_s^{1-} \right), \xi_S^2 \left(z_s^{1-} \right), \psi_S^2 \left(z_s^{1-} \right) \right).$$

Finally, applying the extended method of Poincaré sections for hybrid systems (Theorems 2.2 and 2.5, Chapter 2) completes the proof.   ∎

6.8   SIMULATION RESULTS

This section presents a numerical example for the control scheme proposed in this chapter to exponentially stabilize the desired periodic orbit $\mathcal{O}$ generated by the motion planning algorithm in Section 6.4. The additive Bézier polynomials are chosen with degree $N_s = N_f = 9$. The state matrix of the linearized open-loop restricted Poincaré return map can be obtained numerically as follows:[4]

$$
A_{\text{ol}} =
\begin{bmatrix}
13.1446 & -3.2632 & 1.4352 & -1.7912 & 1.8410 \\
-2.2285 & -0.9782 & 0.0489 & -0.0864 & 0.6449 \\
262.4331 & -62.9310 & 26.4986 & -31.6193 & 29.5938 \\
-0.5343 & -1.3510 & 0.7338 & -1.4694 & 2.5789 \\
5.0767 & -2.1732 & 1.0688 & -1.4461 & 2.3939
\end{bmatrix},
$$

which has the eigenvalues $\mathrm{eig}(A_{\text{ol}}) = \{40.8797, -0.8783 \pm 0.0570i, 0.4726, -0.0061\}$. We remark that without stabilizing parameters, the periodic orbit $\mathcal{O}$ is not stable in the sense of Lyapunov for the closed-loop hybrid model of running. Due to space limitations, the matrices B_s^j, B_f^j for $j = 1, 2$ are not presented here. By using the DLQR design method, the matrix gain of the static stabilizing update laws (i.e., K_s^j, K_f^j for $j = 1, 2$) can be calculated. In this method, the gain matrix K is obtained such that by the static stabilizing feedback law $\delta u = -K \delta z_s^{1-}$, where $\delta z_s^{1-} := z_s^{1-} - z_s^{1-*}$, the cost function

$$
\frac{1}{2} \sum_{k=0}^{\infty} \left\{ \delta z_s^{1-'}[k] \, Q \, \delta z_s^{1-}[k] + \delta u'[k] \, \mathcal{R} \, \delta u[k] \right\}
$$

subject to the linearized system

$$
\delta z_s^{1-}[k+1] = A_{\text{ol}} \, \delta z_s^{1-}[k] + B_{\text{ol}} \, \delta u[k]
$$

is minimized, where $Q = Q' \succeq 0$ and $\mathcal{R} = \mathcal{R}' \succ 0$. For $Q = \mathrm{diag}\{q_i\}_{i=1}^{5}$ and $\mathcal{R} = \mathrm{diag}\{r_i\}_{i=1}^{84}$, where $q_i = 1, i = 1, \ldots, 5$, and $r_i = 10, i = 1, \ldots, 84$, the state matrix

---

[4] For a given smooth function $f : \mathbb{R}^n \to \mathbb{R}^n$, the element (i, j) of the Jacobian matrix evaluated at x^* can be calculated numerically as follows:

$$
Df_{(i, j)}(x^*) \cong \frac{f_i(x_1^*, \ldots, x_j^* + \delta x_j, \ldots, x_n^*) - f_i(x_1^*, \ldots, x_j^* - \delta x_j, \ldots, x_n^*)}{2 \delta x_j},
$$

where δx_j is a scalar perturbation. In this chapter, the perturbations used for calculation of A_{ol} and B_{ol} are assumed to be 10^{-6} and 10^{-5}, respectively.

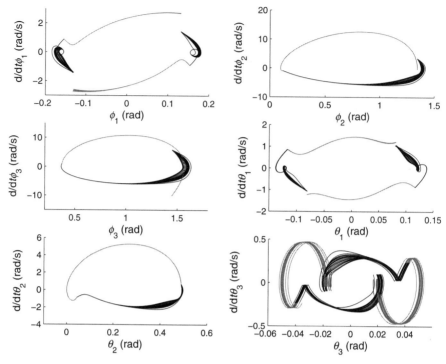

Figure 6.5 Phase portraits of the state trajectories during 40 consecutive steps of running. The stance and flight phases are shown by bold and light curves, respectively. In the figure, the effect of the impact with the ground is illustrated by jumps in the velocity. The convergence to the desired limit cycle $\mathcal{O}$ can be seen.

of the linearized closed-loop Poincaré return map is obtained as follows[5]:

$$A_{\text{cl}} = \begin{bmatrix} -0.5044 & 0.0333 & 0.0107 & -0.0449 & 0.1458 \\ -2.3334 & 0.0639 & -0.0302 & -0.0647 & 0.4142 \\ 0.2268 & -0.1153 & -0.0474 & 0.1162 & -0.2056 \\ -3.1307 & -0.0024 & -0.0190 & -0.1131 & 0.6323 \\ -3.6323 & 0.0440 & 0.0660 & -0.1370 & 0.9969 \end{bmatrix}$$

with the eigenvalues $\text{eig}(A_{\text{cl}}) = \{0.4059, 0.0816 \pm 0.1240i, -0.1735, 0.0004\}$. To illustrate the convergence to the desired periodic orbit $\mathcal{O}$, the simulation of the closed-loop hybrid model of running is started at the end of the stance phase with an initial condition off of this trajectory. Figures 6.5 and 6.6 depict the results of the closed-loop simulation. In these figures, the stance and flight phases are shown by bold and light

---

[5] $A_c l = A_{\text{ol}} - B_{\text{ol}} K.$

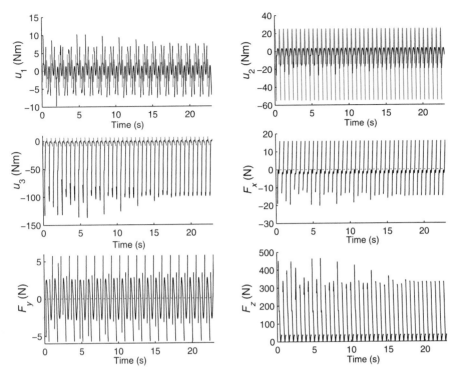

Figure 6.6 Closed-loop control inputs and horizontal and vertical components of the ground reaction force at the leg end during 40 consecutive steps of the monopedal robot. Bold and light curves correspond to stance and flight phases, respectively. Discontinuities are due to discrete transitions between continuous phases. Impulsive ground reaction forces are not presented.

curves, respectively. The phase portraits of the state trajectories during 40 consecutive steps of running are presented in Fig. 6.5 in which the effect of the impact with the ground is illustrated by jumps in the velocity. Figure 6.6 shows closed-loop control inputs and horizontal and vertical components of the ground reaction force during 40 consecutive steps of running.

CHAPTER 7

Stabilization of Periodic Orbits for Walking with Passive Knees

7.1 INTRODUCTION

In this chapter, a motion planning algorithm to generate time trajectories of a periodic walking motion by a five-link, two-actuator planar bipedal robot is presented. In order to reduce the number of actuated joints for walking on a flat ground and restore the walking motion in people with disabilities, we assume that the robot has passive point feet and unactuated knee joints. In other words, only the hip joints of the robot are assumed to be actuated. The motion planning algorithm is developed in terms of a finite-dimensional nonlinear optimization problem with equality and inequality constraints. The equality constraints are necessary and sufficient conditions by which the impulsive model of walking has a period-one orbit, whereas the inequality constraints are introduced to guarantee (i) the feasibility of the periodic motion and (ii) capability of applying the proposed two-level control scheme for stabilization of the orbit. This algorithm is an extension of results developed in the previous chapters.

The main objective of this chapter is to present a time-invariant two-level feedback law based on the concept of *virtual constraints* and *HZD* to exponentially stabilize a desired periodic motion generated by the motion planning algorithm. The mechanical system studied in this chapter has three degrees of underactuation during single support. We present a control methodology for creation of hybrid invariant manifolds and stabilization of a desired periodic orbit for the impulsive model of walking. In particular, for a given integer number $M \geq 2$, we introduce $M - 1$ *within-stride switching hypersurfaces* and thereby split the single support phase into M *within-stride phases*. The within-stride switching hypersurfaces are defined as level sets of a scalar holonomic quantity that is a strictly increasing function of time on the desired walking motion. To stabilize the desired orbit, the overall controller is chosen as a two-level feedback law. At the first level, during a within-stride phase, a parameterized holonomic output function is defined for the dynamical system and imposed to be zero by using a continuous-time feedback law. The output function is

Hybrid Control and Motion Planning of Dynamical Legged Locomotion, Nasser Sadati, Guy A. Dumont, Kaveh Akbari Hamed, and William A. Gruver.
© 2012 by the Institute of Electrical and Electronics Engineers, Inc. Published 2012 by John Wiley & Sons, Inc.

expressed as the difference between the actual values of the angle of hip joints and their desired evolutions, in terms of the latter increasing holonomic quantity. At the second level, the parameters of continuous-time feedback laws are updated during within-stride transitions by event-based update laws. The purpose of updating the parameters is (i) to achieve hybrid invariance, (ii) continuity of continuous-time feedback laws during within-stride transitions, and (iii) stabilization of the desired orbit. From the construction procedure of the parameterized output functions and event-based update laws, intersections of the corresponding zero dynamics manifolds and within-stride switching hypersurfaces are independent of the parameters. Consequently, by choosing one of these common intersections as the Poincaré section, stabilization can be addressed on the basis of a five-dimensional restricted Poincaré return map.

7.2 OPEN-LOOP MODEL OF WALKING

7.2.1 Mechanical Model of the Planar Bipedal Robot

In this chapter, a five-link, two-actuator planar bipedal mechanism with point feet is studied (see Fig. 7.1). The mechanical system consists of a torso link and two identical legs with tibia and femur links. The links are rigid and have mass. They are connected through a set of *body joints* including two actuated revolute hip joints (q_1 and q_2) and two unactuated revolute knee joints (q_3 and q_4). The body joints are relative angles determining the shape of the robot and denoted by $q_b := (q_1, q_2, q_3, q_4)'$, where the subscript "$b$" represents the body joints and prime denotes matrix transpose. The absolute orientation of the torso link with respect to an inertial *world frame* is given by q_5. Assume that the control inputs applied at the hip joints are represented by $u := (u_1, u_2)' \in \mathcal{U}$, where $\mathcal{U}$ is a simply connected and open subset of $\mathbb{R}^2$ containing the origin $u = 0_{2 \times 1}$. It is assumed that bipedal walking can be modeled by a hybrid system

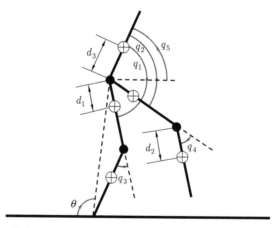

Figure 7.1 A five-link, two-actuator planar bipedal mechanism with point feet during single support. The virtual leg is depicted by the dashed line connecting the stance leg end and the hip joints.

with a continuous single support phase (one leg on the ground) and an instantaneous double support phase (two legs on the ground). During single support, the contacting leg is called the *stance leg* and the other is called the *swing leg*. The *virtual leg* is defined by a line connecting the stance leg end and the hip joint. The angle of the virtual leg with respect to the world frame is given by θ, as shown in Fig. 7.1. An impact occurs when the swing leg end contacts the ground. We assume that during the impact, the swing leg neither slip nor rebound. The instantaneous double support phase is modeled using a coordinate relabeling to swap the role of the legs immediately after impact.

7.2.2 Dynamics of the Single Support Phase

During single support, the mechanical system has five DOF and three degrees of underactuation. Let $q := (q_b', q_5)' \in Q$ denote the generalized coordinates vector of the mechanical system, where Q is the configuration space. The evolution of the mechanical system during single support can be expressed by the following second order equation:

$$D(q_b)\ddot{q} + C(q_b, \dot{q})\dot{q} + G(q) = Bu, \tag{7.1}$$

in which D is a (5×5) mass-inertia matrix, C is a (5×5) matrix containing the Coriolis and centrifugal terms, G is a (5×1) gravity vector, and

$$B := \begin{bmatrix} I_{2\times2} \\ 0_{3\times2} \end{bmatrix}$$

is the input matrix. By introducing the state vector $x := (q', \dot{q}')' \in X$, equation (7.1) can be expressed as $\dot{x} = f(x) + g(x)u$, where X is the single support phase state manifold taken as

$$X := TQ := \{x = (q', \dot{q}')' | q \in Q, \dot{q} \in \mathbb{R}^5\}.$$

Remark 7.1 (Validity of Single Support) *The single support phase model is valid if the ground reaction force at the stance leg end* $F_1 := (F_1^h, F_1^v)'$ *satisfies the unilateral constraints (i)* $F_1^v > 0$ *and (ii)* $|\frac{F_1^h}{F_1^v}| < \mu_s$, *where* μ_s *represents the coefficient of static friction between the leg end and the ground.*

7.2.3 Impact Map

Define the impact switching hypersurface as

$$S := \{x = (q', \dot{q}')' \in X | p_2^v(q) = 0\},$$

in which p_2^v denotes the vertical Cartesian position of the swing leg end with respect to the ground. Following the results of Ref. [18, p. 75], the impact can be modeled by the discrete transition $x^+ = \Delta(x^-)$, in which $x^- = (q^{-\prime}, \dot{q}^{-\prime})'$ and $x^+ = (q^{+\prime}, \dot{q}^{+\prime})'$ represent the state of the mechanical system immediately before and after the impact, respectively. Moreover, $\Delta : \mathcal{S} \to \mathcal{X}$ is the *impact map* given by

$$\Delta(x^-) := \begin{bmatrix} R q^- \\ R \tilde{\Delta}(q^-) \dot{q}^- \end{bmatrix},$$

where R is a nonsingular matrix to swap the role of the legs.

Remark 7.2 (Validity of the Impact Model) *The impact model is valid if (1) the previous stance leg end lifts from the ground without interaction immediately after the impact, that is, $\dot{p}_2^v(q^+, \dot{q}^+) := \frac{\partial p_2}{\partial q}(q^+)\dot{q}^+ \geq 0$ and (2) the intensity of the impulsive ground reaction force $\delta F := (\delta F^h, \delta F^v) \in \mathbb{R}^2$ satisfies the unilateral constraints (i) $\delta F^v > 0$ and (ii) $|\frac{\delta F^h}{\delta F^v}| < \mu_s$.*

7.2.4 Open-Loop Impulsive Model of Walking

By assembling the single support phase model and impact map, the overall open-loop model of walking can be expressed as the impulsive system $\Sigma(\mathcal{X}, \mathcal{S}, \Delta, f, g)$ defined as

$$\Sigma : \begin{cases} \dot{x} = f(x) + g(x)u & x^- \notin \mathcal{S} \\ x^+ = \Delta(x^-) & x^- \in \mathcal{S}. \end{cases} \tag{7.2}$$

We shall assume that for a given initial condition $x_0 \in \mathcal{X} \backslash \mathcal{S}$ (at time $t_0 \in \mathbb{R}$) and the open-loop control input $u \in \mathcal{U}$, the single support phase solution at time $t \geq t_0$ is represented by $\varphi(t; t_0, x_0, u)$. Moreover, as in Ref. [46], the *time to impact function* $T_I : \mathbb{R} \times \mathcal{X} \times \mathcal{U} \to \mathbb{R}$ is defined by

$$T_I(t_0, x_0, u) := \inf\{t > t_0 | \varphi(t; t_0, x_0, u) \in \mathcal{S}\}.$$

Definition 7.1 (Feasible Periodic Orbit) *Assume that there exist $(t_0, x^{-*}) \in \mathbb{R} \times \mathcal{S}$ and the open-loop control $u^* \in \mathcal{U}$ such that (i) $\Delta(x^{-*}) \in \mathcal{X} \backslash \mathcal{S}$, (ii) $T_I^* := T_I(t_0, \Delta(x^{-*}), u^*) < \infty$, and (iii) $\varphi(T_I^*; t_0, \Delta(x^{-*}), u^*) = x^{-*}$. Then, the set*

$$\mathcal{O} := \{x = \varphi(t; t_0, \Delta(x^{-*}), u^*) | t_0 \leq t < T_I^*\}$$

is a period-one orbit for the open-loop impulsive system given in equation (7.2).[1] *Moreover, the orbit $\mathcal{O}$ is said to be feasible if*

> *1. the constraints due to the joint angles and velocities are satisfied on $\mathcal{O}$;*

---

[1] From Definition 2.2 of Chapter 2, it is assumed that the solutions of the impulsive system $\Sigma(\mathcal{X}, \mathcal{S}, \Delta, f, g)$ are right continuous.

2. *the open-loop control input is feasible in the sense that* $\|u^*(t)\|_{\mathcal{L}_\infty} < u_{max}$, *where u_{max} is a positive scalar;*

3. *the ground reaction force experienced at the end of leg-1 satisfies the unilateral constraints of Remark 7.1;*

4. *the impact model satisfies the unilateral constraints of Remark 7.2;*

5. *for every $t \in (t_0, T_I^*)$, $p_2^v(t) > 0$. In addition, the transversality condition is satisfied, that is, the swing leg end contacts the ground with nonzero vertical velocity, $\dot{p}_2^v(T_I^*) < 0$;*

6. *on the orbit $\mathcal{O}$, θ is a strictly increasing function of time, that is,*

$$\min_{t_0 \le t \le T_I^*} \dot{\theta}(t) > 0.$$

7.3 MOTION PLANNING ALGORITHM

In this section, a motion planning algorithm to generate a feasible period-one orbit $\mathcal{O}$ for the open-loop impulsive system of equation (7.2) is presented. The algorithm is developed in terms of a finite-dimensional nonlinear optimization problem with equality and inequality constraints. To describe the motion planning algorithm, we first introduce the generalized coordinates vector

$$\tilde{q} := \begin{bmatrix} q_a \\ q_u \end{bmatrix} := \mathbf{T}\, q,$$

in which $q_a := (q_1, q_2)'$, $q_u := (q_3, q_4, \theta)'$ and the subscripts "a" and "u" denote the actuated and unactuated components, respectively. The matrix $\mathbf{T}$ is also defined as

$$\mathbf{T} := \begin{bmatrix} H_0 \\ \Gamma_0 \end{bmatrix} \in \mathbb{R}^{5 \times 5},$$

where $H_0 := [I_{4 \times 4}\ 0_{4 \times 1}]$ and[2] $\Gamma_0 := [1\ 0\ -\frac{1}{2}\ 0\ -1]$. In coordinates $(\tilde{q}, \dot{\tilde{q}})$, the equation of motions during single support can be expressed as

$$\mathcal{D}(q_b)\, \ddot{\tilde{q}} + \mathcal{H}(\tilde{q}, \dot{\tilde{q}}) = B\, u, \tag{7.3}$$

where

$$\mathcal{D}(q_b) := (\mathbf{T}')^{-1}\, D(q_b)\, \mathbf{T}^{-1}$$

$$\mathcal{H}(\tilde{q}, \dot{\tilde{q}}) := (\mathbf{T}')^{-1} \left(C(q_b, \mathbf{T}^{-1}\dot{\tilde{q}})\, \mathbf{T}^{-1}\dot{\tilde{q}} + G(\mathbf{T}^{-1}\tilde{q}) \right).$$

---

[2] In this chapter, it is assumed that the lengths of the tibia and femur links are identical. Thus, θ can be given by $\theta = \Gamma_0 q = q_1 + \frac{q_3}{2} - q_5$.

Next, assume that

$$\tilde{x}^{-*} := \begin{bmatrix} \tilde{q}^{-*} \\ \dot{\tilde{q}}^{-*} \end{bmatrix}$$

is the state of the mechanical system immediately before the impact on the orbit $\mathcal{O}$. We remark that

$$\tilde{q}^{-*} := \begin{bmatrix} q_a^{-*} \\ q_u^{-*} \end{bmatrix}, \quad \dot{\tilde{q}}^{-*} := \begin{bmatrix} \dot{q}_a^{-*} \\ \dot{q}_u^{-*} \end{bmatrix}.$$

According to the impact map, we obtain the state of the mechanical system at the beginning of the orbit $\mathcal{O}$ (i.e., $\tilde{x}^{+*} := (\tilde{q}^{+*\prime}, \dot{\tilde{q}}^{+*\prime})\prime$) as follows:

$$\tilde{q}^{+*} := \begin{bmatrix} q_a^{+*} \\ q_u^{+*} \end{bmatrix} = \mathbf{T} R \mathbf{T}^{-1} \tilde{q}^{-*}$$

$$\dot{\tilde{q}}^{+*} := \begin{bmatrix} \dot{q}_a^{+*} \\ \dot{q}_u^{+*} \end{bmatrix} = \mathbf{T} R \tilde{\Delta}(\mathbf{T}^{-1} \tilde{q}^{-*}) \mathbf{T}^{-1} \dot{\tilde{q}}^{-*}. \tag{7.4}$$

Since the hip joints are independently actuated, we shall assume that the evolution of the hip joints on the orbit $\mathcal{O}$ can be described by a Bézier polynomial of time. For this purpose, let $N \geq 3$ and $w^* := \mathrm{col}\{w_i^*\}_{i=0}^N \in \mathbb{R}^{2 \times (N+1)}$ denote the degree and coefficient matrix of the Bézier polynomial on $\mathcal{O}$, respectively. In particular, let

$$q_a^*(t) := \mathfrak{B}\left(\frac{t}{T_I^*}, w^*\right), \quad 0 \leq t \leq T_I^*,$$

where $q_a^*(t), 0 \leq t \leq T_I^*$ represents the evolution of the hip joints on the single support phase of $\mathcal{O}$. By properties of the Bézier polynomials in Remark 3.16, the coefficient matrix w^* can be easily adjusted such that the states $(q_a, \dot{q}_a)$ are transferred from the initial conditions $(q_a^{+*}, \dot{q}_a^{+*})$ at time $t = 0$ to the final conditions $(q_a^{-*}, \dot{q}_a^{-*})$ at time $t = T_I^*$. In particular, let

$$w_0^* = q_a^{+*} \qquad\qquad w_N^* = q_a^{-*}$$

$$w_1^* = w_0^* + \frac{T_I^*}{N} \dot{q}_a^{+*} \quad w_{N-1}^* = w_N^* - \frac{T_I^*}{N} \dot{q}_a^{-*}.$$

In addition, from the last three rows of equation (7.3), the evolution of $q_u^*(t), 0 \leq t \leq T_I^*$ can be described by the following ordinary differential equation (ODE):

$$\ddot{q}_u^* = -\mathcal{D}_{uu}^{-1}\left(\mathcal{D}_{ua} \ddot{q}_a^* + \mathcal{H}_u\right), \quad 0 \leq t \leq T_I^*$$

$$q_u^*(0) = q_u^{+*} \tag{7.5}$$

$$\dot{q}_u^*(0) = \dot{q}_u^{+*},$$

in which $\mathcal{D}_{uu}$ and $\mathcal{D}_{ua}$ are the (3×3) and (3×2) lower right and left submatrices of $\mathcal{D}$, respectively. In addition, $\mathcal{H}_u$ consists of the last three rows of $\mathcal{H}$. Thus, for a given $N \geq 3$, the evolution of the mechanical system on the orbit $\mathcal{O}$ can be completely determined by the following vector of parameters:

$$\xi^* := (q_a^{-*'}, q_u^{-*'}, \dot{q}_a^{-*'}, \dot{q}_u^{-*'}, T_I^*, w_2^{*'}, \ldots, w_{N-2}^{*'})' \in \mathbb{R}^{2N+5}.$$

Now we are able to present the following motion planning algorithm.

Algorithm 7.1 *Motion Planning Algorithm*

For a given $N \geq 3$, the motion planning algorithm is expressed as a nonlinear minimization problem in the finite-dimensional parameter space $\Xi \subset \mathbb{R}^{2N+5}$ with the following constraints.

Equality Constraints: The equality constraints are defined as $C_e(\xi^*) = 0_{7 \times 1}$ by $C_e(\xi^*) := (C_e^{1'}(\xi^*), C_e^{2'}(\xi^*), C_e^3(\xi^*))'$, in which

$$C_e^1(\xi^*) := q_u^*(T_I^*; \xi^*) - q_u^{-*}$$
$$C_e^2(\xi^*) := \dot{q}_u^*(T_I^*; \xi^*) - \dot{q}_u^{-*}$$
$$C_e^3(\xi^*) := p_2^v(T_I^*; \xi^*).$$

To emphasize the dependence of solutions on the vector of parameters ξ^* in this latter set of equations, we make use of the notations $q_u^*(.; \xi^*)$ and $p_2^v(.; \xi^*)$. In addition, the equality constraints $C_e^1(\xi^*) = 0_{3 \times 1}$ and $C_e^2(\xi^*) = 0_{3 \times 1}$ are necessary and sufficient conditions for the existence of a period-one orbit $\mathcal{O}$ for the impulsive system Σ and the constraint $C_e^3(\xi^*) = 0$ implies that $x^{-*} \in \mathcal{S}$, where $x^{-*} := (q^{-*'}, \dot{q}^{-*'})'$, $q^{-*} := \mathbf{T}^{-1}\tilde{q}^{-*}$, and $\dot{q}^{-*} := \mathbf{T}^{-1}\dot{\tilde{q}}^{-*}$.

Inequality Constraints: To guarantee the feasibility of $\mathcal{O}$, we can define the vector $C_{ie}(\xi^*) := (C_{ie}^1(\xi^*), \ldots, C_{ie}^p(\xi^*))'$ for some positive integer p such that the inequality constraints $C_{ie}^j(\xi^*) \leq 0$ for $j = 1, \ldots, p - 1$ satisfy items 1–6 of Definition 7.1. Moreover, the last component of the vector $C_{ie}(\xi^*)$ is introduced such that $C_{ie}^p(\xi^*) \leq 0$ implies that

$$\kappa(t; \xi^*) := 1 + e_3' \, \mathcal{D}_{uu}^{-1}(q_b^*(t; \xi^*)) \, \mathcal{D}_{ua}(q_b^*(t; \xi^*)) \frac{\ddot{q}_a^*(t; \xi^*)}{\ddot{\theta}^*(t; \xi^*)} \tag{7.6}$$
$$\neq 0, \quad \forall t \in [0, T_I^*],$$

where $e_3' := [0 \; 0 \; 1]$. In Section 7.5, it will be shown that the condition of equation (7.6) referred to as *invertibility of the decoupling matrix on the orbit $\mathcal{O}$* is a necessary condition by which, our control methodology for stabilization of $\mathcal{O}$ can be applied.

Cost Function: The cost function during the motion planning algorithm is defined as

$$\mathcal{J}(\xi^*) := \frac{1}{L_s(\xi^*)} \int_0^{T_I^*} \|u^*(t; \xi^*)\|_2^2 \, dt,$$

where $L_s(\xi^*)$ represents the step length. To solve the minimization problem by using the fmincon function of the MATLAB's Optimization Toolbox, we employ a two-stage strategy. In the first stage, the cost function is chosen as 1 and we search for a feasible period-one solution of the impulsive system of equation (7.2). To simplify the search process, the components of the equality and inequality constraints can be added in a step-by-step fashion. The solution of the first stage will be used as an initial guess to minimize the cost function $\mathcal{J}(\xi^*)$ in the second stage.

7.4 NUMERICAL EXAMPLE

In this section, a numerical example for the proposed motion planning algorithm is presented. The physical parameters of the biped robot are given in Table 7.1. To obtain an anthropomorphic gait, it is assumed that $q_1, q_2(\text{deg}) \in [110°, 200°], q_3, q_4(\text{deg}) \in [1°, 90°], q_5(\text{deg}) \in [45°, 90°]$, and $\theta(\text{deg}) \in [45°, 135°]$. Moreover, due to actuation limits, we assume that $u_{\max} = 100(\text{Nm})$ and the maximum absolute value of angular velocities at the hip joints is $10(\frac{\text{rad}}{\text{s}})$. Also, suppose that $T_I^*(\text{s}) \in [0.1, 1]$ and $w_i^* \in [-5, 5]^2$ for $i = 2, \ldots, N - 2$. For $N = 5$, a local optimal solution for the motion planning algorithm is obtained. Tables 7.2 and 7.3 show the components of ξ^*. At this point, the optimal motion of the robot has a period of $T_I^* = 0.8586(\text{s})$, a step length of $L_s = 0.3857(\text{m})$, and an average walking speed of $0.4492(\frac{\text{m}}{\text{s}})$.

TABLE 7.1 Physical Parameters of the Biped Robot

	Femur	Tibia	Torso
Length in m	0.5	0.5	0.5
Mass in kg	2	1	4
Mass center in m	0.15	0.2	0.25
Inertia in kgm^2	0.2	0.2	0.5

TABLE 7.2 Components of $q_a^{-*}, q_u^{-*}, \dot{q}_a^{-*}$, and $\dot{q}_u^{-*}$

$q_1^{-*}(\text{rad})$	2.8281	$\dot{q}_1^{-*}(\frac{\text{rad}}{\text{s}})$	6.7814
$q_2^{-*}(\text{rad})$	2.6048	$\dot{q}_2^{-*}(\frac{\text{rad}}{\text{s}})$	6.7929
$q_3^{-*}(\text{rad})$	0.4554	$\dot{q}_3^{-*}(\frac{\text{rad}}{\text{s}})$	-1.9315
$q_4^{-*}(\text{rad})$	0.1160	$\dot{q}_4^{-*}(\frac{\text{rad}}{\text{s}})$	-3.9019
$\theta^{-*}(\text{rad})$	1.7059	$\dot{\theta}^{-*}(\frac{\text{rad}}{\text{s}})$	1.7035

TABLE 7.3   Third and Fourth Columns of the Coefficient Matrix w^*

w_2^*	w_3^*
1.3161	4.6458
0.4362	2.6536

The value of the cost function at this point is also equal to $\mathcal{J}(\xi^*) = 501.7779(\text{N}^2\text{ms})$. On the optimal trajectory, the robot will not slip for a coefficient of friction greater than 0.45. A stick animation of the biped robot taking one step of the optimal motion is depicted in Fig. 7.2. Angular positions of the mechanical system during three steps of the optimal motion are presented in Fig. 7.3. The discontinuities are due to coordinate relabeling for swapping the role of the legs. Figure 7.4 shows the angular velocities of the robot during three steps of walking. The discontinuities are due to impacts and coordinate relabeling. The open-loop control inputs, and the horizonal and vertical components of the ground reaction force at the stance leg end are depicted in Fig. 7.5. Moreover, the absolute ratio of the horizontal component to the vertical component and the path of the swing leg end in the sagittal plane are presented in Fig. 7.5.

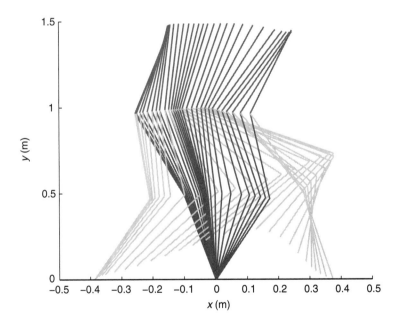

Figure 7.2   Stick animation of the bipedal robot during one step of the optimal motion. (See the color version of this figure in the color plates section.)

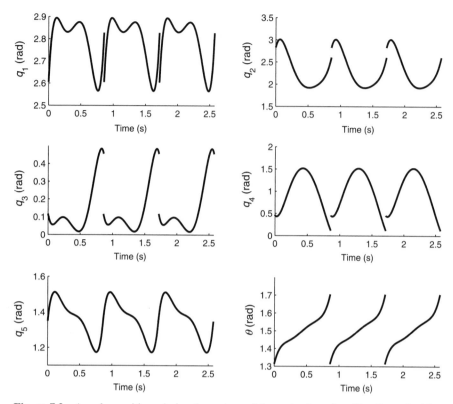

Figure 7.3   Angular positions during three steps of the optimal motion. The discontinuities are due to coordinate relabeling for swapping the role of the legs.

7.5   CONTINUOUS-TIMES CONTROLLERS

To exponentially stabilize a desired period-one orbit $\mathcal{O}$ generated by the motion planning algorithm for the impulsive model of walking, this section presents a time-invariant control scheme that is applied at two levels. In this approach, for a given integer number $M \geq 2$, the state space of the system is split into M subspaces by defining $M - 1$ within-stride switching hypersurfaces. To reduce the dimension of the stabilization problem at the first level of the control scheme, parameterized continuous-time controllers are employed to create finite-time attractive and forward invariant manifolds in the corresponding internal phase. The event-based controllers, which are applied at the second level, update the parameters of the continuous-time controllers during the transitions among the internal phases. This section presents the continuous-time controllers. The event-based update laws will be treated in Sections 7.6 and 7.7.

Assume that θ^{+*} and θ^{-*} are the initial and final values of θ on the orbit $\mathcal{O}$, respectively. From Definition 7.1, we remark that $\theta^{+*} < \theta^{-*}$. Let $M \geq 2$ be an integer number and $\theta^{+*} < \theta_1^* < \theta_2^* < \cdots < \theta_{M-1}^* < \theta^{-*}$ be a partition of the interval

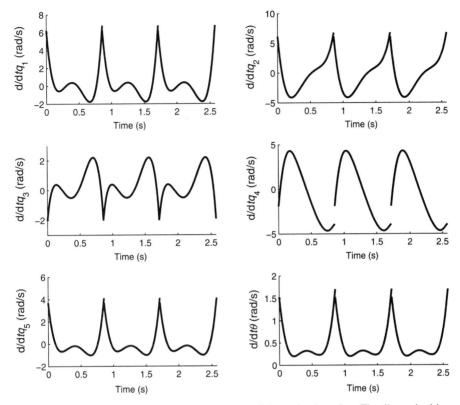

Figure 7.4 Angular velocities during three steps of the optimal motion. The discontinuities are due to impacts and coordinate relabeling for swapping the role of the legs.

$[\theta^{+*}, \theta^{-*}]$. During a step, split the single support phase into M *internal phases*. Denote the index of these phases by j that takes values in the discrete set $\{1, \dots, M\}$. Next introduce the following switching hypersurfaces among the internal phases:

$$\mathcal{S}_1^2 := \{x = (q', \dot{q}')' \in \mathcal{X} | \theta = \theta_1^*\}$$

$$\vdots$$

$$\mathcal{S}_{M-1}^M := \{x = (q', \dot{q}')' \in \mathcal{X} | \theta = \theta_{M-1}^*\}$$

$$\mathcal{S}_M^1 := \mathcal{S}.$$

(7.7)

For simplicity, we define the *index of next phase* function as $\bar{i} : \{1, \dots, M\} \to \{1, \dots, M\}$ by

$$\bar{i}(j) := j + 1, \quad j = 1, \dots, M - 1$$
$$\bar{i}(M) := 1.$$

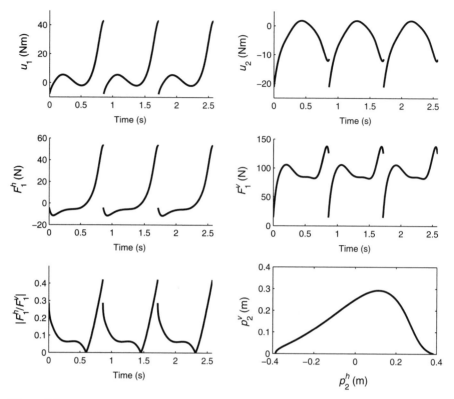

Figure 7.5 Plots of the open-loop control inputs (two top graphs) during three steps of the optimal motion. Two middle graphs represent the horizontal and vertical components of the ground reaction force experienced at the end of leg-1 during three steps. The bottom graphs depict the absolute ratio of the horizontal component to the vertical component of the ground reaction force (during three steps) and the path of the swing leg end in the sagittal plane (during one step).

In our notation, $\bar{S}_j^{\bar{i}(j)}$ represents the switching hypersurface from the jth phase into the $\bar{i}(j)$th phase. Subsequently, the transitions $j \to \bar{i}(j)$ for $j = 1, \ldots, M-1$ are called *within-stride transitions*, whereas the transition $M \to 1$ is called the *impact transition*. The switching maps $\Delta_j^{\bar{i}(j)} : \bar{S}_j^{\bar{i}(j)} \to \mathcal{X}$ are also defined as

$$\Delta_j^{\bar{i}(j)}(x) := \mathrm{Id}_{\bar{S}_j^{\bar{i}(j)}}(x), \quad j = 1, \ldots, M-1$$

$$\Delta_M^1(x) := \Delta(x),$$

in which $\mathrm{Id}_{\bar{S}_j^{\bar{i}(j)}}$ denotes the identity map. In other words, the position and velocity are assumed to be continuous during the within-stride transitions. The switching hypersurfaces in equation (7.7) motivate us to study the stabilization problem of the desired orbit $\mathcal{O}$ for the hybrid system $\Sigma_{\mathcal{H}}(\Sigma^1, \dots, \Sigma^M)$ with the following form:

$$\Sigma^1 : \begin{cases} \dot{x}^1 = f(x^1) + g(x^1)u & x^{1-} \notin \mathcal{S}_1^2 \\ x^{2+} = \Delta_1^2(x^{1-}) & x^{1-} \in \mathcal{S}_1^2 \end{cases}$$

$$\vdots \tag{7.8}$$

$$\Sigma^M : \begin{cases} \dot{x}^M = f(x^M) + g(x^M)u & x^{M-} \notin \mathcal{S}_M^1 \\ x^{1+} = \Delta_M^1(x^{M-}) & x^{M-} \in \mathcal{S}_M^1. \end{cases}$$

In this latter set of equations, x^j, $j = 1, \dots, M$ denotes the state of the mechanical system during the jth internal phase. By the construction procedure of $\Sigma_{\mathcal{H}}$, $\mathcal{O}$ is also a period-one orbit for the hybrid system $\Sigma_{\mathcal{H}}$ and it can be expressed as $\mathcal{O} = \cup_{j=1}^M \mathcal{O}^j$, where $\mathcal{O}^i \cap \mathcal{O}^j = \phi$ for every $i \neq j \in \{1, \dots, M\}$. Moreover, since θ is a strictly increasing function of time on $\mathcal{O}$ (see item 6 of Definition 7.1), the intersections $\overline{\mathcal{O}}^j \cap \mathcal{S}_j^{\bar{i}(j)}$, $j = 1, \dots, M$ are singletons, in which $\overline{\mathcal{O}}^j$ denotes the closure of $\mathcal{O}^j$. In particular, let $\{x^{j-*}\} := \overline{\mathcal{O}}^j \cap \mathcal{S}_j^{\bar{i}(j)}$ for $j = 1, \dots, M$. We observe that $x^{M-*} = x^{-*}$.

Next, assume that $h_{d,\mathcal{O}} : \mathbb{R} \to \mathbb{R}^2$ is at least a C^2 function such that the *nominal holonomic output function* $h_{\mathcal{O}} : \mathcal{Q} \to \mathbb{R}^2$ by

$$h_{\mathcal{O}}(q) := q_a - h_{d,\mathcal{O}}(\theta)$$

vanishes on the single support phase of the periodic orbit $\mathcal{O}$. In particular, $h_{d,\mathcal{O}}$ represents the desired evolution of the hip joints on $\mathcal{O}$ in terms of θ. Now for a given $M \geq 2$ and the sequence $\{\theta_j^*\}_{j=1}^{M-1}$ with the property $\theta^{+*} < \theta_1^* < \theta_2^* < \cdots < \theta_{M-1}^* < \theta^{-*}$, let $\psi^j \in \Psi^j$, $j = 1, \dots, M$ represent the parameter vector of the jth phase controller during a step. In addition, Ψ^j denotes the corresponding parameter space. By adding an augmentation function as a Bézier polynomial to the nominal holonomic output function $h_{\mathcal{O}}$, we define the *parameterized holonomic output function* $y^j : \mathcal{X} \times \Psi^j \to \mathbb{R}^2$ for the jth phase as follows:

$$\begin{aligned} y^j(x^j; \psi^j) &:= h^j(q; \psi^j) \\ &:= q_a - h_{d,\mathcal{O}}(\theta) + \mathcal{B}(s^j; \alpha^j) \\ &= q_a - \left(h_{d,\mathcal{O}}(\theta) - \mathcal{B}(s^j; \alpha^j)\right) \\ &=: q_a - h_d^j(\theta; \psi^j). \end{aligned} \tag{7.9}$$

In equation (7.9), the superscript "j" stands for the jth phase. It is also assumed that the additive Bézier polynomial is of the degree $n \geq 5$ and the coefficient matrix

$\alpha^j := \mathrm{col}\{\alpha_i^j\}_{i=0}^n \in \mathbb{R}^{2\times(n+1)}$. Furthermore, the arguments of the Bézier polynomial are defined as

$$
s^j(\theta) := \begin{cases}
(\theta - \theta^+)/(\theta_1^* - \theta^+), & j = 1 \\[2ex]
(\theta - \theta_{j-1}^*)/(\theta_j^* - \theta_{j-1}^*), & j = 2, \ldots, M - 1 \\[2ex]
(\theta - \theta_{M-1}^*)/(\theta^- - \theta_{M-1}^*), & j = M,
\end{cases}
\tag{7.10}
$$

where θ^+ and θ^- are the parameters of the first and Mth phase controller to be determined in Section 7.6. Consequently, the parameter vector of the jth phase controller can be expressed as

$$
\psi^j := \begin{cases}
\left(\alpha_0^{1'}, \alpha_1^{1'}, \ldots, \alpha_n^{1'}, \theta^+\right)', & j = 1 \\[2ex]
\left(\alpha_0^{j'}, \alpha_1^{j'}, \ldots, \alpha_n^{j'}\right)', & j = 2, \ldots, M - 1 \\[2ex]
\left(\alpha_0^{M'}, \alpha_1^{M'}, \ldots, \alpha_n^{M'}, \theta^-\right)', & j = M.
\end{cases}
$$

By applying the input–output linearization [103] during the jth phase, it can be shown that

$$
\ddot{y}^j(x^j; \psi^j) = L_g L_f y^j(x^j; \psi^j)\, u + L_f^2 y^j(x^j; \psi^j),
$$

where

$$
L_g L_f y^j(x^j; \psi^j) := \frac{\partial h^j}{\partial q}(q; \psi^j)\, D^{-1}(q_b)\, B
$$

$$
L_f^2 y^j(x^j; \psi^j) := \frac{\partial}{\partial q}\left(\frac{\partial h^j}{\partial q}(q; \psi^j)\dot{q}\right)\dot{q} - \frac{\partial h^j}{\partial q}(q; \psi^j)\, D^{-1}(q_b)\left(C(q_b, \dot{q})\dot{q} + G(q)\right).
$$

For the later purposes, $L_g L_f y^j(x^j; \psi^j)$ is called the *decoupling matrix*. The following lemma studies the invertibility of the decoupling matrices on the orbit $\mathcal{O}$.

Lemma 7.1 (Invertibility of the Decoupling Matrices on $\mathcal{O}$) *Assume that $\mathcal{O}$ is a feasible period-one orbit for the impulsive system of equation (7.2) generated by the motion planning algorithm. Let $\psi^{j*}, j = 1, \ldots, M$ represent the nominal parameter vector of the single support phase controller during the jth phase, that is, $\psi^{j*} := 0_{2(n+1)\times 1}$ for $j = 2, \ldots, M - 1$, $\psi^{1*} := (0_{1\times 2(n+1)}, \theta^{+*})'$ and $\psi^{M*} := (0_{1\times 2(n+1)}, \theta^{-*})'$. Then, the invertibility of the decoupling matrices*

$L_g L_f y^j(x^j; \psi^{j*})$, $j = 1, \ldots, M$ on the orbit $\mathcal{O}$ is equivalent to condition of equation (7.6).

Proof. Since $\mathcal{D}^{-1} = \mathbf{T} D^{-1} \mathbf{T}'$ and $(\mathbf{T}')^{-1} B = B$, the decoupling matrix can be rewritten as follows:

$$L_g L_f h^j(q; \psi^j) = \frac{\partial h^j}{\partial q} D^{-1} B = \frac{\partial h^j}{\partial \tilde{q}} \frac{\partial \tilde{q}}{\partial q} D^{-1} B$$

$$= \left[I_{2\times 2} \ 0_{2\times 2} \ -\frac{\partial h^j_d}{\partial \theta} \right] D^{-1} B.$$

By defining

$$\Pi := \mathcal{D}^{-1} = \begin{bmatrix} \Pi_{aa} & \Pi_{au} \\ \Pi'_{au} & \Pi_{uu} \end{bmatrix}$$

and also considering $B = [I_{2\times 2} \ 0_{2\times 3}]'$, the decoupling matrix can be expressed as

$$L_g L_f h^j(q; \psi^j) = \Pi_{aa} - \frac{\partial h^j_d}{\partial \theta} e'_3 \Pi'_{au},$$

together with

$$\Pi_{aa} = \left(\mathcal{D}_{aa} - \mathcal{D}_{au} \mathcal{D}_{uu}^{-1} \mathcal{D}'_{au} \right)^{-1}$$

$$\Pi_{au} = -\Pi_{aa} \mathcal{D}_{au} \mathcal{D}_{uu}^{-1}$$

results in

$$L_g L_f h^j(q; \psi^j) = \left(I_{2\times 2} + \frac{\partial h^j_d}{\partial \theta} e'_3 \mathcal{D}_{uu}^{-1} \mathcal{D}_{ua} \right) \left(\mathcal{D}_{aa} - \mathcal{D}_{au} \mathcal{D}_{uu}^{-1} \mathcal{D}'_{au} \right)^{-1}.$$

Consequently,[3]

$$\det(L_g L_f h^j(q; \psi^j)) = \frac{1 + e'_3 \mathcal{D}_{uu}^{-1} \mathcal{D}_{ua} \frac{\partial h^j_d}{\partial \theta}}{\det(\mathcal{D}_{aa} - \mathcal{D}_{au} \mathcal{D}_{uu}^{-1} \mathcal{D}'_{au})}.$$

---

[3] Note that the matrix $\mathcal{D}$ is positive definite. In addition, we make use of the identity

$$\det(I_{n\times n} + A B) = \det(I_{m\times m} + B A)$$

for every $A \in \mathbb{R}^{n\times m}$ and $B \in \mathbb{R}^{m\times n}$.

Finally, the fact that on the periodic orbit $\mathcal{O}$, $\psi^j = \psi^{j*}$ and thereby $\frac{\partial h_d^j}{\partial \theta} = \frac{\dot{q}_a^*}{\dot{\theta}^*}$, completes the proof. ∎

If the decoupling matrix $L_g L_f y^j(x^j; \psi^{j*})$ is invertible on the orbit $\mathcal{O}^j$, there exists an open neighborhood $\mathcal{N}^j(\mathcal{O}^j \times \psi^{j*}) \subset \mathcal{X} \times \Psi^j$ such that for every $(x^j; \psi^j) \in \mathcal{N}^j$, the control input

$$u_{\text{cl}}^j(x^j; \psi^j) := -(L_g L_f y^j(x^j; \psi^j))^{-1} \left(L_f^2 y^j(x^j; \psi^j) - v^j(y^j, \dot{y}^j) \right) \tag{7.11}$$

is well defined and results in the closed-loop dynamics $\ddot{y}^j = v^j(y^j, \dot{y}^j)$. Assume that $v^j : \mathbb{R}^2 \times \mathbb{R}^2 \to \mathbb{R}^2$ is a continuous function such that the origin for the closed-loop dynamics $\ddot{y}^j = v^j(y^j, \dot{y}^j)$ is globally finite-time stable. For this purpose, the approaches of Refs. [46, 93] can be applied. Next we introduce the *parameterized zero dynamics manifold of the jth phase* as follows:

$$\mathcal{Z}_{\psi^j}^j := \{x^j \in \mathcal{X} | y^j(x^j; \psi^j) = 0_{2\times 1}, L_f y^j(x^j; \psi^j) = 0_{2\times 1}\}.$$

It can be shown that $\mathcal{N}_x^j \cap \mathcal{Z}_{\psi^j}^j$ is a six-dimensional embedded submanifold of $\mathcal{X}$, where $\mathcal{N}_x^j$ is the projection of $\mathcal{N}^j$ onto $\mathcal{X}$. Moreover, from the definition of switching hypersurfaces $\mathcal{S}_j^{\bar{i}(j)}$ in equation (7.7), $\mathcal{N}_x^j \cap \mathcal{Z}_{\psi^j}^j \cap \mathcal{S}_j^{\bar{i}(j)}$ for $j = 1, \ldots, M-1$ is a five-dimensional embedded submanifold of $\mathcal{X}$. We also assume that this is true for $j = M$. The following lemma presents a valid coordinates transformation for the manifolds $\mathcal{N}_x^j \cap \mathcal{Z}_{\psi^j}^j$, $j = 1, \ldots, M$.

Lemma 7.2 (Zero Dynamics)   *Define the following conjugate momenta vector*

$$\sigma_u := \frac{\partial \mathcal{L}'}{\partial \dot{q}_u} = \mathcal{D}_{ua} \dot{q}_a + \mathcal{D}_{uu} \dot{q}_u,$$

in which $\mathcal{L}$ represents the Lagrangian of the single support phase. Then, (q_u, σ_u) is a valid local coordinates transformation for $\mathcal{N}_x^j \cap \mathcal{Z}_{\psi^j}^j$, $j = 1, \ldots, M$. Moreover, in these coordinates, the zero dynamics of the jth phase is given by

$$\dot{q}_u = \left(I_{3\times 3} - \frac{\mathcal{D}_{uu}^{-1} \mathcal{D}_{ua} \frac{\partial h_d^j}{\partial \theta} e_3'}{1 + e_3' \mathcal{D}_{uu}^{-1} \mathcal{D}_{ua} \frac{\partial h_d^j}{\partial \theta}} \right) \mathcal{D}_{uu}^{-1} \sigma_u \tag{7.12}$$

$$\dot{\sigma}_u = \mathcal{H}_u.$$

Proof. Since (i) the distribution generated by the columns of the matrix g (i.e., span$\{g_1, g_2\}$) is involutive and (ii) $L_{g_i}\sigma_u = 0$ for $i = 1, 2$ (this is a consequence of unactuation of q_u), by Ref. [103, p. 222], (q_u, σ_u) is a valid coordinates transformation on $\mathcal{N}_x^j \cap \mathcal{Z}_{\psi^j}^j$ for $j = 1, \ldots, M$. In addition, since the components of q_u are unactuated, the Euler–Lagrange equation immediately implies that $\dot{\sigma}_u = \mathcal{H}_u$. Finally, on the manifold $\mathcal{N}_x^j \cap \mathcal{Z}_{\psi^j}^j$,

$$\dot{q}_u = \left(\mathcal{D}_{uu} + \mathcal{D}_{ua}\frac{\partial h_d^j}{\partial \theta}e_3'\right)^{-1}\sigma_u$$

together with the Matrix Inversion Lemma[4] completes the proof. ∎

Remark 7.3 (Valid Coordinates for $\mathcal{N}_x^j \cap \mathcal{Z}_{\psi^j}^j \cap \mathcal{S}_j^{\bar{i}(j)}$) *From the definition of the within-stride switching hypersurfaces $\mathcal{S}_j^{\bar{i}(j)}$ in equation (7.7), it can be concluded that (q_3, q_4, σ_u) is a valid local coordinates transformation for $\mathcal{N}_x^j \cap \mathcal{Z}_{\psi^j}^j \cap \mathcal{S}_j^{\bar{i}(j)}$, $j = 1, \ldots, M-1$.*

7.6 EVENT-BASED CONTROLLERS

During the transition $j \rightarrow \bar{i}(j)$, $j = 1, \ldots, M$, the *event-based controller* $\pi_j^{\bar{i}(j)}$ updates the parameters of the $\bar{i}(j)$th phase continuous-time controller. The parameter vector $\psi^{\bar{i}(j)}$ remains constant during the $\bar{i}(j)$th phase, that is, $\dot{\psi}^{\bar{i}(j)} = 0$. The purpose of updating the parameters in an event-based manner is (i) to achieve hybrid invariance, (ii) continuity of the continuous-time controllers during the within-stride transitions, and (iii) exponential stabilization of the orbit $\mathcal{O}$ for the system $\Sigma_\mathcal{H}$.

7.6.1 Hybrid Invariance

By applying the continuous-time controllers u_{cl}^j, $j = 1, \ldots, M$, zero dynamics manifolds $\mathcal{Z}_{\psi^j}^j$ are forward invariant.

Definition 7.2 (Hybrid Invariance) *Under the event-based update laws $(\pi_1^2, \pi_2^3, \ldots, \pi_M^1)$, the manifolds $\{\mathcal{Z}_{\psi^j}^j\}_{j=1}^M$ are said to be hybrid invariant for the hybrid system $\Sigma_\mathcal{H}$ if there exist open neighborhoods $\mathcal{V}^j$ of x^{j-*} such that for every $x^{j-} \in \mathcal{V}^j(x^{j-*}) \cap \mathcal{S}_j^{\bar{i}(j)} \cap \mathcal{Z}_{\psi^j}^j$ and $j = 1, \ldots, M$,*

$$\Delta_j^{\bar{i}(j)}(x^{j-}) \in \mathcal{Z}_{\psi^{\bar{i}(j)}}^{\bar{i}(j)},$$

where $\psi^{\bar{i}(j)} := \pi_j^{\bar{i}(j)}(x^{j-})$.

---

[4] The Matrix Inversion Lemma states that for every $A \in \mathbb{R}^{n \times m}$ and $B \in \mathbb{R}^{m \times n}$, if the matrix $(I_{n \times n} + AB)$ is invertible, then

$$(I_{n \times n} + AB)^{-1} = I_{n \times n} - A(I_{m \times m} + BA)^{-1}B.$$

Lemma 7.3 (Hybrid Invariance) *Let $\mathcal{Z}_O$ be the zero dynamics manifold corresponding to the nominal holonomic output function h_O, that is,*

$$\mathcal{Z}_O := \left\{ x = (q', \dot{q}')' \in \mathcal{X} | h_O(q) = 0_{2 \times 1}, \frac{\partial h_O}{\partial q}(q) \dot{q} = 0_{2 \times 1} \right\}.$$

Assume that the event-based update laws $(\pi_1^2, \pi_2^3, \ldots, \pi_M^1)$ are such that for every $j = 2, \ldots, M - 1$,

$$\alpha_0^j = \alpha_1^j = \alpha_{n-1}^j = \alpha_n^j = 0_{2 \times 1} \tag{7.13}$$

and

$$\begin{aligned} \alpha_{n-1}^1 = \alpha_n^1 &= 0_{2 \times 1} \\ \alpha_0^M = \alpha_1^M &= 0_{2 \times 1}. \end{aligned} \tag{7.14}$$

Then, the following statements are true.

(a) *The intersections $\bar{S}_j^{\bar{i}(j)} \cap \mathcal{Z}_{\psi^j}^j$ for $j = 1, \ldots, M - 1$ are independent of ψ^j. In addition, these common intersections are equal to $\bar{S}_j^{\bar{i}(j)} \cap \mathcal{Z}_O$.*

(b) *The sets $\Delta_j^{\bar{i}(j)}(\bar{S}_j^{\bar{i}(j)} \cap \mathcal{Z}_{\psi^j}^j)$ for $j = 1, \ldots, M - 1$ are independent of ψ^j and $\psi^{\bar{i}(j)}$, and equal to $\Delta_j^{\bar{i}(j)}(\bar{S}_j^{\bar{i}(j)} \cap \mathcal{Z}_O)$. Also, for every $x^{j-} \in \bar{S}_j^{\bar{i}(j)} \cap \mathcal{Z}_{\psi^j}^j$, $j = 1, \ldots, M - 1$,*

$$\Delta_j^{\bar{i}(j)}(x^{j-}) \in \mathcal{Z}_{\psi^{\bar{i}(j)}}^{\bar{i}(j)},$$

where $\psi^{\bar{i}(j)} := \pi_j^{\bar{i}(j)}(x^{j-})$.

(c) *Let $x^{M-} = (q^{-'}, \dot{q}^{-'})' \in \bar{S}_M^1$ represent the state of the mechanical system immediately before the impact. Next define*

$$\begin{aligned} \theta^- &:= \theta^{-*} \\ q_a^+ &:= E_1 q^+ = E_1 R q^- \\ \dot{q}_a^+ &:= E_1 \dot{q}^+ = E_1 R \Delta(q^-) \dot{q}^- \\ \theta^+ &:= \Gamma_0 q^+ = \Gamma_0 R q^- \\ \dot{\theta}^+ &:= \Gamma_0 \dot{q}^+ = \Gamma_0 R \Delta(q^-) \dot{q}^-, \end{aligned} \tag{7.15}$$

in which $E_1 := [I_{2\times 2} \; 0_{2\times 3}]$, and update the parameters α_0^1 and α_1^1 as follows:

$$\alpha_0^1 = -q_a^+ + h_{d,\mathcal{O}}(\theta^+)$$

$$\alpha_1^1 = \alpha_0^1 - \frac{\theta_1^* - \theta^+}{n\,\dot\theta^+}\left(\dot{q}_a^+ - \frac{\partial h_{d,\mathcal{O}}}{\partial \theta}(\theta^+)\dot\theta^+\right). \tag{7.16}$$

Then, there exists open neighborhood $\mathcal{V}^M$ of x^{M-*} such that for every $x^{M-} \in \mathcal{V}^M(x^{M-*}) \cap \mathcal{S}_M^1 \cap \mathcal{Z}_{\psi^M}^M$,

$$\Delta_M^1(x^{M-}) \in \mathcal{Z}_{\psi^1}^1,$$

where $\psi^1 := \pi_M^1(x^{M-})$.

Proof. By definition of s^j in equation (7.10), at the end of the jth phase, it follows that

$$s^j = 1, \quad j = 1, \dots, M-1.$$

From properties of the Bézier polynomials given in Remark 3.16, this latter fact together with $\alpha_{n-1}^j = \alpha_n^j = 0_{2\times 1}$ implies part (a). In addition, $\Delta_j^{\bar{i}(j)}(\mathcal{S}_j^{\bar{i}(j)} \cap \mathcal{Z}_{\psi^j}^j)$ is also independent of ψ^j. By an analogous reasoning, it can be shown that $\alpha_0^j = \alpha_1^j = 0_{2\times 1}$ for $j = 2, \dots, M$ together with $s^j = 0$ at the beginning of the jth phase, $j = 2, \dots, M$, results in the first statement of part (b). By part (a), this latter fact and continuity of the position and velocity during the transition $j \to \bar{i}(j)$, $j = 1, \dots$, $M-1$ complete the proof of part (b).

Since the orbit $\mathcal{O}$ is feasible, from item 6 of Definition 7.1 and continuity of the impact map $\Delta_M^1 : \mathcal{S}_M^1 \to \mathcal{X}$, it can be concluded that there exists an open neighborhood $\mathcal{V}^M$ of x^{M-*} such that for every $x^{M-} = (q^{-\prime}, \dot{q}^{-\prime})\prime \in \mathcal{S}_M^1 \cap \mathcal{V}^M$, $\dot\theta^+$ defined in equation (7.15) is positive. Thus, α_1^1 in equation (7.16) is well defined. Moreover, the update laws for α_0^1 and α_1^1 immediately imply that

$$h^1(q^+; \psi^1) = 0_{2\times 1}$$

$$\frac{\partial h^1}{\partial q}(q^+; \psi^1)\dot{q}^+ = 0_{2\times 1},$$

which, in turn, completes the proof of part (c).   ■

The geometry derived from the event-based update laws of Lemma 7.3 and continuous-time controllers of Section 7.5 for $M = 5$ is illustrated in Fig. 7.6.

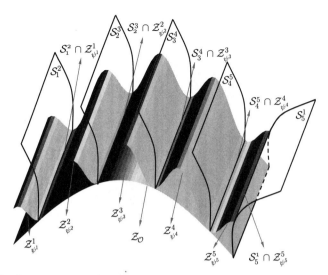

Figure 7.6 Geometry derived from the event-based update laws of Lemma 7.3 and continuous-time controllers of Section 7.5 for $M = 5$. It is seen that the manifolds $\{\mathcal{Z}^j_{\psi^j}\}^5_{j=1}$ are hybrid invariant for the closed-loop hybrid system $\Sigma_{\mathcal{H}}$ under the event-based update laws $(\pi^2_1, \pi^3_2, \pi^4_3, \pi^5_4, \pi^1_5)$. The intersections $\mathcal{S}^{\bar{i}(j)}_j \cap \mathcal{Z}^j_{\psi^j}$ are independent of ψ^j and equal to $\mathcal{S}^{\bar{i}(j)}_j \cap \mathcal{Z}_{\mathcal{O}}$ for $j = 1, 2, 3, 4$. This is not true for the case $j = 5$.

7.6.2 Continuity of the Continuous-Time Controllers During the Within-Stride Transitions

Assume that all assumptions of Lemma 7.3 hold. Then, the event-based update laws $(\pi^2_1, \pi^3_2, \ldots, \pi^1_M)$ given in equations (7.13)–(7.16) result in hybrid invariance. However, we desire that during the within-stride transitions (i.e., $j \to \bar{i}(j)$ for $j = 1, \ldots, M - 1$), the hip torques remain continuous. For this purpose, we present the following lemma by which the acceleration is also imposed to be continuous.

Lemma 7.4 (Continuity of the Continuous-Time Controllers) *Under the assumptions of Lemma 7.3 and the event-based update laws $(\pi^2_1, \pi^3_2, \ldots, \pi^1_M)$, as given in equations (7.13)–(7.16), the additional conditions*

$$\alpha^{\bar{i}(j)}_2 = \alpha^j_{n-2}, \quad j = 1, \ldots, M - 1 \tag{7.17}$$

result in continuity of the continuous-time controllers during the within-stride transitions.

Proof. Since the continuous function v^j vanishes on the zero dynamics manifold $\mathcal{Z}^j_{\psi^j}$, $j = 1, \ldots, M$, the restriction of the continuous-time controller u^j_{cl} to the manifold $\mathcal{Z}^j_{\psi^j}$ can be expressed as

$$u^j_{\text{cl}} = \left(\frac{\partial h^j}{\partial q} D^{-1} B \right)^{-1} \left(\frac{\partial}{\partial q} \left(\frac{\partial h^j}{\partial q} \dot{q} \right) \dot{q} - \frac{\partial h^j}{\partial q} D^{-1} (C\dot{q} + G) \right). \tag{7.18}$$

Equations (7.13) and (7.14) and the fact that the position and velocity remain continuous during the within-stride transition $j \to \bar{i}(j)$, $j = 1, \ldots, M - 1$, imply the continuity of the terms $\frac{\partial h^j}{\partial q} D^{-1} B$ and $\frac{\partial h^j}{\partial q} D^{-1}(C\dot{q} + G)$ during this transition. It can also be easily shown that

$$\frac{\partial}{\partial q} \left(\frac{\partial h^j}{\partial q} (q; \psi^j) \dot{q} \right) \dot{q} = - \frac{\partial^2 h^j_d}{\partial \theta^2} (\theta; \psi^j)(\Gamma_0 \dot{q})^2.$$

Consequently, the following condition

$$\frac{\partial^2 h^j_d}{\partial \theta^2} \left(\theta^*_j; \psi^j \right) = \frac{\partial^2 h^{\bar{i}(j)}_d}{\partial \theta^2} \left(\theta^*_j; \psi^{\bar{i}(j)} \right)$$

for $j = 1, \ldots, M - 1$ results in

$$u^j_{\text{cl}}(x^{j-}; \psi_j) = u^{\bar{i}(j)}_{\text{cl}} \left(\Delta^{\bar{i}(j)}_j (x^-_j), \pi^{\bar{i}(j)}_j (x^-_j) \right),$$

which, in turn, together with properties of Bézier polynomials given in Remark 3.16 completes the proof. ∎

Remark 7.4 (Simplification of the Stabilization Problem) *To simplify the stabilization problem in Section 7.7, we shall assume that*

$$\begin{aligned} \alpha^1_2 = \alpha^2_2 = \cdots = \alpha^M_2 = 0_{2 \times 1} \\ \alpha^1_{n-2} = \alpha^2_{n-2} = \cdots = \alpha^{M-1}_{n-2} = 0_{2 \times 1}. \end{aligned} \tag{7.19}$$

From Lemma 7.4, equation (7.19) results in the continuity of the control inputs during the within-stride transitions. In addition, by equation (7.19), the event-based update laws $\pi^{\bar{i}(j)}_j$, $j = 1, \ldots, M - 1$ can be expressed as a function of x^{j-} and not α^j (see equation (7.17)).

7.7 STABILIZATION PROBLEM

In this section, the event-based update laws proposed by Lemmas 7.3 and 7.4 are modified such that $\mathcal{O}$ is an exponentially stable orbit for the closed-loop hybrid model of walking. Our approach for stabilization of $\mathcal{O}$ is based on the Poincaré sections method developed for systems with impulse effects given in Chapter 2. However, in this chapter, the Poincaré section is chosen to be one of the within-stride switching hypersurfaces. To make this notion precise, we first present some definitions. Since the closed-loop system is time invariant, the solution of the jth internal phase starting from the initial condition x^j (at $t_0 = 0$) can be denoted by

$$\varphi_{\text{cl}}(t; x^j, \psi^j) := \varphi(t; 0, x^j, u^j_{\text{cl}})$$

at time $t \geq 0$. Next for $j = 1, \ldots, M$, define the function $T^j : \mathcal{X} \times \Psi^j \to \mathbb{R}$ by

$$T^j(x^j; \psi^j) := \inf \left\{ t > 0 | \varphi_{\text{cl}}(t; x^j, \psi^j) \in \mathcal{S}^{\bar{i}(j)}_j \right\}.$$

Let the *generalized Poincaré map* of the jth phase, $j = 1, \ldots, M$, given by $\mathcal{P}^j : \mathcal{S}^{\bar{i}(j)}_j \times \Psi^{\bar{i}(j)} \to \mathcal{S}^{\bar{j} \circ \bar{i}(j)}_{\bar{i}(j)}$ be expressed as

$$\mathcal{P}^j(x^{j-}; \psi^{\bar{i}(j)}) := \varphi_{\text{cl}} \left(T^{\bar{i}(j)} \left(\Delta^{\bar{i}(j)}_j(x^{j-}); \psi^{\bar{i}(j)} \right); \Delta^{\bar{i}(j)}_j(x^{j-}), \psi^{\bar{i}(j)} \right).$$

Next by taking the Poincaré section as $\mathcal{S}^2_1$, the Poincaré return map can be expressed as $\mathcal{P} : \mathcal{S}^2_1 \times \Psi^2 \times \cdots \times \Psi^M \times \Psi^1$ by

$$\mathcal{P} := \mathcal{P}^M \circ \cdots \circ \mathcal{P}^1. \tag{7.20}$$

By considering Lemmas 7.3 and 7.4, we denote the remaining parameters of the jth phase controller by ψ^j_{rem}, where

$$\psi^j_{\text{rem}} := \begin{cases} \left(\alpha^{j'}_3, \ldots, \alpha^{j'}_{n-3} \right)', & j = 1, \ldots, M-1 \\ \left(\alpha^{M'}_3, \ldots, \alpha^{M'}_{n-2}, \alpha^{M'}_{n-1}, \alpha^{M'}_n \right)', & j = M. \end{cases}$$

Slightly abusing the notation, we assume that for a given $x^{1-} \in \mathcal{S}^2_1$,

$$\mathcal{P}(x^{1-}; \psi_s)$$

represents the *parameterized Poincaré return map* corresponding to x^{1-}, where

$$\psi_s := \left(\psi^{2'}_{\text{rem}}, \ldots, \psi^{M'}_{\text{rem}}, \psi^{1'}_{\text{rem}} \right)' \in \mathbb{R}^{2(n-5)M+6}$$

denotes the stabilizing parameters vector.

The continuous-time controllers developed in Section 7.5 in combination with the event-based update laws of Lemmas 7.3 and 7.4 reduce the stability analysis of the orbit $\mathcal{O}$ for the full-order hybrid model $\Sigma_{\mathcal{H}}$ to that of the following reduced-order hybrid model:

$$\Sigma^1|_{\mathcal{Z}^1_{\psi^1}} : \begin{cases} \dot{z}^1 = f^1_{\text{zero}}\left(z^1; \psi^1_{\text{rem}}\right) & z^{1-} \notin \mathcal{S}^2_1 \cap \mathcal{Z}^1_{\psi^1} \\ z^{2+} = \Delta^2_1(z^{1-}) & z^{1-} \in \mathcal{S}^2_1 \cap \mathcal{Z}^1_{\psi^1} \end{cases}$$

$$\vdots \tag{7.21}$$

$$\Sigma^M|_{\mathcal{Z}^M_{\psi^M}} : \begin{cases} \dot{z}^M = f^M_{\text{zero}}\left(z^M; \psi^M_{\text{rem}}\right) & z^{M-} \notin \mathcal{S}^1_M \cap \mathcal{Z}^M_{\psi^M} \\ z^{1+} = \Delta^1_M(z^{M-}) & z^{M-} \in \mathcal{S}^1_M \cap \mathcal{Z}^M_{\psi^M}, \end{cases}$$

which is referred to as the *HZD*. The Poincaré return map for the HZD can be expressed as $\rho(z^{1-}; \psi_s)$, where

$$\rho := \rho^M \circ \cdots \circ \rho^1$$

and ρ^j is the restriction of $\mathcal{P}^j$ to $\mathcal{Z}^j_{\psi^j}$, that is, $\rho^j := \mathcal{P}^j|_{\mathcal{Z}^j_{\psi^j}}$. Thus, we can consider the following discrete-time system for stabilization of $\mathcal{O}$:

$$z^{1-}[k+1] = \rho(z^{1-}[k]; \psi_s[k]). \tag{7.22}$$

In equation (7.22), $k \in \{1, 2, \ldots\}$ represents the step number and $\psi_s[k]$ is considered as a control input to be updated in a step-by-step manner on the hypersurface $\mathcal{S}^2_1$. Moreover, by part (a) of Lemma 7.3, the state space for equation (7.22) is taken as the five-dimensional manifold $\mathcal{S}^2_1 \cap \mathcal{Z}_{\mathcal{O}}$ that is independent of the control input ψ_s. Let z^{1-*} be the projection of x^{1-*} onto $\mathcal{S}^2_1 \cap \mathcal{Z}_{\mathcal{O}}$. From the construction procedure, z^{1-*} is an equilibrium point of equation (7.22) when ψ_s is replaced by the nominal parameter $\psi^*_s := 0_{(2(n-5)M+6) \times 1}$. The following theorem presents the main result of this section.

Theorem 7.1 (Exponential Stabilization of the Orbit $\mathcal{O}$) *Define the Jacobian matrices A and B as follows:*

$$A := \frac{\partial \rho}{\partial z^{1-}}(z^{1-}; \psi_s)\Big|_{z^{1-}=z^{1-*}, \psi_s=\psi^*_s} \in \mathbb{R}^{5 \times 5}$$

$$B := \frac{\partial \rho}{\partial \psi_s}(z^{1-}; \psi_s)\Big|_{z^{1-}=z^{1-*}, \psi_s=\psi^*_s} \in \mathbb{R}^{5 \times (2(n-5)M+6)}.$$

If the pair (A, B) is controllable, then there exists the matrix gain

$$K \in \mathbb{R}^{(2(n-5)M+6)\times 5}$$

such that by using the continuous-time controllers given in equation (7.11) and the event-based update laws $(\pi_1^2, \pi_2^3, \ldots, \pi_M^1)$ given in equations (7.13)–(7.17) and

$$\psi_s[k] = -K\left(z^{1^-}[k] - z^{1-*}\right), \tag{7.23}$$

$\mathcal{O}$ *is an exponentially stable period-one orbit for the hybrid systems $\Sigma_{\mathcal{H}}$ and Σ.*

Proof. Linearization of the state equation (7.22) about (z^{1-*}, ψ_s^*) results in

$$\delta z^{1^-}[k+1] = A\,\delta z^{1^-}[k] + B\,\delta\psi_s[k], \tag{7.24}$$

in which $\delta z^{1^-} := z^{1^-} - z^{1-*}$ and $\delta\psi_s := \psi_s - \psi_s^* = \psi_s$. Controllability of (A, B) implies the existence of the matrix gain K such that $|\mathrm{eig}(A_{\mathrm{cl}})| < 1$, where $A_{\mathrm{cl}} := A - BK$. This latter fact implies that z^{1-*} is an exponentially stable equilibrium point for the closed-loop discrete-time system $z^{1^-}[k+1] = \rho_{\mathrm{cl}}(z^{1^-}[k])$, in which

$$\rho_{\mathrm{cl}}(z^{1^-}) := \rho(z^{1^-}, -K(z^{1^-} - z^{1-*})).$$

Finally, applying Theorems 2.2 and 2.5 of Chapter 2 completes the proof. ∎

Remark 7.5 (Event-Based Updating Policy) *We observe that unlike the event-based update laws given in equations (7.13)–(7.17), which are updated during the corresponding within-stride transitions, the stabilizing parameters vector ψ_s is updated only on the switching hypersurface $\mathcal{S}_1^2$ in a step-by-step fashion (see equation (7.23)).*

Remark 7.6 (Assignment of Stabilizing Parameters to Steps) *From equations (7.14) and (7.12), it can be concluded that for a given $z^{1^-} := (q_3, q_4, \sigma_u')' \in \mathcal{S}_1^2 \cap \mathcal{Z}_{\mathcal{O}}$, q_a and $\dot{q}_a$ at the end of the first internal phase are expressed as*

$$q_a = h_d^1(\theta_1^*; \psi^1) = h_{d,\mathcal{O}}(\theta_1^*)$$

$$\dot{q}_a = \frac{\partial h_d^1}{\partial \theta}(\theta_1^*; \psi^1)\dot{\theta} = \frac{\partial h_{d,\mathcal{O}}}{\partial \theta}(\theta_1^*)\dot{\theta}$$

$$= \frac{\partial h_{d,\mathcal{O}}}{\partial \theta}(\theta_1^*)e_3'\left(I_{3\times 3} - \frac{\mathcal{D}_{uu}^{-1}\mathcal{D}_{ua}\frac{\partial h_{d,\mathcal{O}}}{\partial \theta}e_3'}{1 + e_3'\mathcal{D}_{uu}^{-1}\mathcal{D}_{ua}\frac{\partial h_{d,\mathcal{O}}}{\partial \theta}}\right)\mathcal{D}_{uu}^{-1}\sigma_u$$

$$=: \lambda(q_3, q_4)\,\sigma_u.$$

Thus, the state of the mechanical system at the beginning of the second internal phase is independent of α^1. In addition, from equation (7.19), $\alpha_2^2 = 0_{2 \times 1}$ and $\alpha_{n-2}^1 = 0_{2 \times 1}$ (which result in continuity of control inputs during the within-stride transition $1 \to 2$) imply that α^2 is also independent of α^1. Consequently, to calculate the restricted Poincaré return map $\rho(z^{1-}; \psi_s)$, we do not need to know the parameters that have been used by the first internal phase controller to reach the point z^{1-}. We note that the components of ψ_{rem}^1 will be used during the first internal phase of the next step, while the components of ψ_{rem}^j, $j = 2, \ldots, M$ are employed during the jth internal phase of the current step.

7.8 SIMULATION OF THE CLOSED-LOOP HYBRID SYSTEM

In order to confirm the analytical results developed to exponentially stabilize the desired periodic orbit $\mathcal{O}$ generated by the motion planning algorithm, this section presents a numerical example. In this example, the threshold values θ_j^*, $j = 1, \ldots, M - 1$ are generated as follows:

$$\theta_j^* = \theta^{+*} + j \frac{\theta^{-*} - \theta^{+*}}{M}.$$

Furthermore, the gain matrix K in equation (7.23) is obtained by using the DLQR design method subject to the linearized system given in equation (7.24). For this goal, we make use of the `dlqr` function of MATLAB with $Q = I_{5 \times 5}$ and $R = 10 I_{p \times p}$, where $p := 10(n - 5)M + 30$. Tables 7.4 and 7.5 present the spectral radius of the matrices A and A_{cl} for $M \in \{2, \ldots, 5\}$ and $n \in \{5, \ldots, 10\}$, respectively. From these tables, it can be concluded that for a given $M \in \{2, \ldots, 5\}$, by increasing the degree of the Bézier polynomial n, the values of max $|\text{eig}(A)|$ and max $|\text{eig}(A_{\text{cl}})|$ decrease and increase, respectively.

In order to present a numerical example confirming the analytical results obtained in this chapter, the simulation of the closed-loop impulsive model of walking is started at the end of single support for $M = 3$ and $n = 6$. The initial condition of the position vector is assumed to be q^{-*}. However, the initial condition for the velocity vector is

TABLE 7.4 **max $|eig(A)|$ for $M \in \{2, \ldots, 5\}$ and $n \in \{5, \ldots, 10\}$**

	$M = 2$	$M = 3$	$M = 4$	$M = 5$
$n = 5$	461.2236	445.7609	441.6970	440.0632
$n = 6$	453.9132	443.5014	440.5908	439.4399
$n = 7$	449.1912	441.9656	439.8695	439.0794
$n = 8$	446.1504	440.8708	439.4388	438.8435
$n = 9$	444.1028	440.2748	439.1323	438.6916
$n = 10$	442.8971	439.8698	438.9407	438.6029

TABLE 7.5   **max $|eig(A_{cl})|$ for $M \in \{2, \ldots, 5\}$ and $n \in \{5, \ldots, 10\}$**

	$M = 2$	$M = 3$	$M = 4$	$M = 5$
$n = 5$	0.2216	0.2415	0.2470	0.2466
$n = 6$	0.2246	0.2420	0.2474	0.2467
$n = 7$	0.2260	0.2430	0.2474	0.2467
$n = 8$	0.2276	0.2429	0.2474	0.2469
$n = 9$	0.2271	0.2421	0.2476	0.2470
$n = 10$	0.2283	0.2404	0.2477	0.2471

chosen as the value of the velocity vector at the end of single support on $\mathcal{O}$ with an error of $+2(\frac{deg}{s})$ on each of its components. Results of the simulation of the closed-loop system are illustrated in Figs. 7.7 and 7.8. Figure 7.7 presents the phase-plane plots of the state trajectories during four consecutive steps. Discontinuities are due

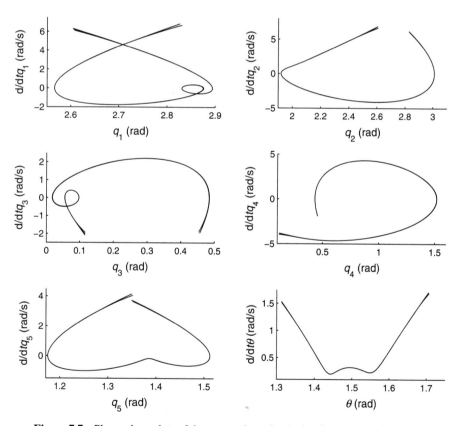

Figure 7.7   Phase-plane plots of the state trajectories during four consecutive steps.

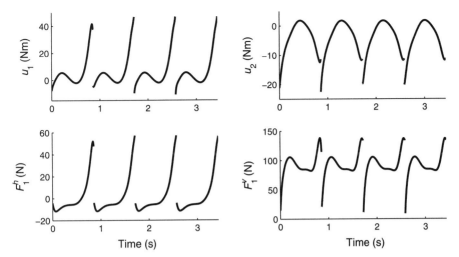

Figure 7.8 Plots of the closed-loop control inputs (two top graphs) during four consecutive steps. Two bottom graphs represent the horizontal and vertical components of the ground reaction force experienced at the end of leg-1 during four steps.

to the impact and coordinate relabeling to swap the role of the legs. The closed-loop control inputs during four consecutive steps are shown in Fig. 7.8. Furthermore, this figure represents the horizontal and vertical components of the ground reaction force experienced at the stance leg end during four steps.

CHAPTER 8

Continuous-Time Update Laws During Continuous Phases of Locomotion

8.1 INTRODUCTION

As studied in the previous chapters, the evolution of a bipedal robot during walking can be described by an impulsive system composed of a single support model and an impact map. The main problem in control of legged locomotion is how to design a feedback law that guarantees the existence of a stable limit cycle for the closed-loop system. The basic tool for analyzing the existence and stability of limit cycles for smooth autonomous dynamical systems is the Poincaré return map. Grizzle et al. extended the method of the Poincaré return map to autonomous systems with impulse effects [46] and made use of virtual constraints to reduce the dimension of the Poincaré return map during bipedal walking. In addition, to improve the convergence rate, the idea of updating the parameters of time-invariant stabilizing controllers by event-based update laws has been described in references [57–59]. The contribution of this chapter is to develop a novel method for designing a class of continuous-time update laws to update the parameters of stabilizing controllers during continuous phases of locomotion such that

(i) a general cost function (such as the energy of the control input over single support) can be minimized in an online manner, and

(ii) the exponential stability behavior of the limit cycle for the closed-loop system is not affected.

In addition, this chapter introduces a class of continuous-time update laws with a *radial basis step length* to minimize a desired cost function in terms of the controller parameters and initial states.

Hybrid Control and Motion Planning of Dynamical Legged Locomotion, Nasser Sadati, Guy A. Dumont, Kaveh Akbari Hamed, and William A. Gruver.
© 2012 by the Institute of Electrical and Electronics Engineers, Inc. Published 2012 by John Wiley & Sons, Inc.

8.2 INVARIANCE OF THE EXPONENTIAL STABILITY BEHAVIOR FOR A CLASS OF IMPULSIVE SYSTEMS

Let us consider the following autonomous impulsive system

$$
\Sigma : \begin{cases} \dot{x} = f(x, \alpha^*) & x^- \notin \mathcal{S} \\ x^+ = \Delta(x^-) & x^- \in \mathcal{S}, \end{cases} \tag{8.1}
$$

where f and α^* are a C^1 vector field defined on the state space $\mathcal{X} \subset \mathbb{R}^n$ and a nominal parameter vector in the parameter space $\mathcal{A} \subset \mathbb{R}^p$, respectively. In addition, $\mathcal{X}$ is a simply connected open subset of $\mathbb{R}^n$ and $\mathcal{S}$ is assumed to be a switching hypersurface taking the form

$$
\mathcal{S} := \{ x \in \mathcal{X} | H(x) = 0 \},
$$

where $H : \mathcal{X} \to \mathbb{R}$ is a smooth function. $\Delta : \mathcal{S} \to \mathcal{X}$ also denotes a C^1 reset map. Assume that the impulsive system Σ has a period-one orbit $\mathcal{O} \subset \mathcal{X}$ that is transversal to $\mathcal{S}$, that is, $\{x^*\} := \overline{\mathcal{O}} \cap \mathcal{S}$ is a singleton and

$$
L_f H(x^*, \alpha^*) := \frac{\partial H}{\partial x}(x^*) f(x^*, \alpha^*) \neq 0,
$$

where $\overline{\mathcal{O}}$ denotes the set closure of $\mathcal{O}$. For the later purposes, let T^* represent the minimum period of the orbit $\mathcal{O}$. Denote the solution of $\dot{x} = f(x, \alpha^*)$ with the initial condition $x(0) = x_0$ by $\varphi(t; x_0, \alpha^*)$ for every $t \geq 0$ in its maximal domain of existence. Throughout this chapter, we assume that the periodic orbit $\mathcal{O}$ satisfies the following key hypothesis.

(H1) The state vector x can be decomposed as

$$
x = \begin{bmatrix} \xi \\ \theta \end{bmatrix},
$$

in which $\xi \in \mathbb{R}^{n-1}$ and $\theta \in \mathbb{R}$ is a strictly increasing function of time on $\overline{\mathcal{O}}$.

Under hypothesis H1, the desired evolution of ξ on the orbit $\mathcal{O}$ can be expressed as a function of θ instead of time, that is, $\xi = \xi_d(\theta)$ for all $x \in \mathcal{O}$. By assuming that on $\overline{\mathcal{O}}$, θ takes values in the closed interval $[\theta_{\min}, \theta_{\max}]$ and solutions of Σ are right continuous (see Definition 2.2, Chapter 2), a description for $\mathcal{O}$ can be expressed as

$$
\mathcal{O} = \{ x = (\xi', \theta)' \in \mathcal{X} | \xi = \xi_d(\theta), \theta_{\min} \leq \theta < \theta_{\max} \}.
$$

By extrapolating the function ξ_d outside the interval $[\theta_{min}, \theta_{max}]$, we define the function $d_a : \mathcal{X}_a \to \mathbb{R}$ by

$$d_a(x, \alpha) := \frac{1}{2}\|\xi - \xi_d(\theta)\|_2^2 + \frac{1}{2}\|\alpha - \alpha^*\|_2^2, \tag{8.2}$$

where $x_a := (x', \alpha')'$ and $\mathcal{X}_a := \mathcal{X} \times \mathcal{A}$ denote the augmented state and augmented state space, respectively. To present the main result of this chapter, we consider the systems

$$\Sigma_a^I : \begin{cases} \begin{bmatrix} \dot{x} \\ \dot{\alpha} \end{bmatrix} = \begin{bmatrix} f(x, \alpha) \\ g(x, \alpha) \end{bmatrix} & x^- \notin \mathcal{S} \\ \\ \begin{bmatrix} x^+ \\ \alpha^+ \end{bmatrix} = \begin{bmatrix} \Delta(x^-) \\ \pi(x^-, \alpha^-) \end{bmatrix} & x^- \in \mathcal{S} \end{cases} \tag{8.3}$$

$$\Sigma_a^{II} : \begin{cases} \begin{bmatrix} \dot{x} \\ \dot{\alpha} \end{bmatrix} = \begin{bmatrix} f(x, \alpha) \\ 0_{p \times 1} \end{bmatrix} & x^- \notin \mathcal{S} \\ \\ \begin{bmatrix} x^+ \\ \alpha^+ \end{bmatrix} = \begin{bmatrix} \Delta(x^-) \\ \pi(x^-, \alpha^-) \end{bmatrix} & x^- \in \mathcal{S}, \end{cases} \tag{8.4}$$

with the following structural hypotheses.

(H2) It is assumed that $\dot{\alpha} = g(x, \alpha)$ is a C^1 continuous-time update law with a radial basis step length, that is,

$$g(x, \alpha) := d_a(x, \alpha)\, \tilde{g}(x, \alpha),$$

where $\tilde{g}$ is C^1 with respect to x_a.

(H3) The function $\pi : \mathcal{S}_a \to \mathcal{A}$ is a C^1 event-based update law with the property $\pi(x^*, \alpha^*) = \alpha^*$, where $\mathcal{S}_a := \mathcal{S} \times \mathcal{A}$ represents the augmented switching hypersurface.

For the later purposes, we define

$$x_a^* := \begin{bmatrix} x^* \\ \alpha^* \end{bmatrix} \in \mathcal{S}_a.$$

Hypotheses H2 and H3 imply that the augmented orbit $\mathcal{O}_a := \mathcal{O} \times \{\alpha^*\}$ is a periodic orbit for the systems Σ_a^I and Σ_a^{II} with the minimum period T^*. In addition, by defining

$$f_a^I(x_a) := \begin{bmatrix} f(x, \alpha) \\ g(x, \alpha) \end{bmatrix}, \qquad f_a^{II}(x_a) := \begin{bmatrix} f(x, \alpha) \\ 0_{p \times 1} \end{bmatrix},$$

and also considering the fact that

$$\mathcal{S}_a = \{x_a \in \mathcal{X}_a | H_a(x_a) = 0\},$$

where $H_a(x_a) := H(x)$, it can be concluded that $\mathcal{O}_a$ is transversal to $\mathcal{S}_a$.

Theorem 8.1 (Invariance of the Exponential Stability Behavior) *Under hypotheses H1–H3, $\mathcal{O}_a$ is exponentially stable for the system Σ_a^I if and only if it is exponentially stable for the system Σ_a^{II}.*

8.3 OUTLINE OF THE PROOF OF THEOREM 8.1

In this section, Theorem 8.1 is proved through a sequence of lemmas. Subsequently, the solution of the augmented system $\dot{x}_a = f_a^j(x_a)$, $j \in \{I, II\}$ with the initial condition $x_{a0} := (x_0', \alpha_0')' \in \mathcal{X}_a$ is denoted by

$$\varphi_a^j(t; x_0, \alpha_0) := \begin{bmatrix} \varphi_x^j(t; x_0, a_0) \\ \varphi_\alpha^j(t; x_0, a_0) \end{bmatrix},$$

where the subscripts "x" and "α" represent the components corresponding to x and α, respectively. In addition,

$$\Phi_a^j(t; x_0, \alpha_0) := \begin{bmatrix} \Phi_{xx}^j(t; x_0, \alpha_0) & \Phi_{x\alpha}^j(t; x_0, \alpha_0) \\ \Phi_{\alpha x}^j(t; x_0, \alpha_0) & \Phi_{\alpha\alpha}^j(t; x_0, \alpha_0) \end{bmatrix}$$

denotes the trajectory sensitivity matrix of $\dot{x}_a^j = f_a^j(x_a)$, $j \in \{I, II\}$. Under hypotheses H2 and H3, for the initial condition $(x_0', \alpha_0')' = (\Delta'(x^*), \alpha^{*'})'$, the solution of $\dot{x}_a = f_a^j(x_a)$, $j \in \{I, II\}$ is denoted by

$$\varphi_a^*(t) := \varphi_a^j(t; \Delta(x^*), \alpha^*) := \begin{bmatrix} \varphi^*(t) \\ \alpha^* \end{bmatrix},$$

where $\varphi^*(t) := \varphi(t; \Delta(x^*), \alpha^*)$.

Lemma 8.1 *Under the assumptions of Theorem 8.1, the components of the trajectory sensitivity matrices* $\Phi_a^j(t; \Delta(x^*), \alpha^*), 0 \leq t \leq T^*$ *for* $j \in \{I, II\}$ *satisfy the following equations:*

$$\dot{\Phi}_{xx}^j(t; \Delta(x^*), \alpha^*) = A^*(t)\, \Phi_{xx}^j(t; \Delta(x^*), \alpha^*)$$

$$\Phi_{xx}^j(0; \Delta(x^*), \alpha^*) = I_{n \times n}$$

$$\dot{\Phi}_{x\alpha}^j(t; \Delta(x^*), \alpha^*) = A^*(t)\, \Phi_{x\alpha}^j(t; \Delta(x^*), \alpha^*) + B^*(t)$$

$$\Phi_{x\alpha}^j(0; \Delta(x^*), \alpha^*) = 0_{n \times p}$$

$$\Phi_{\alpha x}^j(t; \Delta(x^*), \alpha^*) = 0_{p \times n}$$

$$\Phi_{\alpha\alpha}^j(t; \Delta(x^*), \alpha^*) = I_{p \times p},$$

where

$$A^*(t) := D_x f(\varphi^*(t), \alpha^*)$$

$$B^*(t) := D_\alpha f(\varphi^*(t), \alpha^*).$$

Proof. From hypothesis H2 and equation (8.2), it can be concluded that

$$\frac{\partial g}{\partial x}(\varphi^*(t), \alpha^*) = 0_{p \times n}$$

and

$$\frac{\partial g}{\partial \alpha}(\varphi^*(t), \alpha^*) = 0_{p \times p},$$

which, in turn, together with the variational equation [89, Appendix B] completes the proof. ∎

Following the results of Ref. [46], the time-to-impact function for the system $\dot{x}_a = f_a^j(x_a), j \in \{I, II\}$ is defined as $T^j : \mathcal{X}_a \to \mathbb{R}$ by

$$T^j(x_0, \alpha_0) := \inf\{t \geq 0 | \varphi_x^j(t; x_0, \alpha_0) \in \mathcal{S}\}.$$

Under H2 and H3 and results of Ref. [18, pp. 83–84], T^j for $j \in \{I, II\}$ is differentiable at the point $(\Delta'(x^*), \alpha^{*'})'$. Furthermore, equation (C.25) of reference [18, p. 445] together with Lemma 8.1 and

$$\frac{\partial H_a}{\partial x_a}(x_a) = \left[\frac{\partial H}{\partial x}(x)\, 0_{1 \times p} \right]$$

yields

$$
\begin{aligned}
D_1 T^j(\Delta(x^*), \alpha^*) &= -\frac{1}{L_f H(x^*, \alpha^*)} \frac{\partial H}{\partial x}(x^*) \, \Phi^j_{xx}(T^*; \Delta(x^*), \alpha^*) \\
D_2 T^j(\Delta(x^*), \alpha^*) &= -\frac{1}{L_f H(x^*, \alpha^*)} \frac{\partial H}{\partial x}(x^*) \, \Phi^j_{x\alpha}(T^*; \Delta(x^*), \alpha^*).
\end{aligned}
\tag{8.5}
$$

By taking $\mathcal{S}_a$ as the Poincaré section, the augmented Poincaré return map for Σ^j_a, $j \in \{I, II\}$ can be expressed as $\mathcal{P}^j_a : \mathcal{S}_a \to \mathcal{S}_a$ by

$$
\mathcal{P}^j_a(x_a) = \begin{bmatrix} \mathcal{P}^j_x(x, \alpha) \\ \mathcal{P}^j_\alpha(x, \alpha) \end{bmatrix} := \begin{bmatrix} \varphi^j_x \left(T^j(\Delta(x), \pi(x, \alpha)); \Delta(x), \pi(x, \alpha) \right) \\ \varphi^j_\alpha \left(T^j(\Delta(x), \pi(x, \alpha)); \Delta(x), \pi(x, \alpha) \right) \end{bmatrix}.
$$

Lemma 8.2 *Suppose that the assumptions of Theorem 8.1 are satisfied. Then,* $DP^I_a(x^*_a) = DP^{II}_a(x^*_a).$

Proof. Under the assumptions of Theorem 8.1, the Poincaré return map $\mathcal{P}^j_a, j \in \{I, II\}$ is differentiable at the point $x^*_a = (x^{*\prime}, \alpha^{*})'$. Furthermore, using the chain rule, the components of the Jacobian matrix of $\mathcal{P}^j_a$ evaluated at x^*_a can be expressed as

$$
\begin{aligned}
D_x \mathcal{P}^j_p(x^*_a) &= \Big(D_1 \varphi^j_p(T^*; \Delta(x^*), \alpha^*) \, D_1 T^j(\Delta(x^*), \alpha^*) \\
&\quad + D_2 \varphi^j_p(T^*; \Delta(x^*), \alpha^*) \Big) D\Delta(x^*) \\
&\quad + \Big(D_1 \varphi^j_p(T^*; \Delta(x^*), \alpha^*) \, D_2 T^j(\Delta(x^*), \alpha^*) \\
&\quad + D_3 \varphi^j_p(T^*; \Delta(x^*), \alpha^*) \Big) D_1 \pi(x^*, \alpha^*)
\end{aligned}
$$

$$
\begin{aligned}
D_\alpha \mathcal{P}^j_p(x^*_a) &= \Big(D_1 \varphi^j_p(T^*; \Delta(x^*), \alpha^*) \, D_2 T^j(\Delta(x^*), \alpha^*) \\
&\quad + D_3 \varphi^j_p(T^*; \Delta(x^*), \alpha^*) \Big) D_2 \pi(x^*, \alpha^*),
\end{aligned}
$$

where $p \in \{x, \alpha\}$. Since (i) from hypothesis H2,

$$
D_1 \varphi^j_\alpha(T^*; \Delta(x^*), \alpha^*) = 0_{p \times 1},
$$

and (ii) from Lemma 8.1,

$$
D_2 \varphi^j_\alpha(T^*; \Delta(x^*), \alpha^*) = \Phi^j_{\alpha x}(T^*; \Delta(x^*), \alpha^*) = 0_{p \times n}
$$

and

$$D_3\varphi_\alpha^j(T^*; \Delta(x^*), \alpha^*) = \Phi_{\alpha\alpha}^j(T^*; \Delta(x^*), \alpha^*) = I_{p\times p},$$

equation (8.5) in combination with straightforward calculations implies that

$$D_x\mathcal{P}_x^j(x_a^*) = J^j(x^*, \alpha^*) + S^j(x^*, \alpha^*) D_1\pi(x^*, \alpha^*)$$
$$D_\alpha\mathcal{P}_x^j(x_a^*) = S^j(x^*, \alpha^*) D_2\pi(x^*, \alpha^*)$$
$$D_x\mathcal{P}_\alpha^j(x_a^*) = D_1\pi(x^*, \alpha^*)$$
$$D_\alpha\mathcal{P}_\alpha^j(x_a^*) = D_2\pi(x^*, \alpha^*),$$

in which

$$J^j(x^*, \alpha^*) := \Pi\, \Phi_{xx}^j(T^*; \Delta(x^*), \alpha^*) D\Delta(x^*)$$
$$S^j(x^*, \alpha^*) := \Pi\, \Phi_{x\alpha}^j(T^*; \Delta(x^*), \alpha^*)$$

and

$$\Pi := I_{n\times n} - \frac{f(x^*, \alpha^*)\frac{\partial H}{\partial x}(x^*)}{L_f H(x^*, \alpha^*)}.$$

Finally, from Lemma 8.1,

$$\Phi_{xx}^I(T^*; \Delta(x^*), \alpha^*) = \Phi_{xx}^{II}(T^*; \Delta(x^*), \alpha^*)$$
$$\Phi_{x\alpha}^I(T^*; \Delta(x^*), \alpha^*) = \Phi_{x\alpha}^{II}(T^*; \Delta(x^*), \alpha^*),$$

which, in turn, results in $D\mathcal{P}_a^I(x_a^*) = D\mathcal{P}_a^{II}(x_a^*)$. ∎

The proof of Theorem 8.1 is an immediate consequence of Lemma 8.2 and Theorems 2.2 and 2.5.

8.4 APPLICATION TO LEGGED LOCOMOTION

To give an application of Theorem 8.1, assume that $\Gamma(x, \alpha)$ represents a continuous-time stabilizing controller inducing an exponentially stable periodic walking for a bipedal mechanism, where x and α denote the states of the mechanical system and the parameters of the controller, respectively. Moreover, let $\mathcal{J}(x_0, \alpha)$ represent a general cost function to be minimized online, in terms of the initial states of the mechanical system (i.e., x_0) and the controller parameters (i.e., α), such as the energy of controller $\Gamma(x, \alpha)$ over single support,

$$\mathcal{J}(x_0, \alpha) := \int_0^{T^{II}(x_0, \alpha)} \|\Gamma(x, \alpha)\|_2^2 \, dt.$$

Theorem 8.1 states that, under hypotheses H1–H3, the C^1 function $\tilde{g}$ does not affect the exponential stability behavior of the period orbit $\mathcal{O}_a$ for the system Σ_a^I. As a consequence, in order to minimize the cost function $\mathcal{J}$, we can choose $\tilde{g}(x, \alpha) = -\gamma \frac{\partial \mathcal{J}'}{\partial \alpha}(x_0, \alpha)$ or, equivalently,

$$\dot{\alpha} = -\gamma \, d_a(x, \alpha) \frac{\partial \mathcal{J}'}{\partial \alpha}(x_0, \alpha), \tag{8.6}$$

where $\gamma > 0$ is a scalar representing the continuous-time update gain. We remark that equation (8.6) introduces a gradient-based update law in the parameter space $\mathcal{A}$ for which the step length is assumed to change radially.

Proofs Associated with Chapter 3

A.1 PROOF OF LEMMA 3.3

Proof. $\ddot{\bar{x}}_H$ can be given by

$$\ddot{\bar{x}}_H = \frac{\partial \bar{x}_H}{\partial q_i} \ddot{q}_i + \dot{q}_i' \frac{\partial^2 \bar{x}_H}{\partial q_i^2} \dot{q}_i.$$

On the manifold Z_d, the dynamics of double support phase is expressed as equation (3.18), in which (u_3, u_4) is replaced by (u_{3d}^*, u_{4d}^*). Also from equation (3.30), $(u_{3d}^*, u_{4d}^*)'$ can be expressed as

$$\begin{bmatrix} u_{3d}^* \\ u_{4d}^* \end{bmatrix} = -\left(\frac{\partial h_d}{\partial q_i} D_\psi^{-1} \beta_\psi \right)^{-1} \left(\frac{\partial}{\partial q_i} \left(\frac{\partial h_d}{\partial q_i} \dot{q}_i \right) \dot{q}_i - \frac{\partial h_d}{\partial q_i} D_\psi^{-1} \left(C_\psi \dot{q}_i + G_\psi - \frac{\partial \Psi'}{\partial q_i} \begin{bmatrix} u_{1d} \\ u_{2d} \end{bmatrix} \right) \right), \quad \text{(A.1)}$$

which is a quadratic function with respect to $\dot{q}_i$. Since the Coriolis matrix $C_\psi(q_i, \dot{q}_i)$ is linear with respect to $\dot{q}_i$, equations (3.18) and (A.1) together with

$$q_i = \Phi_d^{-1}([0_{1 \times 2}, \bar{x}_H]')$$
$$\dot{q}_i = \lambda_d(q_i) \bar{v}_{xH}$$
$$u_1 = u_{1d}(q_i)$$
$$u_2 = u_{2d}(q_i)$$

yield equation (3.31). ∎

Hybrid Control and Motion Planning of Dynamical Legged Locomotion, Nasser Sadati, Guy A. Dumont, Kaveh Akbari Hamed, and William A. Gruver.
© 2012 by the Institute of Electrical and Electronics Engineers, Inc. Published 2012 by John Wiley & Sons, Inc.

A.2 PROOF OF LEMMA 3.4

Proof. The vertical acceleration of the end of leg-2 in the single support phase, $\ddot{y}_2$, can be given by

$$\ddot{y}_2 = \frac{\partial y_2}{\partial q_s}\ddot{q}_s + \dot{q}_s' \frac{\partial^2 y_2}{\partial q_s^2}\dot{q}_s.$$

On the manifold Z_s, the dynamics of the single support phase can be expressed as $D\ddot{q}_s + C\dot{q}_s + G = Bu_s^*$, where

$$u_s^* = -\left(\frac{\partial h_s}{\partial q_s}D^{-1}B\right)^{-1}\left(\frac{\partial}{\partial q_s}\left(\frac{\partial h_s}{\partial q_s}\dot{q}_s\right)\dot{q}_s - \frac{\partial h_s}{\partial q_s}D^{-1}(C\dot{q}_s + G)\right)$$

is a quadratic function with respect to $\dot{q}_s$. Since

$$\dot{q}_s^+ = \Delta_{\dot{q},d}^s(q_{id}^-)\dot{q}_i^- = \Delta_{\dot{q},d}^s(q_{id}^-)\lambda_d(q_{id}^-)\bar{v}_{xH}^-$$

and the Coriolis matrix $C(q_s, \dot{q}_s)$ is linear with respect to $\dot{q}_s$, it follows that

$$\ddot{y}_2^+ = \bar{\omega}_1(x_{H,d}^-) + \bar{\omega}_2(x_{H,d}^-)\bar{z}_{xH}^-,$$

which completes the proof.   ∎

A.3 PROOF OF LEMMA 3.7

Proof. Equation (3.14) can be decomposed as

$$H_0(D\ddot{q}_d(t) + C\dot{q}_d(t) + G) = u_d(t) + \frac{\partial p_2'}{\partial q_b}F_2(t)$$

$$e_5'(D\ddot{q}_d(t) + C\dot{q}_d(t) + G) = \frac{\partial p_2'}{\partial q_5}F_2(t). \tag{A.2}$$

If rank $\frac{\partial p_2'}{\partial q_5} = 1$, the last row of matrix equation (A.2) yields

$$F_2(t) = \left(\frac{\partial p_2'}{\partial q_5}\right)^+ e_5'(D\ddot{q}_d(t) + C\dot{q}_d(t) + G) + \left(\frac{\partial p_2'}{\partial q_5}\right)^- \Upsilon(t), \tag{A.3}$$

where $\Upsilon(t) := (\Upsilon^h(t), \Upsilon^v(t))' \in \mathbb{R}^2$ is an arbitrary continuously differentiable function. Also $(\frac{\partial p_2'}{\partial q_5})^+$ and $(\frac{\partial p_2'}{\partial q_5})^-$ are the pseudo inverse and projection matrices, respectively. For the bipedal robot, since q_5 decreases in the clockwise direction, using Proposition B.8 of Ref. [18, p. 424],

$$\frac{\partial p_2}{\partial q_5} = \frac{\partial}{\partial q_5} \begin{bmatrix} x_2 \\ y_2 \end{bmatrix} = \begin{bmatrix} -y_2 \\ x_2 \end{bmatrix} = \begin{bmatrix} 0 \\ L_s \end{bmatrix}. \tag{A.4}$$

Therefore, $\operatorname{rank}\frac{\partial p_2'}{\partial q_5} = \operatorname{rank}[0, L_s]' = 1$ for every $t \in [T_s, T)$. Furthermore, from equations (A.3) and (A.4),

$$F_2(t) = \begin{bmatrix} 0 \\ \frac{1}{L_s} \end{bmatrix} e_5'(D\,\ddot{\mathbf{q}}_d(t) + C\,\dot{\mathbf{q}}_d(t) + G) + \begin{bmatrix} 1 & 0 \\ 0 & 0 \end{bmatrix} \Upsilon(t). \tag{A.5}$$

Substituting equation (A.5) into the first four rows of equation (A.2) implies that

$$u_d(t) = u_d^0(t) - \frac{\partial x_2'}{\partial q_b} \Upsilon^h(t).$$

Since $\mathbf{q}_d(t)$ is such that for every $t \in [T_s, T)$,

$$\frac{\partial p_2}{\partial q}(\mathbf{q}_d(t))\,\ddot{\mathbf{q}}_d(t) + \frac{\partial}{\partial q}\left(\frac{\partial p_2}{\partial q}(\mathbf{q}_d(t))\dot{\mathbf{q}}_d(t)\right)\dot{\mathbf{q}}_d(t) = 0_{2\times1}$$

and $\operatorname{rank}\frac{\partial p_2}{\partial q}(\mathbf{q}_d(t)) = 2$, from equation (3.16), knowledge of $u_d(t)$ results in a unique $F_2(t)$. Moreover, $u_d(t)$ satisfies equation (3.17) because equations (3.14) and (3.16) result in equation (3.17), which completes the proof. ∎

Proofs Associated with Chapter 4

B.1 PROOF OF LEMMA 4.2

Proof. Since from Remark 4.3,

$$
x_3^{\max}\left(\tau; x_2^f\right) = \begin{cases} x_3^0 + x_4^0(\tau - t_1^*) + \frac{L_1}{2}(\tau - t_1^*)^2, & t_1^* \leq \tau \leq \bar{\tau}\left(x_2^f\right) \\ \bar{x}_3 + \bar{x}_4\left(\tau - \bar{\tau}\left(x_2^f\right)\right) + \frac{L_2}{2}\left(\tau - \bar{\tau}\left(x_2^f\right)\right)^2, & \bar{\tau}\left(x_2^f\right) \leq \tau \leq t_2^*, \end{cases}
$$

the sensitivity function $\partial x_3^{\max}(\tau; x_2^f)/\partial \bar{\tau}$ can be expressed as follows:

$$
\frac{\partial x_3^{\max}}{\partial \bar{\tau}}\left(\tau; x_2^f\right) = \begin{cases} 0, & t_1^* \leq \tau \leq \bar{\tau}\left(x_2^f\right) \\ (L_1 - L_2)\left(\tau - \bar{\tau}\left(x_2^f\right)\right), & \bar{\tau}\left(x_2^f\right) \leq \tau \leq t_2^*. \end{cases} \tag{B.1}
$$

Moreover, from equation (4.24),

$$
\frac{\partial \bar{\tau}}{\partial x_2^f}\left(x_2^f\right) = \frac{2}{(L_1 - L_2)(\bar{\tau} - t_2^*)^2},
$$

together with equation (B.1) yields

$$
\frac{\partial x_3^{\max}}{\partial x_2^f}\left(\tau; x_2^f\right) = \begin{cases} 0, & t_1^* \leq \tau \leq \bar{\tau}\left(x_2^f\right) \\ \dfrac{2\left(\tau - \bar{\tau}\left(x_2^f\right)\right)}{\left(\bar{\tau}\left(x_2^f\right) - t_2^*\right)^2}, & \bar{\tau}\left(x_2^f\right) \leq \tau \leq t_2^*. \end{cases} \tag{B.2}
$$

Hybrid Control and Motion Planning of Dynamical Legged Locomotion, Nasser Sadati, Guy A. Dumont, Kaveh Akbari Hamed, and William A. Gruver.
© 2012 by the Institute of Electrical and Electronics Engineers, Inc. Published 2012 by John Wiley & Sons, Inc.

Therefore,

$$\frac{\partial x_3^{\max}}{\partial x_2^f}\left(\tau; x_2^f\right) \geq 0$$

for any $(\tau, x_2^f)' \in [t_1^*, t_2^*] \times \Omega_{m,M,L_1,L_2}^{\max}(x_3^0, x_4^0)$, which, in turn, results in

$$x_3^{\max}(\tau; \alpha) \leq x_3^{\max}(\tau; \gamma) \leq x_3^{\max}(\tau; \beta)$$

for any $\tau \in [t_1^*, t_2^*]$. Moreover, from part (a) of Lemma 4.1, $\alpha, \beta \in \Omega_{m,M,L_1,L_2}^{\max}(x_3^0, x_4^0)$ implies that for every $\tau \in [t_1^*, t_2^*]$,

$$m < x_3^{\max}(\tau; \alpha) \leq x_3^{\max}(\tau; \gamma) \leq x_3^{\max}(\tau; \beta) < M,$$

which completes the proof.   ∎

B.2   PROOF OF THEOREM 4.2

Proof. The construction procedure for φ^d and condition (4.34) immediately imply part (a). Next, we show that the trajectory $\varphi^d(t)$, $0 \leq t \leq t_1$ with the initial condition $\theta(0) = \theta_0$ results in $\theta(t_1) = \theta_1$. For this purpose, define

$$\mathcal{C}_1 := \{\psi \in \mathcal{Q}_b | \psi = \varphi^d(t), 0 \leq t \leq t_1\}.$$

Then, by introducing the variable $s = \frac{\sigma_{cm}}{\sigma_{cm}^*}t$,

$$\theta(t_1) = \theta_0 + \int_0^{t_1} \frac{\sigma_{cm}}{A_{3,3}(\varphi^d(t))}\, dt - \int_{\mathcal{C}_1} J(\varphi^d)\, d\varphi^d$$

$$= \theta_0 + \int_0^{t_1^*} \frac{\sigma_{cm}^*}{A_{3,3}(\varphi^*(s))}\, ds - \int_{\mathcal{C}_1^*} J(\varphi^*)\, d\varphi^*$$

$$= \theta_1,$$

where

$$\mathcal{C}_1^* := \{\psi \in \mathcal{Q}_b | \psi = \varphi^*(t), 0 \leq t \leq t_1^*\} = \mathcal{C}_1.$$

Using the proposed reconfiguration algorithm, $\theta(t_1) = \theta_1$ and $\varphi^d(t) = \varphi^*(z_1(t))$, $t_1 \leq t \leq t_2$ imply that $\theta(t_2) = \theta_2$. In addition, the change of variable $s = \frac{\sigma_{cm}}{\sigma_{cm}^*}(t - t_f) + t_f^*$ yields

$$
\theta(t_f) = \theta_2 + \int_{t_2}^{t_f} \frac{\sigma_{cm}}{A_{3,3}(\varphi^d(t))} \, dt - \int_{C_3} J(\varphi^d) \, d\varphi^d
$$

$$
= \theta_2 + \int_{t_2^*}^{t_f^*} \frac{\sigma_{cm}^*}{A_{3,3}(\varphi^*(s))} \, ds - \int_{C_3^*} J(\varphi^*) \, d\varphi^*
$$

$$
= \theta_f,
$$

where

$$
C_3 := \{\psi \in \mathcal{Q}_b | \psi = \varphi^d(t), \, t_2 \leq t \leq t_f\}
$$
$$
C_3^* := \{\psi \in \mathcal{Q}_b | \psi = \varphi^*(t), \, t_2^* \leq t \leq t_f^*\} = C_3.
$$

If $x_3(\tau) \equiv 1$, $t_1^* \leq \tau \leq t_2^*$, the states x_1 and x_2 of the system Σ_a are transferred from the origin at t_1^* to $x_1^{f*} := \int_{t_1^*}^{t_2^*} w(s) ds$ and $x_2^{f*} := l_{max} = t_2^* - t_1^*$ at t_2^*, respectively. On the other hand, $x_3^0 = 1$, $x_4^0 = 0$, $L_2 = -L_1$, and $0 < m < 1 < M$ imply that

$$
\bar{\tau}\left(x_2^{f*}\right) = \underline{\tau}\left(x_2^{f*}\right) = t_2^* - \sqrt[3]{\frac{1}{2}} l_{max}
$$

and $\vartheta = \frac{1}{2}$, which, in turn, follows that $x_3(\tau; x_1^{f*}, x_2^{f*}) \equiv 1$. This completes the proof of part (c). $\blacksquare$

Proofs Associated with Chapter 6

C.1 PROOF OF LEMMA 6.1

Proof. Define $\bar{\mathbf{S}}_s := \text{block diag}\{S_1, S_2\}$. By the symmetrical structure of the monopedal robot, for every $(q_s', \dot{q}_s')' \in T\mathcal{Q}_s$,

$$\mathcal{L}_s(q_s, \dot{q}_s) = \mathcal{L}_s(\bar{\mathbf{S}}_s q_s, \bar{\mathbf{S}}_s \dot{q}_s),$$

which, in turn, results in

$$\mathcal{K}_s(q_s, \dot{q}_s) = \mathcal{K}_s(\bar{\mathbf{S}}_s q_s, \bar{\mathbf{S}}_s \dot{q}_s)$$
$$\mathcal{V}_s(q_s) = \mathcal{V}_s(\bar{\mathbf{S}}_s q_s).$$

Consequently, for every $(q_s', \dot{q}_s')' \in T\mathcal{Q}_s$,

$$D_s(q_s) = \bar{\mathbf{S}}_s' \, D_s(\bar{\mathbf{S}}_s q_s) \, \bar{\mathbf{S}}_s$$
$$C_s(q_s, \dot{q}_s) = \bar{\mathbf{S}}_s' \, C_s(\bar{\mathbf{S}}_s q_s, \bar{\mathbf{S}}_s \dot{q}_s) \, \bar{\mathbf{S}}_s$$
$$G_s(q_s) = \bar{\mathbf{S}}_s' \, G_s(\bar{\mathbf{S}}_s q_s).$$

Next, let $q_s^*(t), 0 \leq t \leq T_s^*$ together with the open-loop control input $u_s^*(t)$ satisfy the differential equation $D_s(q_s^*)\ddot{q}_s^* + C_s(q_s^*, \dot{q}_s^*)\dot{q}_s^* + G_s(q_s^*) = Bu_s^*$. This equation can also be expressed as

$$\bar{\mathbf{S}}_s' \, D_s(\bar{\mathbf{S}}_s q_s^*) \, \bar{\mathbf{S}}_s \, \ddot{q}_s^* + \bar{\mathbf{S}}_s' \, C_s(\bar{\mathbf{S}}_s q_s^*, \bar{\mathbf{S}}_s \dot{q}_s^*) \, \bar{\mathbf{S}}_s \dot{q}_s^* + \bar{\mathbf{S}}_s' \, G_s(\bar{\mathbf{S}}_s q_s^*) = B u_s^*. \qquad (C.1)$$

Hybrid Control and Motion Planning of Dynamical Legged Locomotion, Nasser Sadati, Guy A. Dumont, Kaveh Akbari Hamed, and William A. Gruver.
© 2012 by the Institute of Electrical and Electronics Engineers, Inc. Published 2012 by John Wiley & Sons, Inc.

Premultiplying equation (C.1) by the matrix $(\bar{\mathbf{S}}'_s)^{-1}$ and decomposing the result yields

$$D_{s,\varphi\varphi}(\bar{\mathbf{S}}_s q^*_s)\, S_1\, \ddot{\varphi}^* + D_{s,\varphi\theta}(\bar{\mathbf{S}}_s q^*_s)\, S_2\, \ddot{\theta}^* + H_\varphi(\bar{\mathbf{S}}_s q^*_s, \bar{\mathbf{S}}_s \dot{q}^*_s) = S_1\, u^*_s$$
$$D_{s,\theta\varphi}(\bar{\mathbf{S}}_s q^*_s)\, S_1\, \ddot{\varphi}^* + D_{s,\theta\theta}(\bar{\mathbf{S}}_s q^*_s)\, S_2\, \ddot{\theta}^* + H_\theta(\bar{\mathbf{S}}_s q^*_s, \bar{\mathbf{S}}_s \dot{q}^*_s) = 0_{3\times 1},$$

which completes the proof of part (1). The proof of part (2) is similar. ∎

C.2 PROOF OF LEMMA 6.2

Proof. The first part, that is, $\tilde{q} = T(q) := T_0 q + T_1$ is immediate. For the later purposes, we remark that T_0 is invertible. Since the kinetic energy of the mechanical system is invariant under coordinates transformations, that is, $\mathcal{K}_s(q, \dot{q}) = \mathcal{K}_s(\tilde{q}, \dot{\tilde{q}})$, it can be concluded that

$$\mathcal{D}(\tilde{q}) = (T_0^{-1})'\, D_s(T^{-1}(\tilde{q}))\, T_0^{-1}.$$

By assuming that the decoupling matrix $L_{g_s} L_{f_s} y^j_s(x^j_s; \xi^{*j})$ is invertible on the orbit $\mathcal{O}^j_s$ for $j = 1, 2$, there exists an open neighborhood $\mathcal{N}^j$ of $\mathcal{O}^j_s \times \xi^{j*}$ such that for every $(x^j_s, \xi^j) \in \mathcal{N}^j(\mathcal{O}^j_s \times \xi^{j*})$, the output function $y^j_s(x^j_s; \xi^j)$ has vector relative degree $(2, 2, 2)$. Next, define the new state variable $\tilde{x} := (\tilde{q}', \dot{\tilde{q}}')'$ and introduce the state equation in the new coordinates as $\dot{\tilde{x}} = \tilde{f}_s(\tilde{x}) + \tilde{g}_s(\tilde{x})u$, where

$$\tilde{g}_s(\tilde{x}_s) = \tilde{g}_s(\tilde{q}) := \begin{bmatrix} 0_{6\times 3} \\ \mathcal{D}^{-1}(\tilde{q})\, B \end{bmatrix}.$$

Since (i) the change of coordinates $\tilde{q} = T(q)$ does not change the vector relative degree of the output y^j_s and (ii) the distribution generated by the columns of the matrix $\tilde{g}_s$ is involutive, by Ref. [103, p. 127], $(\tilde{\theta}, \sigma_s)$ is valid local coordinates for the zero dynamics manifold $\mathcal{Z}^j_{s,\xi^j}$ because

$$\sigma_s(\tilde{q}, \dot{\tilde{q}}) = E_2\, \mathcal{D}(\tilde{q})\, \dot{\tilde{q}}$$

and

$$L_{\tilde{g}_s} \sigma_s(\tilde{q}, \dot{\tilde{q}}) = \begin{bmatrix} \frac{\partial \sigma_s}{\partial \tilde{q}}(\tilde{q}, \dot{\tilde{q}}) & \frac{\partial \sigma_s}{\partial \dot{\tilde{q}}}(\tilde{q}, \dot{\tilde{q}}) \end{bmatrix} \begin{bmatrix} 0_{6\times 3} \\ \mathcal{D}^{-1}(\tilde{q})B \end{bmatrix} = E_2\, B = 0_{3\times 3}.$$

In the coordinates $(\tilde{\theta}, \sigma_s)$ for the stance phase zero dynamics, $\varphi = \Phi_{s,d}^j(\gamma; \xi^j)$ and hence, σ_s can be expressed as

$$\sigma_s = E_2 \, \mathcal{D}(\tilde{q}) \, \dot{\tilde{q}} = \mathcal{D}_{\tilde{\theta}\varphi}(\tilde{q}) \, \dot{\varphi} + \mathcal{D}_{\tilde{\theta}\tilde{\theta}}(\tilde{q}) \, \dot{\tilde{\theta}}$$

$$= \left(\mathcal{D}_{\tilde{\theta}\tilde{\theta}}(\tilde{q}) + \mathcal{D}_{\tilde{\theta}\varphi}(\tilde{q}) \, \frac{\partial \Phi_{s,d}^j}{\partial \gamma}(\gamma; \xi^j) \, \mathbf{e}_2' \right) \dot{\tilde{\theta}},$$

where $\mathbf{e}_2 = [0 \; 1 \; 0]'$. In addition, since

$$\sigma_s = \frac{\partial \mathcal{L}_s}{\partial \dot{\tilde{\theta}}} = E_2 \, \mathcal{D}(\tilde{q}) \, \dot{\tilde{q}},$$

the vector $\tilde{\theta}$ is unactuated and θ_3 is the cyclic variable for the stance phase (i.e., $\frac{\partial \mathcal{K}_s}{\partial \theta_3} = 0$), the Euler–Lagrange equations imply that

$$
\begin{aligned}
\dot{\sigma}_{s,1} &= \frac{\partial \mathcal{L}_s}{\partial \theta_1} = \frac{\partial \mathcal{K}_s}{\partial \theta_1} - \frac{\partial \mathcal{V}_s}{\partial \theta_1} \\
\dot{\sigma}_{s,2} &= \frac{\partial \mathcal{L}_s}{\partial \gamma} = \frac{\partial \mathcal{K}_s}{\partial \gamma} - \frac{\partial \mathcal{V}_s}{\partial \gamma} \\
\dot{\sigma}_{s,3} &= \frac{\partial \mathcal{L}_s}{\partial \theta_3} = 0.
\end{aligned}
\tag{C.2}
$$

We remark that θ_3 is the orientation about the z-axis and hence, $\frac{\partial \mathcal{V}_s}{\partial \theta_3} = 0$. Since (i) during stance phases,

$$p_{\text{cm}} = \Gamma_l(q) = R(\theta) \, \Upsilon_l(\varphi)$$

and (ii)

$$\mathcal{V}_s(q) = m_{\text{tot}} g_0 \Gamma_l^v(q) = m_{\text{tot}} g_0 R_3(\theta) \, \Upsilon_l(\varphi),$$

where R_3 denotes the third row of the rotation matrix R, we obtain

$$
\begin{aligned}
\frac{\partial \mathcal{V}_s}{\partial \theta_1} &= \frac{\partial \mathcal{V}_s}{\partial q} \frac{\partial q}{\partial \tilde{q}} \frac{\partial \tilde{q}}{\partial \theta_1} = m_{\text{tot}} \, g_0 \, \frac{\partial R_3}{\partial \theta_1} R^{-1} p_{\text{cm}} \\
&= -m_{\text{tot}} \, g_0 \, \sin(\theta_3) \, x_{\text{cm}} + m_{\text{tot}} \, g_0 \, \cos(\theta_3) \, y_{\text{cm}}
\end{aligned}
\tag{C.3}
$$

$$
\begin{aligned}
\frac{\partial \mathcal{V}_s}{\partial \gamma} &= \frac{\partial \mathcal{V}_s}{\partial q} \frac{\partial q}{\partial \tilde{q}} \frac{\partial \tilde{q}}{\partial \gamma} = \frac{\partial \mathcal{V}_s}{\partial \theta_2} = m_{\text{tot}} \, g_0 \, \frac{\partial R_3}{\partial \theta_2} R^{-1} p_{\text{cm}} \\
&= -m_{\text{tot}} \, g_0 \, \cos(\theta_3) \cos(\theta_1) \, x_{\text{cm}} - m_{\text{tot}} \, g_0 \, \sin(\theta_3) \cos(\theta_1) \, y_{\text{cm}}.
\end{aligned}
$$

Moreover, on the zero dynamics manifold $\mathcal{Z}^j_{s,\xi^j}$,

$$\frac{\partial h^j_s}{\partial q}\dot{q} = \frac{\partial h^j_s}{\partial q}T_0^{-1}\dot{\tilde{q}} = 0_{3\times 1},$$

which together with $\sigma_s = E_2\mathcal{D}(\tilde{q})\dot{\tilde{q}}$ imply that $\dot{\tilde{q}} = \lambda^j(\tilde{\theta};\xi^j)\sigma_s$. Substituting equation (C.3) and $\dot{\tilde{q}} = \lambda^j(\tilde{\theta};\xi^j)\sigma_s$ into equation (C.2) completes the proof. ∎

C.3 INVERTIBILITY OF THE STANCE PHASE DECOUPLING MATRIX ON THE PERIODIC ORBIT

Following the proof of Lemma 6.2 in Appendix C.2, the decoupling matrix in the new coordinates $(\tilde{q}, \dot{\tilde{q}})$ can be expressed as

$$L_{\tilde{g}_s}L_{\tilde{f}_s}h^j_s(q;\xi^j) = \frac{\partial h^j_s}{\partial \tilde{q}}\mathcal{D}^{-1}B = \left(E_1 - \frac{\partial \Phi^j_{s,d}}{\partial \gamma}\mathbf{e}'_2 E_2\right)\mathcal{D}^{-1}\begin{bmatrix} I_{3\times 3} \\ 0_{3\times 3} \end{bmatrix},$$

where $E_1 := [I_{3\times 3}\ 0_{3\times 3}]$, $E_2 := [0_{3\times 3}\ I_{3\times 3}]$, and $\mathbf{e}_2 := [0\ 1\ 0]'$. By defining the symmetric matrix

$$\Lambda := \mathcal{D}^{-1} = \begin{bmatrix} \Lambda_{\varphi\varphi} & \Lambda_{\varphi\tilde{\theta}} \\ \Lambda_{\tilde{\theta}\varphi} & \Lambda_{\tilde{\theta}\tilde{\theta}} \end{bmatrix}$$

and considering the fact that

$$\Lambda_{\tilde{\theta}\varphi} = -\mathcal{D}^{-1}_{\tilde{\theta}\tilde{\theta}}\mathcal{D}_{\tilde{\theta}\varphi}\Lambda_{\varphi\varphi},$$

it can be concluded that

$$L_{\tilde{g}_s}L_{\tilde{f}_s}h^j_s(q;\xi^j) = \Lambda_{\varphi\varphi}\left(I_{3\times 3} + \frac{\partial \Phi^j_{s,d}}{\partial \gamma}\mathbf{e}'_2\mathcal{D}^{-1}_{\tilde{\theta}\tilde{\theta}}\mathcal{D}_{\tilde{\theta}\varphi}\right).$$

Since Λ is positive definite ($\det \Lambda_{\varphi\varphi} \neq 0$) and on the periodic orbit $\Phi^j_{s,d} = \varphi^j_{s,d}$, it can be concluded that $\det L_{\tilde{g}_s}L_{\tilde{f}_s}h^j_s(q;\xi^{j*}) \neq 0$ if and only if[1]

$$\kappa^j(\tilde{\theta}) = 1 + \mathbf{e}'_2\mathcal{D}^{-1}_{\tilde{\theta}\tilde{\theta}}\mathcal{D}_{\tilde{\theta}\varphi}\frac{\partial \varphi^j_{s,d}}{\partial \gamma} = \det\left(I_{3\times 3} + \frac{\partial \varphi^j_{s,d}}{\partial \gamma}\mathbf{e}'_2\mathcal{D}^{-1}_{\tilde{\theta}\tilde{\theta}}\mathcal{D}_{\tilde{\theta}\varphi}\right) \neq 0.$$

---

[1] We remark that for every $A \in \mathbb{R}^{m\times n}$ and $B \in \mathbb{R}^{n\times m}$, $\det(I_{m\times m} + AB) = \det(I_{n\times n} + BA)$.

BIBLIOGRAPHY

[1] J. W. Grizzle, C. H. Moog, and C. Chevallereau, "Nonlinear control of mechanical systems with an unactuated cyclic variable," *IEEE Transactions on Automatic Control*, vol. 30, no. 5, pp. 559–576, May 2005.

[2] K. Hirai, M. Hirose, Y. Haikawa, and T. Takenake, "The development of Honda humanoid robot," *Proceedings of the 1998 IEEE International Conference on Robotics and Automation*, Leuven, Belgium, pp. 1321–1326, May 1998.

[3] M. Vukobratovic and B. Borovac, "Zero-moment point-thirty five years of its life," *International Journal of Humanoid Robotics*, vol. 1, no. 1, pp. 157–173, 2004.

[4] M. Vukobratovic, B. Borovac, D. Surla, and D. Stokic, *Biped Locomotion*, Springer-Verlag, Berlin, 1990.

[5] H. Lim, Y. Yamamoto, and A. Takanishi, "Control to realize human-like walking of a biped humanoid robot," *Proceedings of the IEEE International Conference on Systems, Man and Cybernetics*, Nashville, TN, pp. 3271–3276, June 2000.

[6] J. Yamaguchi, E. Soga, S. Inoue, and A. Takanishi, "Development of a bipedial humanoid robot-control method of whole body cooperative dynamic biped walking," *Proceedings of the 1999 IEEE International Conference on Robotics and Automation*, Detroit, MI, pp. 368–374, May 1999.

[7] K. Loffler, M. Gienger, F. Pfeiffer, and H. Ulbrich, "Sensors and control concept of a biped robot," *IEEE Transactions on Industrial Electronics*, vol. 51, no. 5, pp. 972–980, 2004.

[8] J. H. Park and K. D. Kim, "Biped robot walking using gravity-compensated inverted pendulum mode and computed torque control," *Proceedings of the 1998 IEEE International Conference on Robotics and Automation*, Leuven, Belgium, pp. 3528–3533, May 1998.

[9] K. Erbatur and O. Kurt, "Natural ZMP trajectories for biped robot reference generation," *IEEE Transactions on Industrial Electronics*, vol. 56, no. 3, pp. 835–845, March 2009.

[10] L. Bum-Joo, D. Stonier, K. Yong-Duk, Y. Jeong-Ki, and K. Jong-Hwan, "Modifiable walking pattern of a humanoid robot by using allowable ZMP variation," *IEEE Transactions on Robotics*, vol. 24, no. 4, pp. 917–925, August 2008.

[11] N. Motoi, M. Ikebe, and K. Ohnishi, "Real-time gait planning for pushing motion of humanoid robot," *IEEE Transactions on Industrial Informatics*, vol. 3, no. 2, pp. 154–163, May 2007.

Hybrid Control and Motion Planning of Dynamical Legged Locomotion, Nasser Sadati, Guy A. Dumont, Kaveh Akbari Hamed, and William A. Gruver.
© 2012 by the Institute of Electrical and Electronics Engineers, Inc. Published 2012 by John Wiley & Sons, Inc.

[12] S. Kajita, T. Nagasaki, K. Kaneko, and H. Hirukawa, "ZMP-based biped running control," *IEEE Robotics and Automation Magazine*, vol. 14, no. 2, pp. 63–72, June 2007.

[13] B. Ugurlu and A. Kawamura, "ZMP-based online jumping pattern generation for a one-legged robot," *IEEE Transactions on Industrial Electronics*, vol. 57, no. 5, pp. 1701–1709, May 2010.

[14] T. Sato, S. Sakaino, E. Ohashi, and K. Ohnishi, "Walking trajectory planning on stairs using virtual slope for biped robots," *IEEE Transactions on Industrial Electronics*, vol. 58, no. 4, pp. 1385–1396, April 2011.

[15] P. Sardain and G. Bessonnet, "Forces acting on a biped robot. Center of pressure-zero moment point," *IEEE Transactions on Systems, Man and Cybernetics, Part A: Systems and Humans*, vol. 34, no. 5, pp. 630–637, September 2004.

[16] E. Kim, T. Kim, and J. W. Kim, "Three-dimensional modelling of a humanoid in three planes and a motion scheme of biped turning in standing," *IET (IEE) Control Theory and Applications*, vol. 3, no. 9, pp. 1155–1166, September 2009.

[17] A. Goswami, "Postural stability of biped robots and the foot-rotation indicator (FRI) point," *International Journal of Robotics Research*, vol. 18, no. 6, pp. 523–533, June 1999.

[18] E. R. Westervelt, J. W. Grizzle, C. Chevallereau, J. H. Choi, and B. Morris, *Feedback Control of Dynamic Bipedal Robot Locomotion*, Boca Raton, CRC Press, June 2007.

[19] S. Kajita and K. Tani, "Study of dynamic biped locomotion on rugged terrain-derivation and application of the linear inverted pendulum mode," *Proceedings of the 1991 IEEE International Conference on Robotics and Automation*, pp. 1405–1411, April 1991.

[20] K. Erbatur and U. Seven, "An inverted pendulum based approach to biped trajectory generation with swing leg dynamics," *Proceedings of the 2007 IEEE-RAS International Conference on Humanoid Robots*, pp. 216–221, December 2007.

[21] H. Miura and I. Shimoyama, "Dynamic walk of a biped," *International Journal of Robotics Research*, vol. 3, no. 2, pp. 60–74, June 1984.

[22] R. Katoh and M. Mori, "Control method of biped locomotion giving asymptotic stability of trajectory," *Automatica*, vol. 20, no. 4, pp. 405–414, July 1984.

[23] J. Furusho and M. Masubuchi, "Control of a dynamical biped locomotion system for steady walking," *Journal of Dynamic Systems, Measurement, and Control*, vol. 108, no. 2, pp. 111–118, 1986.

[24] J. Furusho and A. Sano, "Sensor-based control of a nine-link biped," *International Journal of Robotics Research*, vol. 9, no. 2, pp. 83–98, 1990.

[25] A. Sano and J. Furusho, "Realization of natural dynamic walking using the angular momentum information," *Proceedings of the 1990 IEEE International Conference on Robotics and Automation*, Cincinnati, OH, pp. 1476–1481, May 1990.

[26] S. Kajita and K. Tani, "Experimental study of biped dynamic walking," *IEEE Control Systems Magazine*, vol. 16, no. 1, pp. 13–19, February 1996.

[27] S. Kajita, T. Yamaura, and A. Kobayashi, "Dynamic walking control of a biped robot along a potential energy conserving orbit," *IEEE Transactions on Robotics and Automation*, vol. 8, no. 4, pp. 431–438, August 1992.

[28] A. A. Grishin, A. M. Formal'sky, A. V. Lensky, and S. V. Zhitomirsky, "Dynamical walking of a vehicle with two telescopic legs controlled by two drives," *International Journal of Robotics Research*, vol. 13, no. 2, pp.137–147, 1994.

[29] K. Mitobe, N. Mori, K. Aida, and Y. Nasu, "Nonlinear feedback control of a biped walking robot," *Proceedings of the 1995 IEEE International Conference on Robotics and Automation*, Nagoya, Japan, pp. 2865–2870, May 1995.

[30] M. H. Raibert, S. Tzafestas, and C. Tzafestas, "Comparative simulation study of three control techniques applied to a biped robot," *Proceedings of the IEEE International Conference on Systems, Man and Cybernetics Systems Engineering in the Service of Humans*, Le Touquet, France, pp. 494–502, October 1993.

[31] Y. Fujimoto and A. Kawamura, "Simulation of an autonomous biped walking robot including environmental force interaction," *IEEE Robotics and Automation Magazine*, vol. 5, no. 2, pp. 33–42, June 1998.

[32] Y. Fujimoto, S. Obata, and A. Kawamura, "Robust biped walking with active interaction control between foot and ground," *Proceedings of the 1998 IEEE International Conference on Robotics and Automation*, Leuven, Belgium, pp. 2030–2035, May 1998.

[33] S. Kajita, F. Kanehiro, K. Kaneko, K. Fujiwara, K. Yokoi, and H. Hirukawa, "A real-time pattern generator for biped walking," *Proceedings of the 2002 IEEE International Conference on Robotics and Automation*, Washington, D.C., pp. 31–37, 2002.

[34] S. Kajita, F. Kanehiro, K. Kaneko, K. Yokoi, and H. Hirukawa, "The 3D linear inverted pendulum mode: A simple modeling for a biped walking pattern generation," *Proceedings of the 2001 IEEE/RSJ International Conference on Intelligent Robots and Systems*, Maui, HI, pp. 239–246, November 2001.

[35] S. Grillner, "Control of locomotion in bipeds, tetrapods and fish," *Handbook of Physiology II*, Bethesda, MD, American Physiological Society, pp. 1179–1236, 1981.

[36] G. Taga, Y. Yamaguchi, and H. Shimizu, "Self-organized control of bipedal locomotion by neural oscillators," *Biological Cybernetics*, vol. 65, no. 3, pp. 147–159, 1991.

[37] G. Taga, "A model of the neuro-musculo-skeletal system for human locomotion II: Realtime adaptability under various constraints," *Biological Cybernetics*, vol. 73, no. 2, pp. 113–121, 1995.

[38] K. Matsuoka, "Mechanism of frequency and pattern control in the neural rhythm generators," *Biological Cybernetics*, vol. 56, nos. 5–6, pp. 345–353, 1987.

[39] K. Matsuoka, "Sustained oscillations generated by mutually inhibiting neurons with adaptation," *Biological Cybernetics*, vol. 52, no. 6, pp. 367–376, 1985.

[40] M. M. Williamson, "Neural control of rhythmic arm movements," *Neural Networks*, vol. 11, nos. 7–8, pp. 1379–1394, 1998.

[41] N. Sadati, G. A. Dumont, and K. A. Hamed, *Design of a Neural Controller for Walking of a 5-Link Planar Biped Robot via Optimization, Human-Robot Interaction*. D. Chugo (ed.), Rijeka, Croatia, InTech, pp. 267–288, February 2010.

[42] S. Miyakoshi, G. Taga, Y. Kuniyoshi, and A. Nagakubo, "Three dimensional bipedal stepping motion using neural oscillators-towards humanoid motion in the real word," *Proceedings of the 1998 IEEE/RSJ International Conference on Intelligent Robots and Systems*, Victoria, BC, Canada, pp. 84–89, May 1998.

[43] H. Kimura, Y. Fukuoka, Y. Hada, and K. Takase, "Three-dimensional adaptive dynamic walking of a quadruped-rolling motion feedback to CPGs controlling pitching motion," *Proceedings of the IEEE International Conference on Robotics and Automation*, pp. 2228–2233, May 2002.

[44] H. Kimura and Y. Fukuoka, "Adaptive dynamic walking of the quadruped on irregular terrain-autonomous adaptation using neural system model," *Proceedings of the IEEE International Conference on Robotics and Automation*, pp. 436–443, 2000.

[45] J. Or, "A hybrid CPG-ZMP controller for the real-time balance of a simulated flexible spine humanoid robot," *IEEE Transactions on Systems, Man, and Cybernetics, Part C: Applications and Reviews*, vol. 39, no. 5, pp. 547–561, September 2009.

[46] J. W. Grizzle, G. Abba, and F. Plestan, "Asymptotically stable walking for biped robots: Analysis via systems with impulse effects," *IEEE Transactions on Automatic Control*, vol. 46, no. 1, pp. 51–64, January 2001.

[47] C. Chevallereau, G. Abba, Y. Aoustin, F. Plestan, E. R. Westervelt, C. Canudas-DeWit, and J. W. Grizzle, "RABBIT: A testbed for advanced control theory," *IEEE Control Systems Magazine*, vol. 23, no. 5, pp. 57–79, October 2003.

[48] J. W. Grizzle, F. Plestan, and G. Abba, "Poincaré's method for systems with impulse effects: Application to the mechanical biped locomotion," *Proceedings of the 1999 IEEE International Conference on Decision and Control*, Phoenix, AZ, pp. 3869–3876, December 1999.

[49] J. W. Grrizle, G. Abba, and F. Plestan, "Proving asymptotic stability of a walking cycle for a five DOF biped robot model," *Proceedings of the 1999 International Conference on Climbing and Walking Robots*, pp. 69–81, September 1999.

[50] E. R. Westervelt, J. W. Grizzle, and D. E. Koditschek, "Zero dynamics of underactuated planar biped walkers," *Proceedings of the 15th IFAC World Congress*, Barcelona, Spain, pp. 1–6, July 2002.

[51] F. Plestan, J. W. Grizzle, E. R. Westervelt, and G. Abba, "Stable walking of a 7-DOF biped robot," *IEEE Transactions on Robotics and Automation*, vol. 19, no. 4, pp. 653–668, August 2003.

[52] E. R. Westervelt, J. W. Grizzle, and D. E. Koditschek, "Hybrid zero dynamics of planar biped walkers," *IEEE Transactions on Automatic Control*, vol. 48, no. 1, pp. 42–56, January 2003.

[53] C. Chevallereau, J. W. Grizzle, and C. H. Moog, "Nonlinear control of mechanical systems with one degree of underactuation," *Proceedings of the 2004 IEEE International Conference on Robotics and Automation*, New Orleans, LA, pp. 2222–2228, April 2004.

[54] J. H. Choi and J. W. Grizzle, "Planar bipedal walking with foot rotation," *Proceedings of the 2005 American Control Conference*, Portland, OR, pp. 4909–4916, June 2005.

[55] C. Chevallereau, E. R. Westervelt, and J. W. Grizzle, "Asymptotically stable running for a five-link, four-actuator, planar bipedal robot," *International Journal of Robotics Research*, vol. 24, no. 6, pp. 431–464, June 2005.

[56] B. Morris, E. R. Westervelt, C. Chevallereau, G. Buche, and J. W. Grizzle, "Achieving bipedal running with RABBIT: Six steps toward infinity," *Fast Motions Symposium on Biomechanics and Robotics*, Lecture Notes in Control and Information Sciences, Springer-Verlag, Heidelberg, Germany, pp. 277–297, 2006.

[57] J. W. Grizzle, E. R. Westervelt, and C. Canudas, "Event-based PI control of an underactuated biped walker," *Proceedings of the 2003 IEEE International Conference on Decision and Control*, Maui, HI, pp. 3091–3096, December 2003.

[58] E. R. Westervelt, J. W. Grizzle, and C. Canudas, "Switching and PI control of walking motions of planar biped walkers," *IEEE Transactions on Automatic Control*, vol. 48, no. 2, pp. 308–312, February 2003.

[59] J. W. Grizzle, "Remarks on event-based stabilization of periodic orbits in systems with impulse effects," *Second International Symposium on Communications, Control and Signal Processing*, 2006.

[60] B. Morris and J. W. Grizzle, "Hybrid invariant manifolds in systems with impulse effects with application to periodic locomotion in bipedal robots," *IEEE Transactions on Automatic Control*, vol. 54, no. 8, pp. 1751–1764, August 2009.

[61] C. Chevallereau, J. W. Grizzle, and C. L. Shih, "Asymptotically stable walking of a five-link underactuated 3-D bipedal robot," *IEEE Transactions on Robotics*, vol. 25, no. 1, pp. 37–50, February 2009.

[62] C. L. Shih, J. W. Grizzle, and C. Chevallereau, "From stable walking to steering of a 3D bipedal robot with passive point feet," submitted to *Robotica*, January 2010.

[63] H. W. Park, K. Sreenath, J. W. Hurst, and J. W. Grizzle, "Identification of a bipedal robot with a compliant drivetrain," *Control Systems Magazine*, vol. 31, no. 2, pp. 63–88, April 2011.

[64] K. Sreenath, H. W. Park, I. Poulakakis, and J. W. Grizzle, "A compliant hybrid zero dynamics controller for stable, efficient and fast bipedal walking on MABEL," *Internatioanl Journal of Robotics Research*, vol. 30, no. 9, pp. 1170–1193, August 2011.

[65] J. W. Grizzle, J. Hurst, B. Morris, H. W. Park, and K. Sreenath, "MABEL, a new robotic bipedal walker and runner," *Proceedings of the American Control Conference*, St. Louis, MO, pp. 2030–2036, June 2009.

[66] http://mime.oregonstate.edu/research/drl/

[67] Y. Hürmüzlü, "Dynamics of bipedal gait-Part 1: Objective functions and the contact event of a planar five-link biped," *Journal of Applied Mechanics*, vol. 60, no. 2, pp. 331–336, 1998.

[68] K. Akbari Hamed, N. Sadati, W. A. Gruver, and G. A. Dumont, "Stabilization of periodic orbits for planar walking with noninstantaneous double-support phase," *IEEE Transactions on Systems, Man, and Cybernetics, Part A: Systems and Humans*, vol. 42, no. 3, pp. 685–706, May 2012.

[69] N. Sadati, K. Akbari Hamed, W. A. Gruver, and G. A. Dumont, "Radial basis function network for exponential stabilisation of periodic orbits for planar bipedal walking," *IET (IEE) Electronics Letters*, vol. 47, no. 12, pp. 692–694 June 2011.

[70] C. Canudas, B. Siciliano, and G. Bastin, *Theory of Robot Control*, Springer-Verlag, London, 1996.

[71] S. Miossec and Y. Aoustin, "A simplified stability study for a biped walk with underactuated and overactuated phases," *International Journal of Robotics Research*, vol. 24, no. 7, pp. 537–551, June 2005.

[72] D. Djoudi and C. Chevallereau, "Feet can improve the stability property of a control law for a walking robot," *Proceedings of the 2006 IEEE International Conference on Robotics and Automation*, Orlando, FL, pp. 1206–1212, May 2006.

[73] C. Chevallereau and Y. Aoustin, "Optimal reference trajectories for walking and running of a biped robot," *Robotica*, vol. 19, no. 5, pp. 557–569, September 2001.

[74] S. Dubowsky and E. Papadopoulos, "The kinematics, dynamics and control of free-flying and free-floating space robotic systems," *IEEE Transactions on Robotic and Automation*, vol. 9, no. 5, pp. 531–543, October 1993.

[75] C. Fernandes, L. Gurvits, and Z. Li, "Near-optimal nonholonomic motion planning for a system of coupled rigid bodies," *IEEE Transactions on Automatic Control*, vol. 39, no. 3, pp. 450–463, March 1994.

[76] C. Fernandes, L. Gurvits, and Z. Li, "Attitude control of space paltform/manipulator system using internal motion," *Proceedings of the 1992 IEEE International Conference on Robotics and Automation*, Nice, France, pp. 893–898, March 1994.

[77] P. S. Krishnaprasad, "Geometric phases and optimal reconfiguration for multibody systems," *Proceedings of the American Control Conference*, pp. 2440–2444, May 1990.

[78] G. C. Walsh and S. S. Sastry, "On reorienting linked rigid bodies using internal motions," *IEEE Transactions on Robotics and Automation*, vol. 11, no. 1, pp. 139–146, February 1995.

[79] L. Gurvits and Z. X. Li, "Smooth time-periodic feedback solutions for nonholonomic motion planning," *Nonholonomic Motion Planning*. Z. Li and J. F. Canny (eds.), Kluwer, pp. 53–108, 1993.

[80] I. Kolmanovsky, N. McClamroch, and V. Coppola, "New results on control of multibody systems which conserve angular momentum," *Journal of Dynamical and Control Systems*, vol. 1, no. 4, pp. 447–462, 1995.

[81] N. Sadati, G. A. Dumont, K. Akbari Hamed, and W. A. Gruver, "Two-level control scheme for stabilisation of periodic orbits for planar monopedal running," *IET (IEE) Control Theory and Applications*, vol. 5, no. 13, pp. 1528–1543, August 2011.

[82] N. Sadati, K. Akbari Hamed, G. A. Dumont, and W. A. Gruver, "Nonholonomic motion planning based on optimal control for flight phases of planar bipedal running," *IET (IEE) Electronics Letters*, vol. 47, no. 20, pp. 1120–1122, September 2011.

[83] N. Sadati, W. A. Gruver, K. Akbari Hamed, and G. A. Dumont, "Online generation of joint motions during flight phases in running of planar bipedal robots," *IET (IEE) Control Theory and Applications*, under review, January 2012.

[84] C. Chevallereau, Y. Aoustin, and A. Formal'sky, "Optimal walking trajectories for a biped," *Proceedings of the First Workshop on Robot Motion and Control*, Kiekrz, Poland, pp. 171–176, June 1999.

[85] K. Akbari Hamed, N. Sadati, W. A. Gruver, and G. A. Dumont, "Exponential stabilisation of periodic orbits for running of a three-dimensional monopedal robot," *IET (IEE) Control Theory and Applications*, vol. 5, no. 11, pp. 1304–1320, August 2011.

[86] K. Akbari Hamed, N. Sadati, G. A. Dumont, and W. A. Gruver, "Exponential stabilization of periodic motions for a five-link, two-actuator planar bipedal robot," *International Journal of Robotics and Automation*, submitted 15 September 2010, revised 14 April 2011.

[87] K. Akbari Hamed, N. Sadati, W. A. Gruver, and G. A. Dumont, "Continuous-time update laws with radial basis step length for control of bipedal locomotion," *IET (IEE) Electronics Letters*, vol. 46, no. 21, pp. 1431–1433, October 2010.

[88] H. K. Khalil. *Nonlinear Systems - 3rd Edition*, Upper Saddle River, NJ, 2002.

[89] T. S. Parker and L. O. Chua, *Practical Numerical Algorithms for Chaotic Systems*, Springer-Verlag, New York, 1989.

[90] M. W. Spong, S. Hutchinson, and M. Vidyasagar, *Robot Modeling and Control*, Wiley, New York, 2006.

[91] Y. Hürmüzlü and D. B. Marghitu, "Rigid body collisions of planar kinematic chains with multiple contact points," *International Journal of Robotics Research*, vol. 13, no. 1, pp. 82–92, 1994.

[92] X. Mu and Q. Wu, "On impact dynamics and contact events for biped robots via impact effects," *IEEE Transactions on Systems, Man, and Cybernetics-Part B: Cybernetics*, vol. 36, no. 6, pp. 1364–1372, December 2006.

[93] S. P. Bhat and D. S. Bernstein, "Continuous finite-time stabilization of the translational and rotational double integrators," *IEEE Transactions on Automatic Control*, vol. 43, no. 5, pp. 678–682, May 1998.

[94] E. R. Westervelt, B. Morris, and K. D. Farrell, "Sample-based HZD control for robustness and slope invariance of planar passive bipedal gaits," *Proceedings of the 14th Mediterranean Conference on Control and Automation*, pp. 1–6, June 2006.

[95] M. W. Spong, "Energy based control of a class of underactuated mechanical systems," *Proceedings of IFAC World Congress*, San Francisco, CA, pp. 431–435, 1996.

[96] E. R. Westervelt, G. Buche, and J. W. Grizzle, "Experimental validation of a framework for the design of controllers that induce stable walking in planar bipeds," *International Journal of Robotics Research*, vol. 23, no. 6, pp. 559–582, 2004.

[97] X. Mu and Q. Wu, "A complete dynamic model of five-link bipedal walking," *Proceedings of the American Control Conference*, pp. 4926–4931, June 2003.

[98] N. Sreenath, "Nonlinear control of planar multibody systems in shape space," *Mathematics of Control Signals System* vol. 5, no. 4, pp. 343–363, 1992.

[99] A. E. Bryson and Y. C. Ho, *Applied Optimal Control*, Hemisphere, New York, 1975.

[100] D. E. Kirk, *Optimal Control Theory: An Introduction*, Dover, New York, 1970.

[101] M. Athans and P. L. Falb, *Optimal Control: An Introduction to the Theory and Its Applications*, McGraw-Hill, New York, 1966.

[102] A. F. Filippov, *Differential Equations with Discontinuous Righthand Sides*, Kluwer Academic, Dordrecht, 1988.

[103] A. Isidori, *Nonlinear Control Systems-3rd Edition*, Springer-Verlag, Berlin, 1995.

Hybrid Control and Motion Planning of Dynamical Legged Locomotion, Nasser Sadati, Guy A. Dumont,
Kaveh Akbari Hamed, and William A. Gruver.
© 2012 by the Institute of Electrical and Electronics Engineers, Inc. Published 2012 by John Wiley & Sons, Inc.

IEEE PRESS SERIES ON
SYSTEMS SCIENCE AND ENGINEERING

Editor:
MengChu Zhou, *New Jersey Institute of Technology and Tongji University*

Co-Editors:
Han-Xiong Li, *City University of Hong-Kong*
Margot Weijnen, *Delft University of Technology*

The focus of this series is to introduce the advances in theory and applications of systems science and engineering to industrial practitioners, researchers, and students. This series seeks to foster system-of-systems multidisciplinary theory and tools to satisfy the needs of the industrial and academic areas to model, analyze, design, optimize and operate increasingly complex man-made systems ranging from control systems, computer systems, discrete event systems, information systems, networked systems, production systems, robotic systems, service systems, and transportation systems to Internet, sensor networks, smart grid, social network, sustainable infrastructure, and systems biology.

1. *Reinforcement and Systemic Machine Learning for Decision Making*
   Parag Kulkarni
2. *Remote Sensing and Actuation Using Unmanned Vehicles*
   Haiyang Chao, YangQuan Chen
3. *Hybrid Control and Motion Planning of Dynamical Legged Locomotion*
   Nasser Sadati, Guy A. Dumont, Kaveh Akbari Hamed, and William A. Gruver

Forthcoming Titles:

Operator-based Nonlinear Control Systems Design and Applications
Mingcong Deng

Contemporary Issues in Systems Science and Engineering
Mengchu Zhou, Han-Xiong Li and Margot Weijnen

Design of Business and Scientific Workflows: A Web Service-Oriented Approach
Mengchu Zhou and Wei Tan